S0-BQM-203

ADVANCES IN NUTRITION AND CANCER 2

ADVANCES IN EXPERIMENTAL MEDICINE AND BIOLOGY

Editorial Board:

NATHAN BACK, *State University of New York at Buffalo*

IRUN R. COHEN, *The Weizmann Institute of Science*

DAVID KRITCHEVSKY, *Wistar Institute*

ABEL LAJTHA, *N. S. Kline Institute for Psychiatric Research*

RODOLFO PAOLETTI, *University of Milan*

Recent Volumes in this Series

Volume 465
CANCER GENE THERAPY: Past Achievements and Future Challenges
Edited by Nagy A. Habib

Volume 466
CURRENT VIEWS OF FATTY ACID OXIDATION AND KETOGENESIS:
From Organelles to Point Mutations
Edited by Patti A. Quant and Simon Eaton

Volume 467
TRYPTOPHAN, SEROTONIN, AND MELATONIN: Basic Aspects and Applications
Edited by Gerald Heuther, Walter Kochen, Thomas J. Simat, and Hans Steinhart

Volume 468
THE FUNCTIONAL ROLES OF GLIAL CELLS IN HEALTH AND DISEASE:
Dialogue between Glia and Neurons
Edited by Rebecca Matsas and Marco Tsacopoulos

Volume 469
EICOSANOIDS AND OTHER BIOACTIVE LIPIDS IN CANCER, INFLAMMATION, AND RADIATION INJURY, 4
Edited by Kenneth V. Honn, Lawrence J. Marnett, and Santosh Nigam

Volume 470
COLON CANCER PREVENTION:
Dietary Modulation of Cellular and Molecular Mechanisms
Edited under the Auspices of the American Institute for Cancer Research

Volume 471
OXYGEN TRANSPORT TO TISSUE XXI
Edited by Andras Eke and David T. Delpy

Volume 472
ADVANCES IN NUTRITION AND CANCER 2
Edited by Vincenzo Zappia, Fulvio Della Ragione, Alfonso Barbarisi, Gian Luigi Russo, and Rossano Dello Iacovo

Volume 473
MECHANISMS IN THE PATHOGENESIS OF ENTERIC DISEASES 2
Edited by Prem S. Paul and David H. Francis

Volume 474
HYPOXIA: INTO THE NEXT MILLENNIUM
Edited by Robert C. Roach, Peter D. Wagner, and Peter H. Hackett

---

A Continuation Order Plan is available for this series. A continuation order will bring delivery of each new volume immediately upon publication. Volumes are billed only upon actual shipment. For further information please contact the publisher.

ADVANCES IN NUTRITION AND CANCER 2

Edited by

Vincenzo Zappia
Fulvio Della Ragione
Institute of Biochemistry of Macromolecules
Second University of Naples
Naples, Italy

Alfonso Barbarisi
Institute of Clinical Surgery
Second University of Naples
Naples, Italy

Gian Luigi Russo
Institute of Food Science and Technology
National Research Council
Avellino, Italy

and

Rossano Dello Iacovo
National Cancer Institute
Fondazione "Giovanni Pascale"
Naples, Italy

RC268.45
A38
1999

Kluwer Academic / Plenum Publishers
New York, Boston, Dordrecht, London, Moscow

Library of Congress Cataloging-in-Publication Data

Advances in nutrition and cancer 2/edited by Vincenzo Zappia ... [et al.].
p. cm. — (Advances in experimental medicine and biology; v. 472)
"Proceedings of the 2nd International Conference on Nutrition and Cancer, held October 20–23, 1998, in Naples, Italy"—T.p. verso.
Includes bibliographical references and index.
ISBN 0-306-46306-7
1. Cancer—Nutritional aspects—Congresses. I. Zappia, Vincenzo. II. International Conference on Nutrition and Cancer (2nd: 1998: Naples, Italy) III. Series.
RC268.45 .A38 2000
616.99′4071—dc21

99-049491

Proceedings of the 2nd International Conference on Nutrition and Cancer, held October 20–23, 1998, in Naples, Italy

ISBN 0-306-46306-7

©1999 Kluwer Academic / Plenum Publishers, New York
233 Spring Street, New York, N.Y. 10013

http://www.wkap.nl

10 9 8 7 6 5 4 3 2 1

A C.I.P. record for this book is available from the Library of Congress

All rights reserved

No part of this book may be reproduced, stored in a retrieval system, or transmitted in any form or by any means, electronic, mechanical, photocopying, microfilming, recording, or otherwise, without written permission from the Publisher

Printed in the United States of America

PREFACE

This volume includes contributions presented at the Second International Symposium on Nutrition and Cancer, held in Naples, Italy, in October 1998 at the National Tumor Institute "Fondazione Pascale." During the Conference, experts from different disciplines discussed pivotal and timely subjects on the interactions between human nutrition and the development of malignancies.

Comparing the themes of this Meeting with those discussed at the First Symposium in 1992, the major scientific advancements certainly derive from the extensive use of molecular approaches to perform research in nutrition. Moreover, the fundamental observation of R. Doll and R. Peto (1981), which suggested that at least 35% of all cancers (with large differences among different tumors) might be prevented by dietary regimens, has been definitively confirmed by epidemiological studies.

On the other hand, the relationships between diet and cancer are quite intricate and complex; it is difficult, and at the same time not methodologically correct, to reduce them to simple terms. Metabolic and hormonal factors, contaminants and biological agents, and deficiency of specific protective nutrients are all pieces of the same puzzle.

The molecular studies reported here include all the major aspects of investigation on human nutrition and malignant transformation. As discussed during the Conference, in the last decade, a large number of compounds responsible for the biological activity of human foods have been identified and characterized. These molecules not only include important and well-known risk factors but, most promising, compounds which might exert chemopreventive activity. Among them, antioxidants (such as vitamins, phenols, and lycopene) seem to play a critical role in reducing the risk of cancer at different anatomical sites, including colon, breast, and prostate. Other molecules, derived from fiber bacterial intestinal degradation (short fatty acids), are of interest, even if their importance has not been completely unraveled and is still the subject of debate.

The relevance of the individual genetic background as an independent risk factor in the development of diet-related cancer opens new perspectives for molecular epidemiology. The working hypothesis is that genetic differences in the metabolism of nutrients might result in different incidences of malignancies. A further development in the genetic field is the understanding of mechanisms involved in the repair of DNA alterations. Also, in this case, the occurrence of genetic polymorphisms might result in distinct responses to identical dietary risk factors.

A further promising advancement relates to the finding that active dietary molecules might regulate cell transduction pathways and gene expression and, in turn, cell proliferation and differentiation. In recent years, the precise knowledge of the principal protein players of cell division cycle machinery has allowed the elucidation of the biochemistry at growth. Recent data suggest that dietary-derived compounds, including radical scavengers, affect cell proliferation by acting on the cell division cycle engine, and thus might control the process (and the rate) of malignant transformation.

Several of the themes described above are discussed in the present volume, but the final word on their true importance must wait until the dust of the present time has settled. What we can anticipate is that the rapid development of new methodologies and the convergence of different types of approaches (molecular, epidemiological, intervention) to clarify the relationship between nutrition and cancer will require new appraisals, with continuous updating, on this central aspect of human health.

Vincenzo Zappia

ACKNOWLEDGMENTS

The Second International Symposium on Nutrition and Cancer, held in October 1998, was sponsored by the Institute of Biochemistry of Macromolecules of the Second University of Naples, the Institute of Food Science and Technology of the National Research Council in Avellino, the National Tumor Institute "Fondazione Pascale" of Naples, and the Italian Institute for Philosophical Studies of Naples, with the patronage of the Second University of Naples, local and regional governments, and the Italian Health Ministry.

Contributors to the Conference included Air Liquide, Analytical Control, Dia-Chem, Hewlett Packard, IMV Bicef, Libreria Fiorentino, Microglass, Sarstedt, Sigma, and Sigma-tau, as well as the local and regional tourist boards.

Particular thanks are due to Valeria Cucciolla and Valentina Della Pietra, of the Institute of Biochemistry of Macromolecules of the Second University of Naples; Rosanna Palumbo and Giuseppe Iacomino of the Institute of Food Science and Technology, CNR, Avellino; and Annamaria Strollo and Egidio Celentano of the "Pascale" Tumor Institute of Naples, for their help in the scientific and organizational aspects of the Conference. Editorial coordination among editors, authors, and Kluwer Academic/Plenum Publishers was handled by Tricia Reynolds.

The Editors express their gratitude to the authors of the articles and to Kluwer Academic/Plenum Publishers for having made possible the publication of this volume.

CONTENTS

Epidemiological Studies: Risk Factors and Diets

Cell Growth and Differentiation: Molecular Bases of Nutrient Effects

Molecular Epidemiology: An Innovative Trend in Cancer Prevention

1

DIET AND CANCER

Perspectives of Prevention

Peter Greenwald

Division of Cancer Prevention
National Cancer Institute
Building 31, Room 10A52
National Institutes of Health
Bethesda, Maryland 20892

1. INTRODUCTION

Cancer generally has a complex etiology with multiple risk factors that involve an interplay between genetic and environmental influences. Diet, a major environmental factor, has been associated with risk for many types of cancer.[1,2] Evidence from epidemiologic and corroborating experimental studies strongly supports relationships between dietary constituents and cancer risk, suggesting that, in general, vegetables, and fruits, dietary fiber, and certain micronutrients appear to protect against cancer, whereas fat, excessive calories, and alcohol seem to increase cancer risk.[1–4] It follows that a proactive approach to cancer prevention through dietary modification is a prudent choice that may reduce cancer risk and be beneficial to overall good health. Other environmental characteristics, for example, degree of physical activity and obesity, also are lifestyle factors that appear to have major influences on cancer risk.[3,5]

Less is known about the effects of individual genetic susceptibilities on cancer risk. Certain genetic polymorphisms may be important in cancer etiology. For example, one common polymorphism of the gene that codes for methylenetetrahydrofolate reductase—an enzyme critical to the regulation of factors in DNA methylation and synthesis—can influence colorectal cancer risk in affected individuals by altering cellular responses to dietary folate and methionine. Study findings suggest that this polymorphism is inversely associated with colorectal cancer risk,[6] but plays a role only at a late stage of colorectal tumorigenesis—that is, the transformation of adenoma to carcinoma.[7] Until we have a better understanding of gene/nutrient and gene/

Advances in Nutrition and Cancer 2, edited by Zappia *et al.*
Kluwer Academic / Plenum Publishers, New York, 1999.

environment interactions, cancer researchers must continue developing and carrying out strategies and studies aimed at reducing cancer risk that are based on current knowledge.

Discussion in the following sections focuses on several topics—selected from the many encompassed by diet and cancer prevention research—that are currently receiving considerable attention in the scientific community.

2. NATURE VERSUS NURTURE

The most accurate assessment of an individual's risk for developing cancer would be an estimate of dietary and other environmental exposures (nurture) as well as information on interindividual differences in genetic susceptibility, including inheritable variations in carcinogen-metabolizing enzymes, germline mutations in tumor-associated genes, and inherited differences in DNA adduct formation and DNA repair mechanisms (nature).[8,9] Although data on lifestyle factors such as diet, exercise, and smoking can be collected and interpreted with some degree of confidence, genetic assessments for most individuals are not readily available. Links between dietary factors and genetic susceptibility are being investigated through studies in animal models and in human twin pairs.

2.1. Animal Studies

Findings from various animal studies have demonstrated clear interactions between diet and genetics. *p53*-Deficient ($p53^{-/-}$) mice, genetically susceptible animal models that rapidly develop spontaneous tumors, have been used to investigate the potential modulating effects of specific dietary factors on cancer development. To illustrate, findings from a study comparing the effects of caloric restriction (60% of *ad libitum* intake) in $p53^{-/-}$ mice and wild-type mice ($p53^{+/+}$) suggested that caloric restriction inhibits tumor development through a *p53*-independent mechanism(s)[10] and may be related to effects on proliferative pathways.[11] Mice with specific chain-terminating mutations in the *Apc* gene also are particularly useful models for ecogenetic studies, as they develop intestinal tumors in a manner similar to humans with familial adenomatous polyposis (FAP). In one study, $Apc^{\Delta 716}$ knockout mice fed a low-fat, high-fiber diet developed 36% fewer polyps in the small intestine and 64% fewer polyps in the colon, compared with mice fed a high-fat, low-fiber diet.[12] A recent study in *Apc1638* mice fed a long-term (about 32 weeks) Western-style diet that was high in fat and low in calcium and vitamin D reported that a significantly greater number of these mice developed carcinomas than mice fed a standard diet (65% vs. 13%).[13] Only 23% of *Apc1638* mice fed a short-term (about 13 weeks) Western-style diet developed carcinomas, suggesting that diet had a greater influence on promotion and progression—later stages of tumor evolution—than on the initiation of carcinoma development.

2.2. Twin Studies

Concordance rates for cancer for monozygotic (MZ) and dizygotic (DZ) twins—that is, identical and fraternal twins, respectively—provide insight into the relative contributions of environmental and genetic factors to cancer risk. Large differences in concordance between MZ and DZ twins suggest greater genetic influence; low concordance rates among MZ twins indicate greater environmental influences.[14] Overall, findings in twin studies in the United States and Northern Europe suggest that inherited disposition does not explain a large proportion of either specific cancers or all cancer mortality.[15,16] A report of breast and ovarian cancer development in only one of two identical twins who both have an inherited *BRCA1* mutation clearly illustrates the possibility of differences in phenotypic expression of such mutations.[17] One study of concordance in childhood cancer in twins (ages 0–20 years) also suggested that genetic contributions to risk are small.[18] Concordance for all cancers combined (excluding leukemia and retinoblastoma) was 1.2% overall and 2.2% in MZ twins; concordance for leukemia in MZ twins was 5%. Interestingly, a twin study in Sweden found a 4-fold higher concordance rate for prostate cancer in MZ twin pairs than for DZ twin pairs, indicating that genetics might be important in prostate cancer development.[19]

3. FAT AND BREAST CANCER ISSUES

A considerable amount of controversy is present within the scientific community as to whether or not a direct association exists between dietary fat and postmenopausal breast cancer risk. When evaluating findings of research studies on this topic, it is important to recognize that dietary fat is not a completely independent lifestyle risk factor. The effects of other closely related lifestyle risk factors, including caloric intake, weight gain, obesity, and physical activity, have a high probability of being intertwined with those of dietary fat. In fact, it may be unrealistic to expect to separate the independent effects of these diet-related factors on breast cancer risk. Further, individual differences in inherited genetic susceptibility could result in environmental factors being a more important contributor to breast cancer in only some proportion of a study population, limiting the ability to detect associations because of misclassification of outcomes.[20] In addition, methodologic issues related to dietary assessment in epidemiologic studies contribute further uncertainties to data interpretation.

3.1. Epidemiologic Evidence

Although recognizing the limitations of an epidemiologic approach to determining the true association between diet-related factors and breast cancer, epidemiologic evidence, supported by consistent results from animal studies, has clearly suggested direct associations between breast cancer risk and total dietary fat, types of fat, obesity, weight gain, and physical activity.

3.1.1. Dietary Fat. Epidemiologic studies provide conflicting results regarding the association of dietary fat and breast cancer.[2,3] International correlations between fat intake and breast cancer incidence and mortality rates are high, averaging approximately 0.8.[21,22] Migrant and time-trend studies similarly support an increased risk for breast cancer as eating patterns shift from a low-fat, high-fiber diet to a high-fat, low-fiber Western diet.[23,24] Individual case-control and cohort studies, comparing highest versus lowest quintiles/quartiles of fat intake, generally have not found a significant association between fat intake and breast cancer risk.[2,3] One meta-analysis found a relative risk (RR) of 1.21 for breast cancer (premenopausal and postmenopausal women combined), based on fat intake in 16 case-control studies.[25] Two separate meta-analyses investigating the impact of dietary fat on breast cancer development reported RRs of 1.01 for 7 cohort studies (premenopausal and postmenopausal women combined)[25] and 1.05 for 7 cohort studies (similar results for both premenopausal and postmenopausal women);[26] 5 of the same cohort studies were included in both of these meta-analyses.

A growing body of literature suggests that the dietary fat-breast cancer association may be a result of the type of fat consumed rather than or in addition to total fat intake.[3] Using data from the European Community Multicenter Study on Antioxidants, Myocardial Infarction, and Cancer (EURAMIC) breast cancer study, the ratio of long-chain, n-3 polyunsaturated fatty acids (PUFAs)—found in certain fish and fish oils—in adipose tissue to total n-6 PUFAs showed an inverse relationship with breast cancer risk,[27] providing evidence that the balance between n-3 and n-6 PUFAs may be important in breast cancer. These findings are in agreement with ecological mortality data for breast cancer for 24 European countries that show an inverse correlation with fish and fish oil consumption, when expressed as a proportion of animal fat.[28] A comprehensive assessment of evidence from animal, ecologic, and analytic epidemiologic studies linking olive oil (high in oleic acid, a monounsaturated fatty acid) and breast cancer concluded that a modest protective effect of olive oil on breast cancer risk appears likely, but the evidence is not yet conclusive.[29] Findings from a prospective study in Sweden, a country with very low consumption of olive oil, indicated that monounsaturated fat was associated with reduced risk (RR = 0.8) and polyunsaturated fat was associated with increased risk (RR = 1.2).[30] Saturated fat was not associated with breast cancer risk in this study. *Trans* fatty acids also have been investigated in relationship to breast cancer. In the EURAMIC study, data for primarily postmenopausal breast cancer patients supported direct associations between increased breast cancer risk and the concentration of *trans* fatty acids (RR = 1.4) and PUFAs (RR = 1.26) in gluteal adipose tissue. The most significant effect of *trans* fatty acids (RR = 3.65) was seen at the lowest tertile of PUFAs.[31]

3.1.2. Caloric Intake/Weight Gain/Obesity. The roles of factors related to caloric intake—including weight, body size (as measured through body mass index [BMI]), and weight gain—in breast cancer risk, particularly in relationship to menopausal status, have been investigated in numerous epidemiologic studies.[2,3] Obesity prior to menopause appears to protect against breast cancer, whereas postmenopausal obesity is associated with increased risk.[32,33] Gaining weight after age 18 and being overweight during the premenopausal years also appear to increase risk for breast cancer after menopause.[33–35] Some data suggest that high caloric intake and rapid growth during youth may increase breast cancer risk.[36] A recent study reported that women who reached their maximum height when 18 years or older had a 30% lower risk of breast cancer compared with women who reached their maximum height when 13 years or younger.[37]

3.1.3. Physical Activity. Sustained physical activity, by leading to weight loss as well as loss of body fat, helps reduce circulating levels of estrogen and progesterone, and possibly breast cancer risk. A number of case-control and cohort studies have investigated the roles of either occupational, leisure-time, or total physical activity (energy expenditure) in relation to risk of breast cancer.[5,38] Overall, evidence for a protective effect of physical activity is inconsistent; some data, however, are encouraging. A recent comprehensive review of epidemiologic studies on physical activity and breast cancer reported that 11 of 16 studies on the effect of recreational exercise on breast cancer risk reported decreased risk (12%-60%) among both premenopausal and postmenopausal women; Also, higher levels of occupational physical activity were associated with reduced risk in seven of nine studies.[38] Data from a large cohort study in Norway indicated that, compared with a sedentary lifestyle, regular exercise reduces risk in both premenopausal (RR = 0.53) and postmenopausal (RR = 0.67) women.[39] Risk of breast cancer was lowest in lean women (BMI, 22.8, premenopausal and postmenopausal combined) who exercised at least 4 hours per week (RR = 0.28).

3.2. Methodological Considerations

In addition to the unavoidable confounding that is encountered when trying to determine individual contributions to risk from related lifestyle factors, several other factors may contribute to the inability of epidemiologic studies to detect breast cancer risk associations for individual lifestyle components. Measurement error (both random and systematic) in dietary assessment, differences in data collection and analysis methods among studies, insufficient variation in fat intake within a study population, insufficient followup period, failure to distinguish between premenopausal and postmenopausal women, and the importance of diet and other lifestyle components before adulthood all may influence study results.[22,40–42]

Both systematic and random errors can contribute to measurement error in dietary intake data.[40,41,43] One source of random error is the natural variation of day-to-day eating patterns, which can be minimized by investigating dietary intake over an extended period, using diet histories or food frequency questionnaires (FFQs).[41] Use of such measures can introduce systematic error, frequently the result of either underreporting or overreporting of intake, which has possible serious consequences in terms of misclassification in epidemiologic studies. For example, an analysis of self-reported dietary intake data from more than 11,600 adults in the Second National Health and Nutrition Examination Survey (NHANES II) indicated that up to 31% of these individuals underreported their food consumption.[44] Women were five times as likely as men to underreport their intake, and Caucasians more than twice as likely as non-Caucasians. BMI also directly influenced underreporting, with the odds of underreporting increasing by 166% for every 1 unit increase in BMI.

When a measurement error that affects both cases and controls is present in epidemiologic studies, the relative risk will be underestimated such that values will be biased toward 1.0, reducing the chances of finding an existing diet-cancer relationship.[40,45] To investigate the effects of both systematic and random error in epidemiologic studies on the likelihood of finding a diet and breast cancer association, Prentice [43] used data from FFQs and 4-day food records from the original Women's Health Trial to develop a new dietary measurement model. This model links measurement error parameters to BMI and incorporates a random underreporting quantity for self-report instruments. Application of this model to international correlation data on dietary fat and postmenopausal breast cancer showed that adjustment for both systematic and random measurement error in dietary assessment results in a projected RR of about 1.10 for the 90% versus the 10% fat intake percentiles, for either FFQs or 4-day food records. These results highlight the importance of measurement error in epidemiologic studies and clearly suggest that FFQs and diet records may not be adequate to determine a link between dietary fat and postmenopausal breast cancer, even if a strong association exists.[43]

In addition to the uncertainty in epidemiologic results introduced by measurement error, other methodological issues—such as the presence of correlated variables, which can decrease the statistical power of a study—likely contribute to the inconsistent results among studies on dietary fat and breast cancer. Because of the high correlation of total fat with total energy, various methods for energy adjustment have been used in studies investigating the effect of dietary fat on breast cancer risk, including the standard multivariate method,[46,47] the residual analysis method,[48–50] the energy partition method,[51,52] and the nutrient density method.[46,47] Use of different approaches to adjustment for energy intake may contribute to the inconsistency observed among study results and certainly contributes to the difficulty of comparing results across studies. Although the summary RRs of 1.21 from meta-analysis of case-control studies[25] and 1.01[25] and 1.05[26] from meta-analysis of cohort studies on the association between total dietary fat and breast cancer risk suggest only a weak relationship, these results are in agreement with the projected RR of 1.10 from the dietary measurement model proposed by Prentice[43] and do not rule out dietary fat—either alone or combined with other environmental and genetic factors—as a contributor to breast cancer risk. The best hope of clarifying the dietary fat and breast cancer relationship may rest with randomized, controlled clinical trials, such as the Women's Health Initiative[53,54] and the Women's Intervention Nutrition Study,[55,56] both now under way.

4. ANTIOXIDANT TRIALS

Reactive oxygen species such as superoxide, nitric oxide, and hydroxyl radicals—formed continuously as a result of biochemical reactions—and environmental carcinogens such as tobacco smoke, industrial pollution, and food contaminants (e.g., aflatoxins, heterocyclic aromatic amines) can cause significant oxidative damage to DNA, lipids, and proteins in the human body. Such oxidative damage is generally considered to be an important factor in carcinogenesis.[57,58] Antioxidant defenses—for example, enzymes that inhibit carcinogen activation or continually repair DNA damage—help to control the extent of damage, but frequently cannot counteract the oxidative attack completely. Dietary antioxidants, ubiquitous in plant foods where they have evolved to protect the plants against oxidative assault, may be protective to humans also, in terms of reducing cancer risk. Because reactive oxygen species appear to be involved at all stages of cancer development, dietary antioxidants may have potential benefits throughout the carcinogenic process.[59]

Epidemiologic data provide strong evidence of a cancer-protective effect for high intakes of vegetables, fruits, and whole grains. A review summarizing the results from more than 200 case-control and cohort studies found a probable protective effect for cancers of the breast, colon, endometrium, oral cavity and pharynx, pancreas, and bladder, and convincing evidence for inverse associations with cancers of the stomach, esophagus, and lung.[4] The cancer-inhibitory effects reported for these plant foods may be attributed partly to a variety of antioxidant constituents, including micronutrients (e.g., β-carotene [provitamin A], vitamins E and C, selenium) and certain phytochemicals (e.g., polyphenolics, carotenoids).[4,60,61] It is likely, however, that numerous constituents contribute to the overall protective effect.

Based on evidence from epidemiologic, laboratory, and clinical research for compounds that have demonstrated apparent cancer-protective activity, the Chemoprevention Program at the National Cancer Institute (NCI) systematically carries out preclinical and clinical studies on numerous potential chemopreventive agents. Promising chemopreventive agents being investigated include micronutrients, phytochemicals, and synthetics.[62] Continued chemoprevention research efforts have been focused on dietary antioxidants, including carotenoids (e.g., β-carotene, lycopene), selenium, vitamin E, and phytochemicals such as green tea polyphenols. β-Carotene and selenium are highlighted below.

The encouraging epidemiologic evidence, as well as supporting laboratory data, provided a strong scientific rationale for the hypothesis that *β-carotene* can reduce cancer risk. Randomized, controlled clinical chemoprevention trials, the only definitive way to test such hypotheses, were initiated in the 1980s; results of these trials are described below in 4.1. Large-Scale, Randomized Intervention Trials. Early clinical trials on oral cancer showed evidence of benefit from β-carotene. In Filipino Betel nut chewers, who are at high risk for mouth cancer, the percentage of buccal mucosa cells with micronuclei—evidence of genotoxic damage—was significantly lower in people given β-carotene for 9 weeks than in those given canthaxanthin, a carotenoid with no vitamin A activity.[63] One study that examined the effect of β-carotene on leukoplakia, a precancerous oral lesion, found that 17 of 24 patients showed significant reversal of lesions after 6 months of treatment.[64] Trials to prevent colorectal polyps showed no evidence of benefit from β-carotene. In a trial in which individuals, who previously had polyps removed, received either β-carotene, vitamins C and E, all three antioxidants, or placebo for 4 years, no reduction in polyp incidence—and no evidence of harm—was demonstrated for any of the interventions.[65] Similarly, a polyp prevention trial in Australia, in which patients with previous polyps received either β-carotene or placebo together with usual diet, low-fat diet, high-wheat bran diet, or low-fat/high-wheat bran diet, found no significant reduction of polyps.[66] However, after 4 years, persons on the low-fat/high-wheat bran diet developed no large polyps, which have greater malignant potential than small polyps.

The trace mineral *selenium* functions as a cofactor for glutathione peroxidase, an enzyme that may protect against oxidative tissue damage.[4,67] Cancer mortality correlation studies suggest an inverse association between selenium status and cancer incidence. Data from most case-control and cohort studies, however, have not been convincing for those cancer sites investigated, including lung, breast, and stomach cancers.[36,68,69] Data from a recent prospective study that examined the association of serum selenium levels and risk of ovarian cancer indicated that women in the highest tertile, compared with the lowest tertile, were four times less likely to develop ovarian cancer (OR = 0.23).[70] A randomized, controlled clinical intervention trial that tested whether a daily 200 μg selenium supplement (*l*-selenomethionine, as selenium-enriched brewer's yeast) will decrease incidence of basal cell and squamous cell skin cancers found no protective effect against skin cancer. Secondary endpoint analyses, however, showed significant reductions in total cancer mortality (RR = 0.5), total cancer incidence (RR = 0.63), and incidences of lung (RR = 0.54), colorectal (RR = 0.42), and prostate (RR = 0.37) cancers, for individuals who received selenium supplements, compared with controls.[71] These positive findings support the cancer-protective effect of selenium, but must be confirmed in independent clinical intervention trials. Because the effective doses and toxic doses of organoselenium compounds, such as selenomethionine and selenocysteine, are quite close, determining the optimal form of selenium for use in clinical trials is challenging; this is complicated further by the variation in toxicity thresholds among individuals.[72]

4.1. Large-Scale, Randomized Clinical Intervention Trials

This section summarizes several large-scale randomized trials using β-carotene and other antioxidants, as well as certain dietary micronutrients. For some trials, accrual has been closed and the early results published, while long-term followup continues to determine safety as well as efficacy.

4.1.1. Ongoing Trials. The *Harvard Women's Health Study* (WHS) is a chemoprevention trial that was designed to evaluate the risks and benefits of low-dose aspirin, β-carotene, and vitamin E in the primary prevention of cardiovascular disease and cancer in healthy postmenopausal women in the United States.[73,74] Begun in 1992, the WHS has enrolled approximately 40,000 female nurses, ages 45 and older, without a history of either disease. Participants are randomized to treatment or placebo groups for 2 years following a 3-month nonrandomized run-in phase. In response to the lack of benefit for β-carotene seen in closed trials, the WHS removed β-carotene supplementation from its intervention. The study will continue to evaluate aspirin and vitamin E.

4.1.2. Closed Trials. The *Linxian Trials*, conducted by the NCI in collaboration with the Chinese Institute of the Chinese Academy of Medical Sciences, were two randomized, double-blind chemoprevention trials to determine whether daily ingestion of vitamin/mineral supplements would reduce incidence and mortality rates for esophageal cancer in a high-risk population in Linxian, China, where approximately 20% of all deaths result from esophageal cancer. The General Population Trial began in 1986 and randomized more than 30,000 individuals, who received one of four combinations of supplements each day for 5 years, at doses equivalent to one to two times the U.S. Recommended Daily Allowances (RDAs).[75] Combinations included retinal and zinc; riboflavin and niacin; vitamin C and molybdenum; and β-carotene, vitamin E, and selenium. The second study, the Dysplasia Trial, enrolled 3,318 individuals with evidence of severe esophageal dysplasia; subjects were randomized to receive either a placebo or a daily supplement of 14 vitamins and 12 minerals, at two to three times the U.S. RDAs, for 6 years.[76]

Results of the General Population Study indicated a significant benefit for those receiving the β-carotene/vitamin E/selenium combination—a 13% reduction in cancer mortality, largely the result of a 21% drop in stomach cancer mortality.[75] Also, this group experienced a 9% reduction in deaths from all causes, a 10% decrease in deaths from strokes, and a 4% decrease in deaths from esophageal cancer. The effects of the β-carotene/vitamin E/selenium combination began to appear within 1 to 2 years after the intervention began and continued throughout the study; the three other combinations did not affect cancer risk.

A nonsignificant, 16% reduction in mortality from esophageal cancer was reported for the Dysplasia Trial.[76] Analysis of esophageal dysplasia data showed that supplementation had a significant beneficial effect; individuals receiving supplements were 1.2 times as likely to have no dysplasia after 30 and 72 months of intervention, compared with individuals receiving the placebo.[77] Postintervention followup is continuing. The results of these trials are encouraging but may not be directly applicable to Western cultures, which tend to be well nourished and not deficient in multiple micronutrients, in contrast to the Linxian community. Even so, valuable information might be gained by determining cancer incidence for the 5 years since the trials ended, including analysis according to baseline nutrient levels.[78]

The *Physicians' Health Study* (PHS), a general population trial in 22,000 U.S. physicians that evaluated the effect of aspirin and β-carotene supplementation on the primary prevention of cardiovascular disease and cancer, began in 1982. The aspirin component of PHS ended in 1987, because a benefit of aspirin on risk of first heart attack (44% reduction) was found. The treatment period for β-carotene continued until December 1995; data showed no significant evidence of benefit or harm from β-carotene on either cardiovascular disease or cancer.[79]

The *Alpha-Tocopherol, Beta-Carotene Lung Cancer (ATBC) Prevention Study*, conducted in Finland, and the *Beta-Carotene and Retinol Efficacy Trial (CARET)* both were carried out in populations at high risk for lung cancer. The ATBC study investigated the efficacy of vitamin E (α-tocopherol) alone, β-carotene alone, or a combination of the two compounds in preventing lung cancer among more than 29,000 male cigarette smokers ages 50 to 69, with an average treatment/followup of 6 years. Unexpectedly, this study showed an 16% higher incidence of lung cancer in the β-carotene group. However, 34% fewer cases of prostate cancer and 16% fewer cases of colorectal cancer were diagnosed among men who received vitamin E.[80,81] Further, recent analysis of ATBC followup data found a 36% decrease in prostate cancer incidence and a 41% decrease in mortality from prostate cancer among men receiving vitamin E.[82] In the ATBC trial, the adverse effects of β-carotene were observed at the highest two quartiles of ethanol intake, indicating that alcohol consumption may enhance the actions of β-carotene.[81] Also, followup data analysis showed a 32% lower incidence of prostate cancer among nondrinkers, whereas the risk among drinkers decreased 25%, 42%, and 40% by tertiles.[82] It is not clear how alcohol might influence the protective effect of β-carotene. CARET tested the efficacy of a combination of β-carotene and retinol (as retinyl palmitate) in male and female former heavy smokers and in men with extensive occupational asbestos exposure. This trial was terminated in January 1996 after 4 years of treatment, when data showed an overall 28% higher incidence of lung cancer in participants receiving the β-carotene/retinyl palmitate combination.[83] Male current smokers in CARET, excluding those exposed to asbestos, showed a 39% higher incidence,[84] compared with the 16% higher incidence in the ATBC study,[81] suggesting a possible adverse effect for supplemental retinol.

Several possible explanations for the unanticipated outcomes of these β-carotene trials have been considered. The median followup of 6 years may have been too short, either to show any effect on carcinogenesis or to reverse or overcome lung cancer risk factors, particularly in active smokers. Posttrial followup will help to clarify this issue.[83,85] Also, many heavy smokers and asbestos-exposed individuals may have developed the initial stages of lung cancer prior to supplementation. Considering that the effect of β-carotene appeared after only 2 years of supplementation, the observed effect likely was related to the growth of cells that had already undergone malignant transformation.[85,86] Such suppositions, however, do not fully explain the excess risk observed in ATBC and CARET in individuals receiving supplements. Another consideration is that, at the high doses of β-carotene used, direct oxidative attack of β-carotene by extremely reactive constituents of high-intensity cigarette smoke in the lungs of heavy smokers may induce the formation of β-carotene products that have pro-oxidant activity.[85–87] A recent study in ferrets reported that the formation of β-apo-carotenals (β-carotene oxidative metabolites structurally similar to retinoids) was threefold higher in ferrets exposed to smoke.[88] Study data indicated that cell proliferation and squamous metaplasia were observed in all β-carotene-supplemented animals and were enhanced by exposure to tobacco smoke. The authors postulated that formation of these metabolites might interfere with retinoid signaling, leading to decreased retinoic acid synthesis and, consequently, a decrease in expression of the RARβ gene (encodes retinoic acid receptor β) and an increase in the expression of c-fos and c-jun (encode activator protein-1). A decrease in RARβ may lead to squamous metaplasia and an increase in AP-1 may lead to hyperproliferation.[88,89]

A further possibility for the observed results in ATBC and CARET is competitive inhibition by supplemental β-carotene of the antioxidant activity of other dietary carotenoids, such as α-carotene and lutein. α-Carotene, for example, has been reported to show higher potency than β-carotene in suppressing tumorigenesis in animal lung and skin models.[90] Epidemiologic data support these results. Reanalyses of dietary intake data from New Jersey[91] and Finland[92] indicated stronger inverse associations with lung cancer for α-carotene than for β-carotene. A recent study, however, reported that supplementation with α-carotene, β-carotene, lutein, or lycopene in human volunteers had no significant effect, protective or otherwise, on endogenous oxidative damage, as measured by strand breaks, oxidized pyrimidines, and altered purines in the DNA of lymphocytes.[93] It is worthy of mention that participants with higher serum β-carotene concentrations at entry into the ATBC study[81] and CARET[84] developed fewer lung cancers during the course of the trials, even among those who received β-carotene supplements. Baseline serum concentrations of β-carotene reflect total intake of vegetables and fruits, which contain numerous other antioxidants, as well as many naturally occurring potential anticarcinogens that may exert their effects through diverse mechanisms. Thus, β-carotene serum concentrations, as well as concentrations of other carotenoids, may simply be markers for the actual protective agents, reaffirming the importance of including an abundance of plant foods in our diets.[81,87]

5. DIETARY GUIDELINES

Advances in diet and cancer research, however exciting and innovative, will not result in health benefits to the public unless the knowledge gained about diet and cancer is effectively translated into practical applications and usable guidelines that can promote and encourage diet-related behaviors important to good health. Since the mid-1970s, various organizations—in the United States as well as Japan and a number of European countries—have formulated dietary recommendations, based on the available evidence linking diet and cancer. A comprehensive review and summary of these recommendations, which include both numerical goals and more general nonquantified advice, is presented in the recent report *Food, Nutrition, and the Prevention of Cancer: A Global Perspective*, developed by the World Cancer Research Fund (WCRF) in association with the American Institute for Cancer Research (AICR).[3] In the United States, the National Academy of Sciences (NAS) published *Diet, Nutrition, and Cancer*, a report commissioned by the NCI that summarized the scientific evidence on diet and cancer and concluded that cancer risk for most major sites is influenced by dietary patterns.[94] The NAS also formulated interim dietary guidelines for the general public that were both consistent with good nutrition and, based on the available evidence, likely to reduce cancer risk. The NAS report, in part, served as the basis of subsequent dietary

Table 1. American Cancer Society guidelines on diet, nutrition, and cancer prevention[97]

1. Choose most of the foods you eat from plant sources.
   Eat five or more servings of fruits and vegetables each day.
   Eat other foods from plant sources, such as breads, cereals, grain products, rice, pasta, or beans several times each day.
2. Limit your intake of high-fat foods, particularly from animal sources.
   Choose foods low in fat.
   Limit consumption of meats, especially high-fat meats.
3. Be physically active; achieve and maintain a healthy weight.
   Be at least moderately active for 30 minutes or more on most days of the week.
   Stay within your healthy weight range.
4. Limit consumption of alcoholic beverages, if you drink at all.

guidelines for cancer prevention from the American Cancer Society (ACS)[95] and the NCI.[96] The NCI dietary guidelines, which still stand, are: (1) reduce fat intake to ≤30% or less of calories; (2) increase fiber intake to 20–30 g/d with an upper limit of 35 g; (3) include a variety of vegetables and fruits in the daily diet; (4) avoid obesity; (5) consume alcoholic beverages in moderation, if at all; and (6) minimize consumption of salt-cured, salt-pickled, or smoked foods.[96] These guidelines are consistent with those presented in the 1988 *Surgeon General's Report on Nutrition and Health*[1] and the report by the National Research Council (NRC) in 1989, *Diet and Health: Implications for Reducing Chronic Disease Risk*.[2]

On the basis of its recent review of the scientific evidence, the ACS 1996 Advisory Committee on Diet, Nutrition, and Cancer Prevention reaffirmed the previous conclusions of the ACS in 1984 that dietary practices and physical activity are important factors in cancer prevention.[97] The report of the committee presented four broad guidelines to reduce cancer risk among people aged 2 years and older. These guidelines, summarized in Table 1, are notable in that they recommend limiting consumption of meat, especially red meat, which has been linked to cancer at several sites.[3,97] The recent WCRF/AICR report, published in 1997, presents 14 dietary recommendations—either derived from or generally supported by available evidence on food, nutrition, and cancer prevention—that are specified both as goals for policymakers and as advice to individuals (the general public).[3] The recommendations, which are quantified wherever practical, cover foods and drinks, eating patterns, dietary supplementation, physical activity, and obesity. The advice to individuals is summarized in Table 2. Although more indepth research is needed to understand the mechanisms that underlie the evidence on which the current recommendations of various organizations are based, all of the recommendations are consistent with general good health and may reduce cancer risk.

6. BIOENGINEERING AND THE FOOD SUPPLY

New technologies have made it possible for the food industry to modify existing foods and develop new foods, for the purposes of benefitting overall health and reducing risk for chronic diseases. Such foods have various names—engineered foods, functional foods, designer foods, formulated foods, neutraceuticals, and pharmafoods; they are referred to here as engineered foods. Bioengineering can take various routes in the development of new food products. Genetic engineering, begun in the early 1980s, has

Table 2. WCRF/AICR advice to individuals[3]

1. **Food supply and eating**—Choose predominantly plant-based diets rich in a variety of vegetables and fruits, pulses (legumes) and minimally processed starchy staple foods.
2. **Maintaining body weight**—Avoid being underweight or overweight and limit weight gain during adulthood to less than 5 kg (11 pounds).
3. **Maintaining physical activity**—If occupational activity is low or moderate, take an hour's brisk walk or similar exercise daily; also exercise vigorously for at least one hour (total) in a week.
4. **Vegetables and fruits**—Eat 400–800 grams (15–30 ounces) or five or more servings a day of a variety of fruits and vegetables, all year round.
5. **Other plant foods**—Eat 600–800 grams (20–30 ounces) or more than seven servings a day of a variety of cereals (grains), pulses (legumes), roots, tubers, and plantains. Prefer minimally processed foods. Limit consumption of refined sugar.
6. **Alcoholic drinks**—Alcohol consumption is not recommended. If consumed at all, limit alcoholic drinks to less than two drinks a day for men and one for women.
7. **Meat**—If eaten at all, limit intake of red meat to less than 80 grams (3 ounces) daily. It is preferable to choose fish, poultry, or meat from nondomesticated animals.
8. **Total fats and oils**—Limit consumption of fatty foods, particularly those of animal origin. Choose modest amounts of appropriate vegetable oils.
9. **Salt and salting**—Limit consumption of salted foods and use of cooking and table salt. Use herbs and spices to season foods.
10. **Storage**—Do not eat food that, as a result of prolonged storage at ambient temperatures, is susceptible to contamination with mycotoxins.
11. **Preservation**—Use refrigeration and other appropriate methods to preserve perishable food.
12. **Additives and preservatives**—When levels of additives, contaminants, and other residues are properly regulated, their presence in food and drink is not known to be harmful. However, unregulated or improper use can be a health hazard, particularly in economically developing countries.
13. **Preparation**—Do not eat charred food. Avoid burning meat and fish juices. Consume only occasionally: meat and fish grilled (broiled) in direct flame; cured and smoked meats.
14. **Dietary supplements**—For those who follow the recommendations presented here, dietary supplements are probably unnecessary, and possibly unhelpful, for reducing cancer risk.

resulted in development of a number of modified plant products that are now reaching the marketplace.[98–100] Foods also can be developed by other biotechnological approaches, such as using novel ingredients—for example, sugar and fat substitutes—to either formulate new foods or reformulating existing foods.[101]

6.1. Genetic Engineering

Genetically engineered foods have been altered by the addition of genes from plants or other organisms to increase resistance to pests, to retard spoilage, or to improve flavor, nutrient composition, or other desirable qualities. Products of genetic engineering available in the retail marketplace include virus-resistant squash and papayas, insect-resistant cotton plants, herbicide-tolerant corn, nonbruisable potatoes, and canola and soybean oils with modified fatty acid chain lengths (low-saturated oils and high-monounsaturated oils) for improving both human health and oxidative stability.[98–100] Specific examples include two of the most commonly consumed produce items—potatoes and tomatoes. A gene from a bacterium has been inserted into the Russet Burbank potato to increase its starch content, which, in turn, reduces the oil absorbed during frying and results in a french fry or potato chip with a lower fat content.[102] Also, a genetic modification in the Flavr Savr™ tomato, approved by the FDA in 1994, that reduces the expression of polygalacturonase—a cell wall-degrading enzyme—delays the softening of the fruit and allows longer time on the vine for greater flavor development; this may result in a product that has greater consumer appeal, leading to increased consumption.[98] Following the release of data indicating that the tomato-based carotenoid lycopene was associated with reduced risk for prostate cancer,[103] a group of tomato processors initiated the formation of a scientific advisory panel to help guide tomato product development.[104] Such collaboration between the food industry and the scientific community is essential to translate research findings into practical applications in terms of new and modified food products. A recent comprehensive review of the epidemiologic literature regarding intake of tomatoes and tomato-based products, as well as blood lycopene levels, in relation to cancer risk adds support for the anticancer properties of tomatoes.[105] The review found that, of 72 studies identified, 57 reported inverse associations (35 of these statistically significant) between either tomato intake or blood lycopene level for a specific type of cancer.

6.2. Biotechnology

Using both established and recently developed biological, chemical, and physical techniques, food technologists can create novel molecular structures that simulate existing structures found in food. Technological advances in development and large-scale production of new ingredients have provided an extensive range of choices that can enhance nutritional value and improve flavor and texture of food, as well as reduce costs and facilitate product handling.

Currently, there is great interest in sugar and fat substitutes, particularly in low-calorie/calorie-free substitutes for fat. The formulation of low-fat and fat-free foods using fat substitutes—without altering flavor, mouthfeel, or overall product acceptability—has received major attention from the food industry because of the consumer demand for these products and their perceived health benefits. Reduced-fat products account for about one-third of the leading new food products, based on recent sales;[104] it has been estimated that three-quarters of the U.S. population consume low-fat foods and one-half consume no-fat products.[106]

Most fat substitutes are protein-based, carbohydrate-based, or synthetic compounds. Simplesse®, a protein-based fat substitute that contains 4 kcal/g (dry weight), is produced from milk and/or egg protein by a heating and blending process called microparticulation. Precise blending and shearing shapes the protein gel into small spheroidal particles that the tongue perceives to be fluid. This imparts the rich and creamy mouthfeel associated with fat. In addition to frozen desserts, Simplesse® has applications in other dairy products and in oil-based products such as mayonnaise and salad dressings. Although Simplesse® is not sufficiently heat stable to withstand frying, it can be used in foods that may undergo cooking or heat processing, such as baked goods, soups, and sauces.[101]

Carbohydrate-based molecules (e.g., gums, dextrins) have been used to partially or totally replace fats and oils in a variety of foods for more than a decade.[101] Olestra (marketed as Olean®), the most current carbohydrate-based, calorie-free fat substitute, is prepared by esterifying sucrose with long-chain fatty acids isolated from edible fats and oils—forming a sucrose polyester. This nonabsorbable synthetic fat maintains all the properties—appearance, flavor, heat stability, flash point, and shelf life—of a conventional edible fat and can be used in both high- and low-temperature applications.[101] In 1996, the FDA approved olestra for use in snack foods,[107] and these products now appear on store shelves around the country. Although indigestible, olestra has properties similar to those of dietary fats; therefore, the fat-soluble vitamins A, D, E, and K are soluble in this compound. Because olestra is not absorbed into the bloodstream, neither are the solubilized vitamins. This effect, however, has been offset by adding fat-soluble vitamins to olestra-containing foods. Reduced absorption also is observed for carotenoids, including β-carotene, but not for water-soluble micronutrients.[108] Gastrointestinal (GI) symptoms experienced by some people eating large amounts of olestra appear to present no health risks.[107] A postmarketing surveillance program and observational epidemiologic studies are being carried out to monitor nutritional and GI effects of olestra under free-living conditions.[107]

7. FUTURE DIRECTIONS

The existence of an association between diet and nutrition and cancer risk is not in question. Recognizing the existence of such an association, however, is just the first step in reaching a better understanding of the specific roles that diet- and nutrition-related factors play in the etiology and subsequent development of cancer. Achieving this goal will require a well integrated, multidisciplinary approach, with research efforts ranging from laboratory- to population-based studies, as well as application of innovative techniques in molecular biology and genetics. An impressive number of questions require answers—for example, questions about the fundamental mechanisms-of-action of dietary patterns and dietary constituents in the initiation, promotion, progression, and prevention of cancer, particularly in the context of complex gene/nutrient and nutrient/environment interactions. Many gaps in knowledge exist regarding dietary nutrient and nonnutrient microconstituent interactions with and effects on cellular metabolism, genomic stability, intracellular signal transduction, expression of growth-promoting and growth-inhibiting genes, transcription factors, and other modulators of gene expression and carcinogenesis. Further, additional clinical-metabolic information is needed about transport of nutrients to target tissues, metabolism of nutrients, and individual variability in susceptibility and response to nutrient effects. Knowledge of the metabolic characteristics and basic mechanisms-of-action of dietary components in cancer development is necessary to make informed recommendations about the optimal range of intake of specific dietary constituents and to refine and individualize dietary guidance for cancer prevention. Although the scope of existing research questions appears daunting, it nonetheless presents an extraordinarily exciting opportunity for the scientific community to expand the knowledge base for diet and nutrition cancer research and to establish research in this very important area as solid, state-of-the-art, science.

REFERENCES

1. U.S. Department of Health and Human Services. The Surgeon General's Report on Nutrition and Health, NIH Publication No. 88-50210, Washington, D.C. Public Health Service, U.S. Govt. Printing Office; 1988.
2. National Academy of Sciences, National Research Council, Commission on Life Sciences, Food and Nutrition Board. Diet and Health. Implications for Reducing Chronic Disease Risk, Washington D.C. National Academy Press; 1989.
3. World Cancer Research Fund. Food, Nutrition and the Prevention of Cancer: A Global Perspective, Washington, D.C., American Institute for Cancer Research; 1997.
4. Steinmetz K.A. and Potter J.D. Vegetables, fruit, and cancer prevention a review. J Am Diet Assoc 1996;96:1027–1039.

5. U.S. Department of Health and Human Services. Physical Activity and Health. A Report of the Surgeon General, Atlanta, G.A., U.S. Department of Health and Human Services, Centers for Disease Control and Prevention, National Centers for Chronic Disease Prevention and Health Promotion; 1996.
6. Chen J., Giovannucci E., Kelsey K., Rimm E.B., Stampfer M.J., Colditz G.A., Spiegleman D., Willett W.C., and Hunter D.J. A methylenetetrahydrofolate reductase polymorphism and the risk of colorectal cancer. Cancer Res 1996;56:4862–4864.
7. Chen J., Giovannucci E., Hankinson S.E., Ma J., Willett W.C., Spiegelman D., Kelsey K.T., and Hunter D.J. A prospective study of methylenetetrahydrofolate reductase and methionine synthase gene polymorphisms, and risk of colorectal adenoma. Carcinogenesis 1998;19:2129–2132.
8. Spitz M.R. Risk factors and genetic susceptibility. Cancer Treat Res 1995;74:73–87.
9. Ishibe N. and Kelsey K.T. Genetic susceptibility to environmental and occupational cancers. Cancer Causes Control 1997;8:504–513.
10. Hursting S.D., Perkins S.N., Brown C.C., Haines D.C., and Phang J.M. Calorie restriction induces a p53-independent delay of spontaneous carcinogenesis in p53-deficient and wild-type mice. Cancer Res 1997;57:2843–2846.
11. Wang T.T.Y., Hursting S.D., Perkins S.N., and Phang J.M. Effects of dehydroepiandrosterone and calorie restriction on the Bcl-2/Bax-mediated apoptotic pathway in p53-deficient mice. Cancer Lett 1997;116:61–69.
12 Hioki K., Shivapurkar N., Oshima H., Alabaster O., Oshima M., and Taketo M.M. Suppression of intestinal polyp development by low-fat and high-fiber diet in APC^{716} knockout mice. Carcinogenesis 1997;18:1863–1865.
13. Yang K., Edelmann W., Fan K., Lau K., Leung D., Newmark H., Kucherlapati R., and Lipkin M. Dietary modulation of carcinoma development in a mouse model for human familial adenomatous polyposis. Cancer Res 1998;58:5713–5717.
14. Deapen D., Escalante A., Weinrib L., Horwitz D., Bachman B., Roy-Burman P., Walker A., and Mack T.M. A revised estimate of twin concordance in systemic lupus erythematosus. Arthritis Rheum 1992;35(3):311–318.
15. Braun M.M., Caporaso N.E., Page W.F., and Hoover R.N. A cohort study of twins and cancer. Cancer Epidemiol Biomark Prev 1995;4:469–473.
16. Carmelli D. and Page W.F. Twenty-four year mortality in World War II U.S. male veteran twins discordant for cigarette smoking. Int J Epidemiol 1996;25:554–559.
17. Diez O., Brunet J., Sanz J., del Rio E., Alonso M.C., and Baiget M. Differences in phenotypic expression of a new *BRCA1* mutation in identical twins. Lancet 1997;350:713.
18. Buckley J.D., Buckley C.M., Breslow N.E., Draper G.J., Roberson P.K., and Mack T.M. Concordance for childhood cancer in twins. Med Pediatr Oncol 1996;26:223–229.
19. Gronberg H., Damber L., and Damber J.-E. Studies of genetic factors in prostate cancer in a twin population. J Urol 1994;152:1484–1489.
20. Slattery M.L., O'Brien E., and Mori M. Disease heterogeneity does it impact our ability to detect dietary associations with breast cancer? Nutr Cancer 1995;24:213–220.
21. Rose D.P., Boyar A.P., and Wynder E.L. International comparisons of mortality rates for cancer of the breast, ovary, prostate, and colon and per capita food consumption. Cancer 1986;58:2363–2371.
22. Wynder E.L., Cohen L.A., Muscat J.E., Winters B., Dwyer J.T., and Blackburn G. Breast cancer: weighing the evidence for a promoting role of dietary fat. J Natl Cancer Inst 1997;89:766–775.
23. McMichael A.J., Monett A., and Roder D. Cancer incidence among migrant populations in South Australia. The Medical Journal of Australia 1989;150:417–420.
24. Ziegler R.G., Hoover R.N., Hildesheim A., Nomura A.M.Y., Pike M.C., West D., Wu-Williams A.H., Kolonel L.N., Horn-Ross P.L., Rosenthal J.F., and Hyer M.B. Migration patterns and breast cancer risk in Asian-American women. J Natl Cancer Inst 1993;85:1819–1827.
25. Boyd N.F., Martin L.J., Noffel M., Lockwood G.A., and Tritchler D.L. A meta-analysis of studies of dietary fat and breast cancer risk. Br J Cancer 1993;68:627–636.
26. Hunter D.J., Spiegelman D., Adami H.-O., Beeson L., Van den Brandt P.A., Folsom A.R., Fraser G.E., Goldbohm R.A., Graham S., Howe G.R., Kushi L.H., Marshall J.R., McDermott A., Miller A.B., Speizer F.E., Wolk A., Yaun S.-S., and Willett W. Cohort studies of fat intake and the risk of breast cancer—a pooled analysis. N Engl J Med 1996;334:356–361.
27. Simonsen N., van't Veer P., Strain J.J., Martin-Moreno J.M., Huttunen J.K., Navajas J.F.-C., Martin B.C., Thamm M., Kardinaal A.F.M., Kok F.J., and Kohlmeier L. Adipose tissue omega-3 and omega-6 fatty acid content and breast cancer in the EURAMIC study. Am J Epidemiol 1998;147:342–352.
28. Caygill C.P.J., Charlett A., and Hill M.J. Fat, fish, fish oil, and cancer. Br J Cancer 1996;74:159–164.

29. Lipworth L., Martinez M.E., Angell J., Hsieh C.-C., and Trichopoulos D. Olive oil and human cancer: an assessment of the evidence. Prev Med 1997;26:181–190.
30. Wolk A., Bergstrom R., Hunter D., Willett W., Ljung H., Holmberg L., Berkvist L., Bruce A., and Adami H.-O. A prospective study of association of monounsaturated fat and other types of fat with risk of breast cancer. Arch Intern Med 1998;158:41–45.
31. Kohlmeier L., Simonsen N., van't Veer P., Straain J.J., Martin-Moreno J.M., Margolin B., Huttenen J.K., Navajas J.F.-C., Martin B.C., Thamm M., Kardinaal A.F.M., and Kok F.J. Adipose tissue *trans* fatty acids and breast cancer in the European Community Multicenter Study on Antioxidants, Myocardial Infarction, and Breast Cancer. Cancer Epidemiol Biomark Prev 1997;6:705–710.
32. La Vecchia C., Negri E., Franceschi S., Talamini R., Bruzzi P., Palli D., and Decarli A. Body mass index and post-menopausal breast cancer: an age-specific analysis. Br J Cancer 1997;75:441–444.
33. Trentham-Dietz A., Newcomb P.A., Storer B.E., Longnecker M.P., Baron J., Greenberg E.R., and Willett W.C. Body size and risk of breast cancer. Am J Epidemiol 1997;145:1011–1019.
34. Brinton L.A. and Swanson C.A. Height and weight at various ages and risk of breast cancer. Ann Epidemiol 1992;2:597–609.
35. Radimer K., Siskind V., Bain C., and Schofield F. Relation between anthropometric indicators and risk of breast cancer among Australian women. Am J Epidemiol 1993;138(2):77–89.
36. Hunter D.J. and Willett W.C. Nutrition and breast cancer. Cancer Causes Control 1996;7:56–68.
37. Li C.I., Malone K.E., White E., and Daling J.R. Age when maximum height is reached as a risk factor for breast cancer among young U.S. women. Epidemiology 1997;8:559–565.
38. Gammon M.D., John E.M., and Britton J.A. Recreational and occupational physical activities and risk of breast cancer. J Natl Cancer Inst 1998;90:100–117.
39. Thune I., Benn T., Lund E., and Gaard M. Physical activity and the risk of breast cancer. N Engl J Med 1997;336:1269–1275.
40. Byar D.P. and Freedman L.S. Clinical trials in diet and cancer. Prev Med 1989;18:203–219.
41. Beaton G.H. Approaches to analysis of dietary data: relationship between planned analyses and choice of methodology. Am J Clin Nutr 1994;59:253–261.
42. Beaton G.H., Burema J., and Ritenbaugh C. Errors in the interpretation of dietary assessments. Am J Clin Nutr 1997;65:1100–1107.
43. Prentice R.L. Measurement error and results from analytic epidemiology: dietary fat and breast cancer. J Natl Cancer Inst 1996;88:1738–1747.
44. Klesges R.C., Eck L.H., and Ray J.W. Who underreports dietary intake in a dietary recall? Evidence from the second National Health and Nutrition Examination Survey. J Consult Clin Psychol 1995;63:438–444.
45. Kipnis V., Freedman L.S., Brown C.C., Hartman A.M., Schatzkin A., and Wacholder S. Effect of measurement error on energy-adjustment models in nutritional epidemiology. Am J Epidemiol 1997;146:842–855.
46. Knekt P., Albanes D., Seppanen R., Aromaa A., Jarvinen R., Hyvonen L., Teppo L., and Pukkala E. Dietary fat and risk of breast cancer. Am J Clin Nutr 1990;52:903–908.
47. Kushi L.H., Sellers T.A., Potter J.D., Nelson C.L., Munger R.G., Kaye S.A., and Folsom A.R. Dietary fat and postmenopausal breast cancer. J Natl Cancer Inst 1992;84:1092–1099.
48. Jones D.Y., Schatzkin A., Green S.B., Block G., Brinton L.A., Ziegler R.G., Hoover R.N., and Taylor P.R. Dietary fat and breast cancer in the National Health and Nutrition Examination Survey I Epidemiologic Follow-up Study. J Natl Cancer Inst 1987;79:465–471.
49. Willett W.C., Hunter D.J., Stampfer M.J., Colditz G.A., Manson J.E., Spiegelman D., Rosner B.A., Hennekens C.H., and Speizer F.E. Dietary fat and fiber in relation to risk of breast cancer—an 8-year follow-up. JAMA 1992;268:2037–2044.
50. Van den Brandt P.A., van't Veer P., Goldbohm R.A., Dorant E., Volovics A., Hermus R.J.J., and Sturmans F. A prospective cohort study on dietary fat and the risk of postmenopausal breast cancer. Cancer Res 1993;53:75–82.
51. Graham S., Zielezny M., Marshall J., Priore R., Freudenheim J., Brasure J., Haughey B., Nasca P., and Zdeb M. Diet in the epidemiology of postmenopausal breast cancer in the New York state cohort. Am J Epidemiol 1992;136:1327–1337.
52. Howe G.R., Friedenreich C.M., Jain M., and Miller A.B. A cohort study of fat intake and risk of breast cancer. J Natl Cancer Inst 1991;83:336–340.
53. Rossouw J.E., Finnegan L.P., Harlan W.R., Pinn V.W., Clifford C., and McGowan J.A. The evolution of the Women's Health Initiative: perspectives from the NIH. JAMWA 1995;50:50–55.
54. The Women's Health Initiative Study Group. Design of the Women's Health Initiative clinical trial and observational study. Controlled Clin Trials 1998;19:61–109.

55. Chlebowski R.T., Blackburn G.L., Buzzard I.M., Rose D.P., Martino S., Khandekar J.D., York R.M., Jeffrey R.W., Elashoff R.M., and Wynder E.L. Adherence to a dietary fat intake reduction program in postmenopausal women receiving therapy for early breast cancer. J Clin Oncol 1993;11:2072–2080.
56. Buzzard I.M., Faucett C.L., Jeffery R.W., McBane L., McGovern P., Baxter J.S., Shapiro A.C., Blackburn G.L., Chlebowski R.T., Elashoff R.M., and Wynder E.L. Monitoring dietary change in a low-fat diet intervention study: advantages of using 24-hour dietary recalls vs food records. J Am Diet Assoc 1996;96:574–579.
57. Loft S. and Poulsen H.E. Cancer risk and oxidative DNA damage in man. J Mol Med 1996;74:297–312.
58. Jacob R.A. and Burri B.J. Oxidative damage and defense. Am J Clin Nutr 1996;63:985S–990S.
59. Diplock A.T. Antioxidants and disease prevention. Food Chem Toxicol 1996;34:1013–1023.
60. van Poppel G. and van den Berg H. Vitamins and cancer. Cancer Lett 1997;114:195–202.
61. Steinmetz K.A. and Potter J.D. Vegetables, fruit, and cancer. II. Mechanisms. Cancer Causes Control 1991;2:427–442.
62. Kelloff G.J., Boone C.W., Crowell J.A., Nayfield S.G., Hawk E., Malone W.F., Steele V.E., Lubet R.A., and Sigman C.C. Risk biomarkers and current strategies for cancer chemoprevention. J Cell Biochem 1996;25S:1–14.
63. Stich H.F., Stich W., Rosin M.P., and Vallejera M.O. Use of the micronucleus test to monitor the effect of vitamin A., beta-carotene and canthaxanthin on the buccal mucosa of betel nut/tobacco chewers. Int J Cancer 1984;34:745–750.
64. Garewal H.S., Meyskens F.L., Jr., Killen D., Reeves D., Kiersch T.A., Elletson H., Strosberg A., King D., and Steinbronn K. Response of oral leukoplakia to beta-carotene. J Clin Oncol 1990; 8:1715–1720.
65. Greenberg E.R., Baron J.A., Tosteson T.D., Freeman D.H., Jr., Beck G.J., Bond J.H., Colacchio T.A., Coller J.A., Frankl H.D., Haile R.W., and Mandel J.S. A clinical trial of antioxidant vitamins to prevent colorectal adenoma. N Engl J Med 1994;331:141–147.
66. MacClennan R., Macrae F., Bain C., Battistutta D., Chapuis P., Gratten H., Lambert J., Newland R.C., Ngu M., Russell A., Ward M., and Wahlqvist M.L. Randomized trial of intake of fat, fiber, and beta carotene to prevent colorectal adenomas. J Natl Cancer Inst 1995;87:1760–1766.
67. Rock C.L., Jacob R.A., and Bowen P.E. Update on the biological characteristics of the antioxidant micronutrients: vitamin C, vitamin E, and the carotenoids. J Am Diet Assoc 1996;96:693–702.
68. Ziegler R.G., Mayne S.T., and Swanson C.A. Nutrition and lung cancer. Cancer Causes Control 1996;7:157–177.
69. Kono S. and Hirohata T. Nutrition and stomach cancer. Cancer Causes Control 1996;7:41–55.
70. Helzlsouer K.J., Alberg A.J., Norkus E.P., Morris J.S., Hoffman S.C., and Comstock G.W. Prospective study of serum micronutrients and ovarian cancer. J Natl Cancer Inst 1996;88:32–37.
71. Clark L.C., Combs G.F., Jr., Turnbull B.W., Slate E.H., Chalker D.K., Chow J., Davis L.S., Glover R.A., Graham G.F., Gross E.G., Krongrad A., Lesher J.L., Jr., Park H.K., Sanders B.B., Jr., Smith C.L., and Taylor J.R. Effects of selenium supplementation for cancer prevention in patients with carcinoma of the skin. JAMA 1996;276:1957–1963.
72. Patterson B.H. and Levander O.A. Naturally occurring selenium compounds in cancer chemoprevention trials: a workshop summary. Cancer Epidemiol Biomark Prev 1997;6:63–69.
73. Buring J.E. and Hennekens C.H. The Women's Health Study: summary of the study design. J Myocardial Ischemia 1992;4:27–29.
74. Buring J.E. and Hennekens C.H. The Women's Health Study: rationale and background. J Myocardial Ischemia 1992;4:30–40.
75. Blot W.J., Li J.-Y., Taylor P.R., Guo W., Dawsey S.M., Wang G.-Q., Yang C.S., Zheng S.-F., Gail M.H., Li G.-Y., Yu Y., Liu B.-Q., Tangrea J.A., Sun Y.-H., Liu F., Fraumeni J.F., Jr., Zhang Y.-H., and Li B. Nutrition intervention trials in Linxian, China: supplementation with specific vitamin/mineral combinations, cancer incidence, and disease-specific mortality in the general population. J Natl Cancer Inst 1993;85:1483–1492.
76. Li J.-Y., Taylor P.R., Li B., Wang G.-Q., Ershow A.G., Guo W., Liu S.-F., Yang C.S., Shen Q., Wang W., Mark S.D., Zou X.-N., Greenwald P., Wu Y.-P., and Blot W.J. Nutrition intervention trials in Linxian, China: multiple vitamin/mineral supplementation, cancer incidence, and disease-specific mortality among adults with esophageal dysplasia. J Natl Cancer Inst 1993;85:1492–1498.
77. Mark S.D., Liu S.-F., Li J.-Y., Gail M.H., Shen Q., Dawsey S.M., Liu F., Taylor P.R., Li B., and Blot W.J. The effect of vitamin and mineral supplementation on esophageal cytology: results from the Linxian Dysplasia Trial. Int J Cancer 1994;57:162–166.
78. Omenn G.S. Interpretations of the Linxian vitamin supplement chemoprevention trials [editorial]. Epidemiology 1998;9:1–3.

79. Hennekens C.H., Buring J.E., Manson J.E., Stampfer M., Rosner B., Cook N.R., Belanger C., LaMotte F., Gaziano J.M., Ridker P.M., Willett W., and Peto R. Lack of effect of long-term supplementation with beta carotene on the incidence of malignant neoplasms and cardiovascular disease. N Engl J Med 1996;334:1145–1149.
80. Alpha-Tocopherol Beta-Carotene Cancer Prevention Study Group, Heinonen O.P., Huttunen J.K., and Albanes D. The effect of vitamin E and beta carotene on the incidence of lung cancer and other cancers in male smokers. N Engl J Med 1994;330(15):1029–1035.
81. Albanes D., Heinonen O.P., Taylor P.R., Virtamo J., Edwards B.K., Rautalahti M., Hartman A.M., Palmgren J., Freedman L.S., Haapakoski J., Barrett M.J., Pietinen P., Malila N., Tala E., Liippo K., Salomaa E.-R., Tangrea J.A., Teppo L., Askin F.B., Taskinen E., Erozan Y., Greenwald P., and Huttunen J.K. α-tocopherol and β-carotene supplements and lung cancer incidence in the Alpha-Tocopherol, Beta-Carotene Cancer Prevention Study: effects of base-line characteristics and study compliance. J Natl Cancer Inst 1996;88:1560–1570.
82. Heinonen O.P., Albanes D., Virtamo J., Taylor P.R., Huttenen J.K., Hartman A.M., Haapakoski J., Malila N., Rautalahti M., Ripatti S., Mäenpää H., Teerenhovi L., Koss L., Virolainen M., and Edwards B.K. Prostate cancer and supplemantation with α-tocopherol and β-carotene: incidence and mortality in a closed trial. J Natl Cancer Inst 1998;90:440–446.
83. Omenn G.S., Goodman G.E., Thornquist M.D., Balmes J., Culler M.R., Glass A., Keogh J.P., Meyskens F.L., Jr., Valanis B., Williams J.H., Jr., Barnhart S., and Hammar S. Effects of a combination of beta carotene and vitamin A on lung cancer and cardiovascular disease. N Engl J Med 1996;334:1150–1155.
84. Omenn G.S., Goodman G.E., Thornquist M.D., Balmes J., Cullen M.R., Glass A., Keogh J.P., Meyskens F.L., Jr., Valanis B., Williams J.H., Jr., Barnhart S., Cherniack M.G., Brodkin C.A., and Hammar S. Risk factors for lung cancer and for intervention effects in CARET, the Beta-Carotene and Retinol Efficacy Trial. J Natl Cancer Inst 1996;88:1550–1559.
85. Rautalahti M., Albanes D., Virtamo J., Taylor P.R., Huttunen J.K., and Heinonen O.P. Beta-carotene did not work: aftermath of the ATBC study. Cancer Lett 1997;114:235–236.
86. Erdman J.W., Jr., Russell R.M., Mayer J., Rock C.L., Barua A.B., Bowen P.E., Burri B.J., Curran-Celentano J., Furr H., Mayne S.T., and Stacewicz-Sapuntzakis M. Beta-carotene and the carotenoids: beyond the intervention trials. Nutr Rev 1996;54:185–188.
87. Mayne S.T., Handelman G.J., and Beecher G. β-carotene and lung cancer promotion in heavy smokers—a plausible relationship? J Natl Cancer Inst 1996;88:1513–1515.
88. Wang X.-D., Liu C., Bronson R.T., Smith D.E., Krinsky N.I., and Russell R.M. Retinoid signaling and activator protein-1 expression in ferrets given β-carotene supplements and exposed to tobacco smoke. J Natl Cancer Inst 1999;91:60–66.
89. Lotan R. Lung cancer promotion by β-carotene and tobacco smoke: relationship to suppression of retinoic acid receptor-β and increased activator protein-1? J Natl Cancer Inst 1999;91:7–9.
90. Nishino H. Cancer chemoprevention by natural carotenoids and their related compounds. J Cell Biochem Suppl Supplement 1995;22:231–235.
91. Ziegler R.G., Colavito E.A., Hartge P., Mc Adams M.J., Schoenberg J.B., Mason T.J., and Fraumeni J.F., Jr. Importance of α-carotene, β-carotene, and other phytochemicals in the etiology of lung cancer. J Natl Cancer Inst 1996;88:612–615.
92. Knekt P., Jarvinen R., Teppo L., Aromaa A., and Seppanen R. Role of various carotenoids in lung cancer prevention. J Natl Cancer Inst 1999;91:182–184.
93. Collins A.R., Olmedilla B., Southon S., Granado F., and Duthie S.J. Serum carotenoids and oxidative DNA damage in human lymphocytes. Carcinogenesis 1998;19:2159–2162.
94. National Academy of Sciences, National Research Council, Committee on Diet, Nutrition, and Cancer. Diet, Nutrition, and Cancer, Washington, D.C., National Academy Press; 1982.
95. American Cancer Society. Nutrition and cancer: cause and prevention: American Cancer Society special report. CA Cancer J Clin 1984;34:121–126.
96. Butrum R.R., Clifford C.K., and Lanza E. NCI dietary guidelines: rationale. Am J Clin Nutr 1988;48:888–895.
97. American Cancer Society 1996 Advisory Committee. Guidelines on diet, nutrition, and cancer prevention: reducing the risk of cancer with healthy food choices and physical activity. CA Cancer J Clin 1996;46:325–341.
98. Wilkinson J.Q. Biotech plants: from lab bench to supermarket shelf. Food Tech 1997;51(12):37–42.
99. Katz F. Biotechnology—new tools in food technology's toolbox. Food Tech 1996;50(11):63–65.
100. Liu K. and Brown E.A. Enhancing vegetable oil quality through plant breeding and genetic engineering. Food Tech 1996;50(11):67–71.
101. Akoh C.C. Fat replacers. Food Tech 1998;52(3):47–53.

102. Brewer M.S. and Kendall P. Position of the American Dietetic Association: biotechnology and the future of food. J Am Diet Assoc 1995;95:1429–1432.
103. Giovannucci E., Ascherio A., Rimm E.B., Stampfer M.J., Colditz G.A., and Willett W.C. Intake of carotenoids and retinol in relation to risk of prostate cancer. J Natl Cancer Inst 1995;87:1767–1776.
104. Hollingsworth P. Mainstreaming healthy foods. Food Tech 1997;51:55–58.
105. Giovannucci E. Tomatoes, tomato-based products, lycopene, and cancer: review of the epidemiologic literature. J Natl Cancer Inst 1999;91:317–331.
106. Katz F. Fat-free & reduced-fat reach maturity. Food Tech 1998;52:54–56.
107. Peters J.C., Lawson K.D., Middleton S.J., and Triebwasser K.C. Assessment of the nutritional effects of olestra, a nonabsorbed fat replacement: introduction and overview. J Nutr 1997;127:1539S–1546S.
108. Peters J.C., Lawson K.D., Middleton S.J., and Triebwasser K.C. Assessment of the nutritional effects of olestra, a nonabsorbed fat replacement: summary. J Nutr 1997;127:1719S–1728S.

2

EPIC-ITALY

A Molecular Epidemiology Project on Diet and Cancer

Domenico Palli,[1] Vittorio Krogh,[2] Antonio Russo,[1] Franco Berrino,[2]
Salvatore Panico,[3] Rosario Tumino,[4] and Paolo Vineis[5]
on behalf of EPIC-Italy*

[1] Sez. Epidemiologia Analitica
U.O. Epidemiologia
CSPO—Azienda Ospedaliera Careggi
Firenze
[2] Divisione di Epidemiologia
Istituto Nazionale per lo Studio e la Cura dei Tumori
Milano
[3] Dipartimento di Medicina Clinica e Sperimentale
Università Federico II, Napoli
[4] Registro dei Tumori della Provincia di Ragusa
Azienda Ospedaliera Civile M.P. Arezzo
Ragusa
[5] Servizio di Epidemiologia dei Tumori
Dipartimento di Scienze Biomediche e Oncologia Umana
Università di Torino, Torino

1. BACKGROUND

Examination of epidemiological and experimental studies concerning dietary habits/constituents and cancer risk reveals a substantial body of evidence supporting the hypothesis that diet plays an important role in the occurrence of some types of cancer.[1,2,3,4] Research on nutrition and cancer has developed substantially over the past 20-years, initially stimulated by a number of ecological studies which drew attention to the large world-wide variations in cancer incidence, suggesting that these variations could be related to differences in diet and lifestyle among populations.[5] The most

* For the full list of EPIC-Italy collaborators see the end of this chapter

Advances in Nutrition and Cancer 2, edited by Zappia *et al.*
Kluwer Academic / Plenum Publishers, New York, 1999.

consistent result so far is that a diet rich in vegetables and fruit is associated with lower cancer risk.[6,7,8,9] Various hypotheses have been put forward to explain why consumption of vegetables and fruit is associated with a reduced risk of cancer. Despite the current difficulty of identifying active anti-carcinogens, the accumulation of epidemiological studies lends strength to the observed association between vegetables and fruit intake and lower cancer risk.

However, during the 1980s new horizons for cancer research were opened by progress made in the field of the molecular biology of cancer and by the development of research on biochemical markers for a variety of environmental exposures and host factors.

The idea of setting up very large multi-centre prospective studies on nutrition and cancer in Europe was conceived during the 1980s. The design of such a study evolved in light of results of a series of pilot and methodological studies which were conducted to test the validity and feasibility of collecting data through different types of questionnaires, of taking anthropometric measurements and drawing, aliquoting and storing blood samples. A large project known as EPIC (European Prospective Investigation into Cancer and Nutrition), a prospective study on diet and cancer based on healthy, middle-aged subjects who agreed, following an active invitation, to participate in the study and to have their health status followed up for the rest of their lives, was designed in 1989–90 and enrollment started in 1992–93. The study is based on 22 collaborating centres in nine European countries and, overall, recruited approximately 460,000 subjects.[10]

2. EPIC-ITALY

2.1. Study Design

EPIC-Italy is the Italian section of EPIC (European Prospective Investigation into Cancer and Nutrition) and involved four areas covered by long-established cancer registries: the provinces of Florence, Ragusa, Varese, and the city of Turin. An associate centre has joined the project in Naples (Fig. 1).

The recruitment of EPIC-Italy volunteers, according to a common protocol, has been in operation in the period 1993–98; 48,025 volunteers of both sexes have been enrolled in the age interval 35–64 years in 5 participating centres, across different areas of the country: Varese (12,095 volunteers) and Turin (10,867) in the Northern part of the country; Florence (13,597) and Ragusa (6397) in Central and Southern Italy, respectively. The associate centre in Naples has enrolled 5069 women in a study known as ATENA mainly devoted to cardiovascular disease but with many features in common with EPIC.

2.2. Methodological Considerations

Prospective cohort studies offer major methodological advantages for investigating the relation between diet, biochemical markers of nutritional status or metabolic status, and cancer risk. The advantage of the prospective or cohort study design is the possibility of studying dietary and environmental exposure markers in samples collected a long time before cancer onset; these measurements will not be influenced by the disease process itself (Fig. 2). Data on dietary and life-style habits have also been

Figure 1. EPIC study areas in Italy.

collected well before the occurrence of disease: comparability of reported information between cases and controls is therefore much higher than in population based or hospital-based case-control studies. Sampling of a conveniently matched group of controls in a prospective study offers the additional advantage of avoiding the selection bias which can typically affect the interpretation of results of hospital-based case-control studies. Therefore even traditional (questionnaire-based) epidemiological analyses on diet and cancer will provide more reliable results in the frame of a prospective study design: a cohort study will measure cancer incidence at all sites in relation to consumption of foods and intake of several nutrients.

A major strength of the EPIC study is that dietary habits differ substantially among the centres involved in the study, enhancing the capacity of the study to assess the effects of these dietary differences on cancer risk. The recruitment of subjects in Ragusa and Naples (populations with typical Mediterranean dietary habits), in Florence (where a high consumption of fresh and cured meats and wine is combined with olive oil as the main cooking and dressing fat) and in Varese and Turin (the latter two

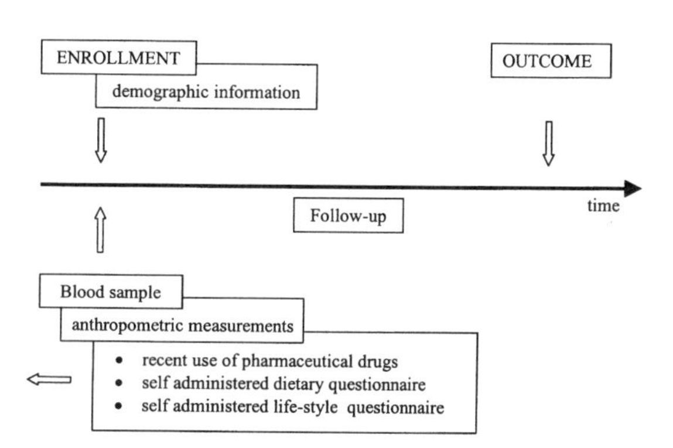

Figure 2. The study design of EPIC is a prospective cohort: information is collected at enrollment for all volunteers who are then followed until outcomes are identified.

Table 1. Age-standardized (*) Cancer Incidence Rates (×100,000) for all sites and a selected group of major sites provided by the Cancer Registries covering EPIC-Italy study areas

	Varese (1988–1992)	Turin (1988–1991)	Florence (1988–1991)	Ragusa (1988–1992)
All sites				
Males	492.9	420.5	439.1	246.6
Females	365.3	336.3	361.2	229.3
Stomach				
Males	38.7	24.9	50.8	19.3
Females	16.3	25.3	31.8	12.8
Colon & rectum				
Males	62.0	53.6	58.8	28.6
Females	48.9	45.6	53.6	28.8
Lung				
Males	108.4	89.9	86.8	53.7
Females	14.5	18.6	18.4	8.4
Breast				
Females	109.4	98.8	97.2	64.2

*Reference population: Italy 1981.

populations share more continental habits, high consumption of butter and dairy products, but also include a large proportion of migrants from Southern Italy) will allow the investigation of a wide range of dietary patterns, representative of the major areas of the country.

At the same time, a high variability (Table 1) in cancer incidence and mortality is present within Italy, ranging from nearly 2-fold differences for cancers of the colon-rectum, breast, and lung to a 3-fold ratio for upper aereo-digestive tract cancers in males. Cancer risk is generally higher in Northern and Central Italy, with, however, a few noteworthy exceptions (e.g., high mortality rates for cancer of the liver, uterus, and bladder in Southern regions.[11]

2.3. Individual Information

For each volunteer the following information and measurements have been collected: 1) demographic information; 2) short questionnaire on recent use of pharmaceutical drugs, current smoking, and occupation; 3) anthropometric measurements (weight, height, sitting height, waist, and hip circumference) and blood pressure according to a standardised protocol with periodical quality control procedures (including between- and within- observer variability); 4) self administered dietary questionnaire, with pictures of different portion sizes of a series of selected foods; 5) self administered life-style questionnaire (smoking, reproductive, and medical history).

The food frequency questionnaire had been specifically developed for the Italian dietary pattern and tested in a pilot phase.[12] Dietary information on the frequency of consumption of more than 120 foods and beverages obtained by this questionnaire, has been checked, coded, computerised by optical reading and then transformed into estimates of intake of 40 nutrients according to specifically developed Italian Food Tables.[13]

A standardised life style questionnaire (specifically printed in two versions for men and women) has also been filled by each participant; detailed information has been collected on reproductive history, physical activity, smoking history, alcohol drinking, medical history, occupation, education level, and other socio-economic variables. The questionnaire is the Italian translation of a common English version developed at the European level.

A computerised data-base with the dietary and life-style information has been implemented and completed.

A detailed 24-hour dietary recall interview about foods and beverages consumed during a specific 24-hour period (usually the day prior to the blood sample) has been carried out in an 8% subgroup of the whole cohort (3932 subjects), with a PC-assisted innovative technique. Between 3000 and 4000 subjects have been interviewed following a standardised protocol in each country in order to "calibrate" the dietary information provided by each individual participant through the country-specific self administered food frequency questionnaire.[14]

2.4. Biological Samples

The collection and storage at low temperature of blood and other biological samples constitutes a major characteristic of the new generation of prospective studies on cancer and diet-related hormonal and nutritional factors.

Following a common protocol and using identical equipment, 30 ml of blood were collected for all 48,000 participants. The blood samples have been processed and centrifuged in a dedicated laboratory in each centre, in the same day of collection, divided in 28 aliquots of 0.5 ml each (12-plasma, 8-serum, 4-concentrated red blood cells, and 4-buffy coat) using an automatic aliquoting and sealing machine specifically developed by BICEF (Cryo-Bio Straw). The aliquots, after a slow freezing process at –80°C, have been divided into two series and stored in liquid nitrogen tanks at –196°C. One series is in a local biological bank specifically developed in each centre in Italy (Florence, Milan, Turin, and Ragusa); the other one has been shipped on dry ice to the central EPIC repository at IARC in Lyon. A specific computer programme has been developed to automatically assign the position in the liquid nitrogen container to each subject's aliquots. This programme will also allow the management of the bank for the retrieval of specimens in this project and in the near future.

2.5. Follow-up

In addition to the variations in risk and dietary habits, a crucial element of the statistical power, which was central in the design of EPIC, is the study size. In a prospective study, this is best expressed by the number of expected cases of the disease of interest. Estimations based on the incidence of cancer at specific sites led to the conclusion that for a total European cohort, it would provide sufficient power to detect statistically significant relative risk as low as 1.2 for all major sites. The total expected number of incident cases during the period 1993–2006, based on the age and sex distribution of the European cohort is 23,460. Standardised procedures have been developed in order to identify newly diagnosed cases of cancer at all sites and other outcome of interest among participants (Fig. 3).

2.6. EPIC Italy Cohort

In Italy, enrollment has been completed in March 1998: 48,025 volunteers have been included in five participating centres (Table 2). Table 3 shows the distribution by age and sex for each centre: Varese (12,095 volunteers) and Turin (10,867) in

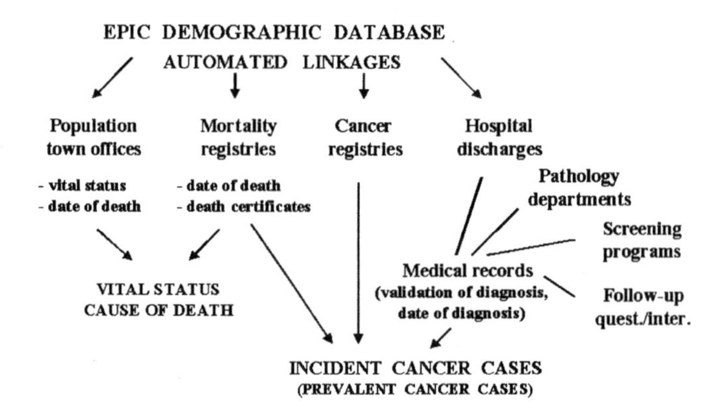

Figure 3. Flow-chart of follow-up procedures developed in the frame of EPIC-Italy.

the Northern part of the country; Florence (13,597) and Ragusa (6397) in Central and Southern Italy, respectively. The associate centre in Naples has enrolled 5069 women.

In each EPIC-Italy centre follow up procedures have been tested and are periodically implemented for the identification of incident cases and the retrieval of medical records (Fig. 3): overall more than 1000 newly diagnosed cancer cases are expected in the period between enrollment and December 1999 and will be soon available for analyses at the national level (286 breast, 135 colon, 130 lung) (Table 4).

3. CONCLUSIONS

EPIC represents the largest multicentre molecular epidemiology study ever planned and carried out in Italy. The successful enrollment phase of EPIC-Italy creates a unique opportunity for co-operation between epidemiologists and laboratory scientists in order to test emerging hypotheses in samples obtained from a large population of well-characterised subjects followed over a long period of time.

Cancer incidence/mortality will be related to reported dietary habits and biochemical markers of food consumption and individual susceptibility, in order to test the role of diet-related exposures in the aetiology of cancer and its interaction with other environmental or genetic determinants.

The availability of such a large biological bank linked to individual data on dietary

Table 2. Age and sex distribution of the 48,025 volunteers enrolled in the EPIC-Italy cohort (1993–1998)

	Males		Females		Total	
Age	N	(%)	N	(%)	N	(%)
≤40	997	6.5	3,087	9.4	4,084	8.5
40–44	3,042	19.7	5,206	16.0	8,248	17.2
45–49	3,507	22.8	6,093	18.7	9,600	20.0
50–54	2,890	18.8	7,074	21.7	9,964	20.7
55–59	2,873	18.7	6,216	19.0	9,089	19.0
60+ years	2,080	13.5	4,960	15.2	7,040	14.6
Total	15,389	32.0	32,636	68.0	48,025	100

Table 3. Age and sex distribution of the 48,025 EPIC volunteers enrolled in each center (Florence, Varese, Turin, Ragusa and Naples)

	Florence		Varese		Turin		Ragusa		Naples
Age	M	F	M	F	M	F	M	F	F
≤39	194	856	28	678	696	369	79	826	358
40–44	648	1,203	328	1,654	1,119	709	947	760	880
45–49	833	1,218	519	2,023	1,317	944	838	706	1,202
50–54	675	2,600	497	1,829	1,224	990	494	494	1,161
55–59	677	2,469	608	1,658	1,169	938	419	329	822
60+	487	1,737	576	1,697	743	649	274	231	646
Total	3,514	10,083	2,556	9,539	6,268	4,599	3,051	3,346	5,069

Table 4. EPIC-Italy: absolute numbers of expected cancer cases among participants in the follow-up period between enrollment and 31/12/99 by site and sex

Cancer sites:	Colon Rectum	Stomach	Lung	Breast	Total (all sites)
Sex					
Females	88	32	35	286	765
Males	47	26	95	—	390
Total	135	58	130	286	1155

and life-style exposures will also provide the unique opportunity to evaluate the prevalence of selected genotypes involved in the metabolism of chemical compounds potentially related to the risk of cancer, in samples of the Italian population resident in geographical areas characterised by specific cancer risk and different dietary patterns.

ACKNOWLEDGMENTS

The authors would like to thank all participants for their cooperation and Chiara Zappitello for editorial assistance. EPIC Italy was financially supported by a generous grant from the Associazione Italiana per la Ricerca sul Cancro (AIRC, Milan) and the European Union (97 SOC 200302 05F02). The international study coordinator of EPIC is Dr. Elio Riboli (IARC, Lyon). EPIC-Italy local investigators and collaborators are the following:

Florence	Epidemiology Unit, CSPO, A.O. Careggi	*Palli D.*, Masala G., Russo A., Saieva C., Cordopatri G., Corsi A.M., Ermini I., Martinez M., Rigacci M.
Milan	National Cancer Institute	*Berrino F.*, Krogh V., Sieri S., Pala V., Bellegotti M., Evangelista A., Villa S.
Turin	Cancer Epidemiology Unit, Turin University	*Vineis P.*, Davico L., Sacerdote C., Fiorini L., Magnino A., Faggiano F., Leo N.

Ragusa	Ragusa Cancer Registry	*Tumino R.*, Gafà L., Ruschena A.M., Lauria C.
Naples	Federico II University and National Cancer Institute Fondazione Pascale	*Panico S.*, Celentano E., Galasso R., Dello Iacovo R.

REFERENCES

1. Doll, R. and Peto, R. The causes of cancer: quantitative estimates of avoidable risk of cancer in US today, JNCI 66:1191–1308, 1981.
2. Henderson, B.E., Ross, R.K., and Pike, M.C. Toward the primary prevention of cancer, Science. 254:1131–1138, 1991.
3. Willett, W.C., Stampfer, M.J., Colditz, A.G., Rosner, B.A., and Speizer, F.E. Relation of meat, fat and fiber intake to the risk of colon cancer in a prospective study among women. New Engl. J. Med. 323:1664–1672, 1990.
4. Willett, W.C. Diet and Nutrition. In: Cancer Epidemiology and Prevention, 2nd edition (eds Schottenfeld D, Fraumeni JF), pp. 438–461, WB Saunders, 1996.
5. Armstrong, B. and Doll, R. Environmental factors and cancer incidence and mortality in different countries, with special reference to dietary practice, Int. J. Cancer. 15:617–631, 1975.
6. Steinmetz, K.A. and Potter, J.D. Vegetables fruit and cancer. I. Epidemiology. Cancer Causes Control 2:325–357, 1991a.
7. Steinmetz, K.A. and Potter, J.D. Vegetables fruit and cancer. I. Mechanisms, Cancer Causes Control 2:427–442, 1991b.
8. Block, G., Patterson, B., and Subar, A. Fruit, vegetables, and cancer prevention: a review of the epidemiological evidence, Nut. Cancer 18:1–20, 1992.
9. Miller, A.B., Berrino, F., Hill, M., Pietinen, P., Riboli, E., and Wahrendorf, J. Diet in the aetiology of cancer: a review, Eur. J. Cancer. 30A:207–220, 1994.
10. Riboli, E. and Kaaks, R. The EPIC Project: rationale and study design, Int. J. Epidemiol. 26 (Suppl. 1):S6–S14, 1997.
11. Parkin, D.M., Whelan, S.L., Ferlay, J., Raymond, L., and Young, J. (eds) Cancer Incidence In Five Continents, IARC Scientific Publications, VII (143), Lyon (1997).
12. Pisani, P., Faggiano, F., Krogh, V., Palli, D., Vineis, P., and Berrino, F. Relative validity and reproducibility of a food-frequency dietary questionnaire for use in the Italian EPIC centres, Int. J. Epidemiol. 26 (Suppl. 1):S152–S160, 1997.
13. Salvini, S., Parpinel, M., Gnagnarella, P., Maisonneuve, P., and Turrini, A. Banca Dati di Composizione degli Alimenti per Studi Epidemiologici in Italia, IEO, Milan 1998.
14. Kaaks, R. and Riboli, E. Validation and calibration of dietary intake measurements in the EPIC project: methodological considerations, Int. J. Epidemiol. 26 (Suppl. 1):S15–S25, 1997.

3

NUTRITIONAL FACTORS IN HUMAN CANCERS

Edward Giovannucci

Channing Laboratory, Department of Medicine
Brigham and Women's Hospital and Harvard Medical School
Departments of Nutrition and Epidemiology
Harvard School of Public Health
Boston, Massachusetts 02115

ABSTRACT

A variety of external factors interacting with genetic susceptibility influence the carcinogenesis process. External factors including oxidative compounds, electrophilic agents, and chronic infections may enhance genetic damage. In addition, various hormonal factors which influence growth and differentiation are critically important in the carcinogenic process. Diet and nutrition can influence these processes directly in the gastrointestinal tract by providing bioactive compounds to specific tissues via the circulatory system, or by modulating hormone levels. Differences in certain dietary patterns among populations explain a substantial proportion of cancers of the colon, prostate and breast. These malignancies are largely influenced by a combination of factors related to diet and nutrition. Their causes are multifactorial and complex, but a major influence is the widespread availability of energy-dense, highly processed and refined foods that are also deplete in fiber. These dietary patterns in combination with physical inactivity contribute to obesity and metabolic consequences such as increased levels of IGF-1, insulin, estrogen, and possibly testosterone. These hormones tend to promote cellular growth. For prostate cancer, epidemiologic studies consistently show a positive association with high consumption of milk, dairy products, and meats. These dietary factors tend to decrease $1,25(OH)_2$ vitamin D, a cell differentiator, and low levels of this hormone may enhance prostate carcinogenesis. While the nutritional modulation of growth-enhancing and differentiating hormones is likely to contribute to the high prevalence of breast, colorectal, prostate, and several other cancers in the Western world, these cancers are relatively rare in less economically developed countries, where malignancies of the upper gastrointestinal tract are quite common. The major causes

Advances in Nutrition and Cancer 2, edited by Zappia *et al.*
Kluwer Academic / Plenum Publishers, New York, 1999.

of upper gastrointestinal tract cancers are likely related to various food practices or preservation methods other than refrigeration, which increase mucosal exposure to irritants or carcinogens.

INTRODUCTION

The wide variation in incidence and mortality rates for most cancer sites across populations worldwide[1] indicates that cancer is a disease influenced primarily by external factors. While genetic differences among populations may contribute to this variation, there are many examples for specific cancer sites for which age-adjusted incidence rates within a geographically restricted population changed substantially over a relatively short period. Moreover, migrants generally acquire the cancer rates of the new country following their assimilation to new customs, lifestyle practices, and dietary patterns.[2–7] Of course, on an individual basis, genetic susceptibility is also important. Cancer victims may have an unfortunate combination of inherited factors in addition to environmental factors.

Many external factors have been documented as contributing to risk for specific cancers. Some examples are tobacco use, alcohol consumption, and infectious agents. Diet and nutrition are also likely to be important. Dietary practices, nutritional status, and physical activity vary substantially throughout different populations of the world as a result of economics, culture, geography, and history. Differences in dietary patterns are likely to contribute to the vast variation in cancer rates.

While most cancer investigators acknowledge an important role of nutrition, researchers have found it challenging to separate the roles of specific dietary factors. Even assuming we can measure long term intakes of single nutrients reasonably well, diet represents a set of highly intercorrelated and interacting factors. Diet is not easily characterized into a single quantifiable variable, such as the average number of cigarettes smoked per day.

While appreciating the difficulties of implicating specific factors, such as a single nutrient in causing or preventing cancers, we should not ignore the wealth of available epidemiologic, clinical, and laboratory evidence that supports the role of general dietary patterns. Thus, this report will examine general dietary patterns, specifically, how eight aspects of diet and nutrition may contribute to different patterns of cancer worldwide. These include energy balance, the contribution to total energy by different proportion of macronutrients, the sources of macronutrients, food processing methods, nutrients affecting DNA processes, the influence on activating and detoxifying metabolic enzymes, oxidation, and direct "toxic" effects of diet through carcinogens or by modulation inflammatory processes. The focus will be on how these factors contribute to the burden of specific cancers in populations across the world.

ENERGY BALANCE

The role of energy balance is critical for carcinogenesis. Animal studies demonstrate that a moderate restriction of energy intake reduces the incidence of various experimentally-induced tumors.[8,9] Direct evidence is difficult in human studies because use of direct measures of energy balance is infeasible in large epidemiologic studies. However, various anthropometric measures provide relevant information regarding chronic energy balance in humans. Height reflects a combination of genetic and exogenous influences, of which energy balance during the period of growth is clearly important. To some extent, short stature reflects insufficient energy intake during the growth period. Body mass index and other anthropometric variables, such as waist circumference and weight gain over a specified period, indicate energy balance during adulthood. In addition, physical activity influences energy balance, and at least vigorous activity level can be assessed via questionnaires or diaries,[10] or inferred through occupational history.

Energy imbalances are likely to contribute substantially to cancer causation, particularly in economically developed countries. Some of the most common human cancers in North America and Northern Europe, including those of the large bowel, breast and prostate gland, are closely correlated with indicators of energy balance. Taller height increases risk of prostate, breast, and colon cancers, indicating that energy restriction severe enough to restrict maximal height attainment reduces occurrence of these cancers.[11–13] While interesting from an etiologic perspective, the public health relevance of limiting energy intake during the growth period is unclear. At least for colon and postmenopausal breast cancer,[14] adult obesity clearly increases risk. The relation between body mass and prostate cancer risk is presently unclear.[15] Moreover, adult physical inactivity is a consistent risk factor for colon cancer,[16] and possibly for cancers of the breast[17] and prostate as well.[18] Thus, minimizing excess energy balance by limiting intake and increasing physical activity in middle-aged individuals is likely to confer benefits for some of the major cancers in economically developed areas.

Excess energy intake and deficient physical activity increase the levels of various growth-promoting hormones, including insulin, insulin-like growth factors (IGFs), and steroid hormones. Chronic exposure to elevated levels of these hormones, singularly or in combination, may enhance carcinogenesis, possibly by enhancing mitogenicity. This concept is supported by laboratory studies.[8,9,15] More recently, correlations between high levels of IGF-1 and prostate cancer[19] and premenopausal breast cancer,[20] estrogens and breast cancer[21] and testosterone and prostate cancer[22] have been found.

PROPORTIONS OF MACRONUTRIENTS IN CONTRIBUTING TO TOTAL ENERGY INTAKE

The three major sources of energy, excluding alcohol, are fat, protein, and carbohydrates. As described above, energy balance is a clear determinant of risk of various malignancies. To the extent that overconsumption of energy increases the incidence of a particularly malignancy, it could be argued that either of the sources, such as fat, also increase risk. However, attributing an effect to a specific macronutrient first requires establishing that this influence is independent of energy balance. For example, if fat intake is a risk factor for a cancer solely because of its energy content, then the replacement an equivalent amount of energy from fat by that from carbohydrate should not have an impact on the cancer risk.

Because carbohydrates, fats, or protein do not represent homogenous entities, recommendations for substituting one energy component for the other, such as carbohydrate for fat, are simplistic, and could have unintended deleterious effects. For example, replacing the equivalent energy level of monounsaturated fat with highly refined carbohydrates is likely to have a deleterious impact on cardiovascular risk factor profile (hyperglycemia, hyperinsulinemia, hypertriglyceridemia),[23] and possibly on risk of some cancers.[24] Use of the terms "fat" and "carbohydrates" may be inadequate; replacing saturated fat with complex carbohydrates in fiber-rich foods may be quite different from replacing monounsaturated fat with simple carbohydrates.

The amount of dietary protein may play a role in carcinogenesis, though this has not been demonstrated definitively in humans. Circulating IGF-1 levels are reduced by lowering protein intake in controlled studies in humans.[25–27] Whether this relationship is important in carcinogenesis within normal dietary ranges in free-living populations remains to be established.

SOURCES OF MACRONUTRIENTS

The major energy-providing foods among populations can be quite varied. In most populations, the major energy sources are cereals or grains, starchy foods such as roots and tubers, meats, milk, and dairy products, added sugar and fats, alcohol, fruits, and vegetables, and other foods such as pulses, nuts, and seeds. In most economically developing areas, the major proportion of energy intake is derived from cereals and starchy foods and other sources (pulses, nuts, seeds); meat and milk and dairy products may be trivial sources of energy in these populations. In the West, in contrast, the majority of energy is provided by meat, milk, and dairy products, and added sugar and fats.

The fact that energy in Western diets is provided primarily by animal products and highly processed foods has some consequences. First, as discussed above, the combination of energy-dense, palatable diets in combination with general lack of energy expenditure may contribute to energy imbalances and high prevalences of obesity. In addition, such diets tend to be low in potentially beneficial compounds, such as fiber and micronutrients which are considered in the following sections. Finally, considered in this section, are potentially deleterious patterns directly related to the high animal fat and animal protein content of these diets.

Diets that are based on meats, dairy foods and added oils and fats will tend to be high in fat and protein as percentages of energy, and the lipid composition will consist primarily of saturated and monounsaturated fats, rather than polyunsaturated fats. The type of fat may impact carcinogenesis. In many animal models, linoleic acid (the major polyunsaturated fat) promotes tumor growth, whereas long-chain omega-3 (n-3) fatty acids may be inhibitory.[28] However, the relevance of this finding for humans is unclear. In fact, little is known about the impact of fatty acids in human carcinogenesis. Because fatty acids may have very specific effects that theoretically have a role in cancer, for example on prostaglandin synthesis, further study is warranted.

While much research emphasis has been on the influence of nutrients, some relatively consistent risk patterns have emerged from analyses based on food groups. For example, a quite consistent finding from ecologic studies, case-control and cohorts studies is a positive association between milk, dairy products, and meats and eggs to a lesser extent, with risk of prostate cancer.[29] For colorectal cancer, in contrast, positive associations have been more consistently related to higher consumption of red meat.[1,30–45] Some recent cohort studies indicate the association may be specific for red meat, and may not exist for other sources of total fat or animal fat,[46–49] suggesting non-lipid components of red meat may account for the relationship between red meat intake and colon cancer. The specific factors have not been conclusively identified to date. Although the mechanisms underlying these associations for prostate and colorectal cancer are not established, the strong empirical evidence of a deleterious effect cannot be ignored when dietary recommendations are made.

Carbohydrates represent the major source of energy in most diets, though generally to a lesser extent in economically developed areas. The amount, type, and source of carbohydrate, plus modifying factors, are important in health and disease. Regardless of source, carbohydrates are ultimately broken down to several simple sugars, mainly glucose, fructose, and galactose. It is unknown whether the proportion of simple sugars that form the basis of carbohydrates in the diet are important. Probably, a more useful approach is to consider the influence of carbohydrate source on physiologic and metabolic parameters, such as the degree of the postprandial rise in glucose, insulin, and triglycerides, which potentially could impact on cancer risk. How carbohydrate load and food processing methods may contribute to certain cancers is discussed in the following section.

FOOD PROCESSING—INFLUENCE ON FIBER, MICRONUTRIENTS

The ways certain foods are processed have had a huge impact on health, including cancer. In general, processing methods in economically developed areas have produced energy-dense foods that are deplete in micronutrients and fiber. The palatability and availability of these foods, in combination with an environment requiring little energy expenditure, tend to lead to high rates of obesity. In addition, individuals obtaining much of their required energy from highly processed foods tend to consume only a small amount of micronutrients or other potentially beneficial phytochemicals.

The general effects of micronutrient depletion extends to many cancer sites, but perhaps the most important effect of processing has been on colon cancer. Highly refined diets may have a direct impact on colon cancer through influencing the lumenal characteristics of the large bowel. Dietary fiber,[50] as well as physical activity, could be important modulators of risk by reducing colonic transit time, thus limiting the exposure of the colonic mucosa to carcinogens. Another relevant hypothesis is based on the differentiation effects of butyrate, which is formed during fermentation of fiber in the large bowel.[51] In the more refined diets that are depleted of fiber in the West, little undigested carbohydrates reach the distal regions of the bowel, thus limiting the exposure to potentially beneficial effects of butyrate.

A more recent hypothesis focuses on the systemic influences of highly refined diets, most importantly on hyperinsulinemia.[24,52,53] If hyperinsulinemia is cancer-enhancing, the total carbohydrate load and composition are likely to be important. In general, slower rates of carbohydrate digestion will flatten the blood glucose and insulin responses.[54] Food processing methods in Western countries tend to produce foods with a pronounced glycemic effect. Such methods include grinding, extruding, flaking, and popping, and increased starch gelatinization, as opposed to more traditional methods of food processing such as parboiling of wheat and rice, use of whole grains in rye breads and cold extrusion in the making of pasta.[55]

Traditionally, much of the focus has been on the lumenal consequences of a highly refined diet. However, the postprandial physiologic systemic effects cannot be ignored, and may be of especial importance because the deleterious effects may extend to other organs beside the intestines.

NUTRIENTS INFLUENCING DNA SYNTHESIS, REPAIR, AND METHYLATION

Specific nutritional factors may influence carcinogenesis through effects on DNA synthesis and repair, and through postsynthetic modification of DNA (DNA methylation). DNA synthesis and repair processes are dependent on various nutritional factors, including folic acid. 5,10-methylene tetrahydrofolate is the cellular form of folate required for synthesis of thymidine. When levels of 5,10-methylene tetrahydrofolate are low, uracil may be misincorporated for thymidine into DNA during synthesis.[56] Experiments show that low levels of available thymidine may cause an increase in spontaneous DNA mutation rates,[57] an enhanced sensitivity to DNA-damaging agents,[58] a higher frequency of chromosomal aberrations,[59,60] and an increase in the error rate during DNA replication.[60–62] Recently, Blount et al.[63] demonstrated that folate deficiency caused massive incorporation of uracil into human DNA and to increased frequency of chromosomal breaks. These were reversed when folate-deficient individuals were treated with folic acid supplementation.

The methylation of DNA cytosine at specific cytosine-phospho-guanine (CpG) sites inhibits gene transcription and, thus, has a role in the regulation of gene expression.[64] DNA hypomethylation is among the earliest events observed in colon carcinogenesis,[65–70] although it remains unproven whether DNA methylation abnormalities directly influence carcinogenesis. Three nutrients, folate, vitamin B_{12}, and methionine, may be important in DNA methylation. When methionine intake is low, levels of S-adenosylmethionine, the ultimate methyl donor, decrease. To compensate, cells endogenously convert 5,10-methylene tetrahydrofolate to 5-methyl tetrahydrofolate. The transfer of the methyl group from 5-methyl tetrahydrofolate to homocysteine forms methionine, a reaction also requiring cobalamin. If compensatory production of methionine is hindered by insufficient 5-methyl tetrahydrofolate level, and possibly cobalmine levels, the methyl supply for DNA methylation may be inadequate.

Because these processes would appear to be important for cell types in general, it is possible that folate deficiency could be a risk factor for numerous cancer types. In fact, some supportive though inconclusive evidence exists for various cancer sites.[71] To date, the strongest epidemiologic evidence regarding folate appears to be for colon cancer. Four case-control studies[72–75] and four prospective studies[76–79] and five studies of adenomas, cancer precursors,[80–84] have found a higher risk of colon cancer among individuals with low folate intakes or blood levels. The major dietary sources of folate include fruits and vegetables, and these have been consistently related to a lower risk of colon cancer. Multivitamin supplements, also a major source of folate, have also been related to a reduced risk of colon cancer.[73,77,79,82]

INDUCTION OF ACTIVATING OR DETOXIFYING METABOLIC ENZYMES

External carcinogens may play an important role for at least some human cancers. The metabolism of many potential carcinogens is mediated in part by cytochromes P450 1A1 and P450 1A2. Some of the compounds generated are highly reactive metabolites that can damage DNA. In addition to enzymes that have the capability to transform compounds into carcinogens, some enzymes may have a “detoxifying” effect. Examples include glutathione transferases, epoxide hydroxylase, quinone reductase, and glucurosyltransferases. These are often called phase 2 detoxification enzymes. Higher activity of these phase 2 enzymes may help protect against cancer by reducing the impact of reactive compounds that could damage DNA.

A potential role for diet in preventing cancer is through the induction of phase 2 enzymatic detoxification systems. Specifically, food groups such as allium (for example, onions, garlic) and cruciferous vegetables (for example, broccoli, Brussels sprouts, cauliflower, kale) contain a wide variety of minor metabolites, some of which appear to be potent inducers of the phase 2 enzymes.[85] For example, cruciferous vegetables contain large amounts of isothiocyanates, some of which induce these enzymes. Repeated exposures of these through diet may keep the detoxifying enzymes "primed" for periodic exposures of potentially genotoxic compounds. Relatively high consumption of garlic and cruciferous vegetables has been associated with reduced risk of colon cancer in some studies.

OXIDANTS AND ANTIOXIDANTS

DNA damage may result from highly reactive oxidative compounds, such as superoxide, hydrogen peroxide, singlet oxygen, and hydroxyl radical. Oxidative damage induced by these is believed to contribute to the causation of various human cancers.[86,87] Numerous antioxidant defenses may help limit damage to DNA, as well as to lipids and protein. Enzymatic antioxidant systems include superoxide dismutase, catalase, and glutathione peroxidase. In addition, macromolecules such as albumin and small molecules such as ascorbic acid, α-tocopherol, and carotenoids, as well as numerous others may help quench these highly reactive compounds or free radicals. Despite these cellular defenses, the potential for DNA damage remains substantial because these protective systems are imperfect. Ames and colleagues have estimated that the number of oxidative hits to DNA per human cell is about 10,000.[88]

Diet provides an important source of antioxidants. The antioxidant enzyme glutathione peroxidase is selenium-dependent and selenium deficiency appears to predispose one to at least some cancers, although selenium may act through other mechanisms.[89] Vitamin E, or α-tocopherol, is an essential nutrient that has an established antioxidant function. A recent randomized clinical trial among heavy smokers showed a benefit of α-tocopherol on prostate cancer.[90] Fruits and vegetables may be rich sources of various antioxidants. One study[91] found a large capacity of antioxidants in individual fruits based on an *in vitro* assay (oxygen radical absorbence capacity, or ORAC). On a dry weight basis, strawberries had the highest ORAC capacity of the various fruits tested. Interestingly, vitamin C represented only a small proportion of the antioxidant capacity. Part of the antioxidant capacity may be contained in flavonoids, but largely, the source of the antioxidant capacity was uncharacterized. The carotenoid lycopene,[92] a potent antioxidant, may be particularly effective against prostate cancer.[93,94]

Although the epidemiologic evidence suggests that fruits and vegetables are protective for many cancers, the individual role of the various antioxidants and other biologically active compounds is difficult to assess. The complexity is enormous, and little is known whether these compounds are absorbed intestinally or how they distribute into various tissues, influence oxidative processes in the cell, and interact with each other.

"TOXIC" EFFECTS OF DIET: CARCINOGENS AND INFLAMMATION

Dietary factors may have a direct impact on the gastrointestinal lumen, including the oral cavity, esophagus, stomach, and intestines. For example, diet can be a direct supply of genotoxic compounds, or may cause chronic irritation or inflammation. These pathologic processes are associated with increased cellular proliferation, local production of growth factors, and oxidative stress,[95] all of which may enhance carcinogenesis.

Cancer of the mouth and pharynx is the sixth most common cancer worldwide, and nearly 80% of cases occur in economically developing countries.[96] Besides alcohol and tobacco, some dietary factors, through direct exposure, may also influence oral carcinogenesis. Limited evidence has found a higher risk of oral cancer associated with increased intake of salt-preserved meat and fish,[97] smoked foods,[98] and charcoal grilled meat.[99] Intake of various beverages drunk very hot, such as maté drunk as tea, has been associated with about a two-fold excess in risk,[99–101] possibly through a direct irritant effect on the mucosa.

As for the oral cavity, alcohol and tobacco use are important risk factors for esophageal cancers. Also, the frequency of drinking hot drinks has been related to a higher risk of esophageal cancer in numerous studies.[102–105] A role for hot drinks is supported by the finding in laboratory animals that drinking hot beverages was associated with development of premalignant lesions of the esophagus.[106] Food processing practices may also provide direct exposure from N-nitrosoamines,[107] and heterocyclic amines formed from cooking meat at high temperatures have also been speculated to be important for cancers of the esophagus.[100,105]

Though becoming rarer in economically-developed countries, cancers of the stomach remain the second most common cancer worldwide. Helicobacter pylori infection, widely prevalent in developing countries, predisposes for stomach cancer.[108] Possibly, diet may affect stomach cancer risk by altering the bacterial milieu. Some compounds in garlic and onions have a strong antibacterial effect against Helicobacter pylori infection, possibly accounting for an apparent protective effect of these items against stomach cancer.[109] Perhaps related to a direct irritant effect, high salt intake may increase risk of stomach cancer. More limited evidence also suggests grilling, curing and smoking meats increases risk of stomach cancer.

Diet may also be a source of carcinogens. For example, frying or broiling of meat may produce potential carcinogens including polynuclear aromatic hydrocarbons (PAHs) and heterocyclic amines such as 2-amino-1-methyl-6-phenylimidazol (4,5-β) pyridine (PhIP).[110–112] *In vitro* studies and animal models suggest that heterocyclic amines increase risk of colorectal cancer.[113] Whether these are consumed by humans at sufficient quantities to be carcinogenic is an area of active study.

SUMMARY

Macronutrients, micronutrients, nutritional imbalances, and "nonessential" dietary components affect carcinogenesis. Certain dietary patterns and practices explain a substantial proportion of the international variation of rates of cancers. Cancers of the colon, prostate, and breast may be considered cancers of the Western world. The causes of these malignancies are multifactorial and complex, but a major influence is the widespread availability of energy-dense, highly processed and refined foods. In addition, the modern types of processing techniques tend to lead to foods that are deplete in fiber and have a high glycemic index, which enhances insulin levels. The dietary patterns in combination with physical inactivity contribute to obesity and metabolic consequences such as increased levels of IGF-1, insulin, estrogen, and testosterone, which tend to promote cellular growth. High consumption of milk, dairy products, and meats tend to decrease 1,25$(OH)_2$vitamin D, a cell differentiator in the prostate and possibly other tissues.

Nutritionally influenced growth-enhancing and differentiating hormones are likely to contribute to the high prevalence of breast, colorectal, prostate, and several other cancers in the Western world, but in less economically developed countries, these cancers are relatively rare, whereas malignancies of the upper gastrointestinal tract are quite common. Upper gastrointestinal tract cancers account for about 20% of the cancer load worldwide, and are proportionally much higher in developing countries. The major causes of these cancers are likely related to various food practices or preservation methods that increase mucosal exposure to irritants or carcinogens. Among non-smokers and non-abusers of alcohol in affluent populations, these upper gastrointestinal malignancies are relatively rare, and as such, appear largely preventable through prevention of smoking and alcohol abuse.

While general overconsumption and pattern of macronutrients causes various cancers in the Western world and diet-related carcinogens and irritants contribute to cancer causation in developing countries, various micronutrients consumed primarily from fruits and vegetables play an important inhibitory role against many types of cancer. It is likely that nutritional deficiency interacts with other etiologic factors. For example, many populations in developing areas have marginal folate status, but nonetheless have low overall rates of colon cancer. However, in economically developed countries where factors that increase risk of colon cancer are very prevalent, poor folate status substantially enhances risk. As an additional example, heavy smokers appear to benefit from some unidentified anti-cancer factor(s) in fruits and vegetables. Avoiding energy excess, relying less on highly-processed foods and animal meats and fats as major sources of energy, and increasing the diversity of fruits and vegetables in the diet are steps likely to minimize one's risk of numerous cancers.

REFERENCES

1. Armstrong B. and Doll R. Environmental factors and cancer incidence and mortality in different countries, with special reference to dietary practices. Int J Cancer. 1975;15:617–631.
2. Staszewski J. and Haenszel W. Cancer mortality among the Polish-born in the United States. J Natl Cancer Inst. 1965;35:291–297.
3. Adelstein A.M., Staszewski J., and Muir C.S. Cancer mortality in 1970–1972 among Polish-born migrants to England and Wales. Br J Cancer. 1979;40:464–475.
4. McMichael A.J. and Giles G.G. Cancer in migrants to Australia: extending the descriptive epidemiological data. Cancer Res. 1988;48:751–756.
5. Shimizu H., Ross R.K., Bernstein L., Yatani R., Henderson B.E., and Mack T.M. Cancers of the prostate and breast among Japanese and white immigrants in Los Angeles County. Br J Cancer. 1991;63:963–966.
6. Ziegler R.G., Hoover R.N., Pike M.C., Hildesheim A., Nomura A.M.Y., West D.W., Wu-Williams A.H., Kolonel L.N., Horn-Ross P.L., Rosenthal J.F., and Hyer M.B. Migration patterns and breast cancer risk in Asian-American women. J Natl Cancer Inst. 1993;85:1819–1827.

7. Haenszel W. and Kurihara M. Studies of Japanese migrants. I. Mortality from cancer and other diseases among Japanese in the United States. J Natl Cancer Inst. 1968;40:43–68.
8. Ruggeri B.A., Klurfeld D.M., Kritchevsky D., and Furlanetto R.W. Caloric restriction and 7,12-dimethylbenz(a)anthracene-induced mammary tumor growth in rats: alterations in circulating insulin, insulin-like growth factors I and II, and epidermal growth factor. Cancer Res. 1989;49:4130–4134.
9. Klurfeld D.M., Lloyd L.M., Welch C.B., Davis M.J., Tulp O.L., and Kritchevsky D. Reduction of enhanced mammary carcinogenesis in LA/N-cp (corpulent) rats by energy restriction. Proc Soc Exp Biol Med. 1991;196:381–384.
10. Chasan-Taber S., Rimm E.B., Stampfer M.J., Spiegelman D., Colditz G.A., Giovannucci E., Ascherio A., and Willett W.C. Reproducibility and validity of a self-administered physical activity questionnaire for male health professionals. Epidemiology. 1996;7:81–86.
11. Giovannucci E., Ascherio A., Rimm E.B., Colditz G.A., Stampfer M.J., and Willett W.C. Physical activity, obesity, and risk for colon cancer and adenoma in men. Ann Intern Med. 1995;122:327–334.
12. Albanes D., Jones D.Y., Schatzkin A., Micozzi M.S., and Taylor P.R. Adult stature and risk of cancer. Cancer Res. 1988;48:1658–1662.
13. Hebert P.R., Ajani U., Cook N.R., Lee I.-M., Chan K.S., and Hennekens C.H. Adult height and incidence of cancer in male physicians (United States). Cancer Causes Control. 1997;8:591–597.
14. Huang Z., Hankinson S.E., Colditz G.A., Stampfer M.J., Hunter D.J., Manson J.E., Hennekens C.H., Rosner B., Speizer F.E., and Willett W.C. Dual effects of weight and weight gain on breast cancer risk. J Am Med Assoc. 1997;278:1407–1411.
15. Giovannucci E., Rimm E.B., Stampfer M.J., Colditz G.A., and Willett W.C. Height, body weight, and risk of prostate cancer. Cancer Epidemiol Biomarkers Prev. 1997;6:557–563.
16. Colditz G., Cannuscio C., and Frazier A. Physical activity and reduced risk of colon cancer: implications for prevention. Cancer Causes Control. 1997;8:649–667.
17. Bernstein L., Henderson B.E., Hanisch R., Sullivan-Halley J., and Ross R.K. Physical exercise and reduced risk of breast cancer in young women. J Natl Cancer Inst. 1994;86:1403–1408.
18. Lee I.M., Paffenbarger R.S., Jr., and Hsieh C.C. Physical activity and risk of prostatic cancer among college alumni. Am J Epidemiol. 1992;135:169–175.
19. Chan J.M., Stampfer M.J., Giovannucci E., Gann P.H., Ma J., Wilkinson P., Hennekens C.H., and Pollak M. Plasma insulin-like growth factor-I and prostate cancer risk: a prospective study. Science. 1998; 279:563–566.
20. Hankinson S.E., Willett W.C., Colditz G.A., Hunter D.J., Michaud D.S., Deroo B., Rosner B., Speizer F.E., and Pollak M. Circulating concentrations of insulin-like growth factor-I and risk of breast cancer. Lancet. 1998;351:1393–1396.
21. Hankinson S.E., Willett W.C., Manson J.E., Colditz G.A., Hunter D.J., Spiegelman D., Barbieri R.L., and Speizer F.E. Plasma sex steriod hormone levels of breast cancer in postmenopausal women. J Natl Cancer Inst. 1998;••:(in press).
22. Gann P.H., Hennekens C.H., Ma J., Longcope C., and Stampfer M.J. A prospective study of sex hormone levels and risk of prostate cancer. J Natl Cancer Inst. 1996;88:1118–1126.
23. Reaven G.M. Do high carbohydrate diets prevent the development or attenuate the manifestations (or both) of syndrome X? A viewpoint strongly against. Current Opinion In Lipidology. 1997;8:23–27.
24. Giovannucci E. Insulin and colon cancer. Cancer Causes Control. 1995;6:164–179.
25. Isley W.L., Underwood L.E., and Clemmons D.R. Dietary components that regulate serum somatomedin-C concentrations in humans. J Clin Invest. 1983;71:175–182.
26. Clemmons D.R., Underwood L.E., Dickerson R.N., Brown R.O., Hak L.J., MacPhee R.D., and Heizer W.D. Use of plasma somatomedin-C/insulin-like growth factor I measurements to monitor the response to nutritional repletion in malnourished patients. Am J Clin Nutr. 1985;41:191–198.
27. Unterman T.G., Vazquez R.M., Slas A.J., Martyn P.A., and Phillips L.S. Nutrition and somatomedin. XIII. Usefulness of somatomedin-C in nutritional assessment. Am J Med. 1985;78:228–234.
28. Rose D.P. and Connolly J.M. Dietary fat, fatty acids, and prostate cancer. Lipids. 1992;27:798–803.
29. Giovannucci E. How is individual risk for prostate cancer assessed? Hematology/Oncology Clinics of North America. 1996;10:537–548.
30. Rose D.P., Boyar A.P., and Wynder E.L. International comparisons of mortality rates for cancer of the breast, ovary, prostate, and colon, and per capita food consumption. Cancer. 1986;58:2263–2271.
31. Jain M., Cook G.M., Davis F.G., Grace M.G., Howe G.R., and Miller A.B. A case-control study of diet and colo-rectal cancer. Int J Cancer. 1980;26:757–768.
32. Potter J.D. and McMichael A.J. Diet and cancer of the colon and rectum: a case-control study. J Natl Cancer Inst. 1986;76:557–569.

33. Lyon J.L., Mahoney A.W., West D.W., Gardner J.W., Smith K.R., Sorenson A.W., and Stanish W. Energy intake: its relationship to colon cancer risk. J Natl Cancer Inst. 1987;78:853–861.
34. Graham S., Marshall J., Haughey B., Mittelman A., Swanson M., Zielezny M., Byers T., Wilkinson G., and West D. Dietary epidemiology of cancer of the colon in western New York. Am J Epidemiol. 1988;128:490–503.
35. Bristol J.B., Emmett P.M., Heaton K.W., and Williamson R.C. Sugar, fat, and the risk of colorectal cancer. Br Med J (Clin Res Ed). 1985;291:1467–1470.
36. Kune G.A. and Kune S. The nutritional causes of colorectal cancer: an introduction to the Melbourne study. Nutr Cancer. 1987;9:1–4.
37. West D.W., Slattery M.L., Robison L.M., Schuman K.L., Ford M.H., Mahoney A.W., Lyon J.L., and Sorensen A.W. Dietary intake and colon cancer: sex- and anatomic site-specific associations. Am J Epidemiol. 1989;130:883–894.
38. Peters R.K., Pike M.C., Garabrandt D., and Mack T.M. Diet and colon cancer in Los Angeles County, California. Cancer Causes Control. 1992;3:457–473.
39. Gerhardsson de Verdier M., Hagman U., Peters R.K., Steineck G., and Overik E. Meat, cooking methods, and colorectal cancer: A case-referent study in Stockholm. Int J Cancer. 1991;49:520–525.
40. Manousos O., Day N.E., Trichopoulos D., Gerovassilis F., Tzanou A., and Polychronopoulou A. Diet and colorectal cancer: a case-control study in Greece. Int J Cancer. 1983;32:1–5.
41. La Vecchia C., Negri E., Decarli A., D'Avanzo B., Gallotti L., Gentile A., and Franceschi S. A case-control study of diet and colo-rectal cancer in northern Italy. Int J Cancer. 1988;41:492–498.
42. Miller A.B., Howe G.R., Jain M., Craib K.J.P., and Harrison L. Food items and food groups as risk factors in a case-control study of diet and colo-rectal cancer. Int J Cancer. 1983;32:155–161.
43. Young T.B. and Wolf D.A. Case-control study of proximal and distal colon cancer and diet in Wisconsin. Int J Cancer. 1988;42:167–175.
44. Benito E., Obrador A., Stiggelbout A., Bosch F.X., Mulet M., Muñoz N., and Kaldor J. A population-based case-control study of colorectal cancer in Majorca. I. Dietary factors. Int J Cancer. 1990;45:69–76.
45. Lee H.P., Gourley L., Duffy S.W., Esteve J., Lee J., and Day N.E. Colorectal cancer and diet in an Asian population–A case-control study among Singapore Chinese. Int J Cancer. 1989;43:1007–1016.
46. Willett W.C., Stampfer M.J., Colditz G.A., Rosner B.A., and Speizer F.E. Relation of meat, fat, and fiber intake to the risk of colon cancer in a prospective study among women. N Engl J Med. 1990;323:1664–1672.
47. Goldbohm R.A., van den Brandt P.A., van't Veer P., Brants H.A.M., Dorant E., Sturmans F., and Hermus R.J.J. A prospective cohort study on the relation between meat consumption and the risk of colon cancer. Cancer Res. 1994;54:718–723.
48. Bostick R.M., Potter J.D., Kushi L.H., Sellers T.A., Steinmetz K.A., McKenzie D.R., Gapstur S.M., and Folsom A.R. Sugar, meat, and fat intake, and non-dietary risk factors for colon cancer incidence in Iowa women (United States). Cancer Causes Control. 1994;5:38–52.
49. Giovannucci E., Rimm E.B., Stampfer M.J., Colditz G.A., Ascherio A., and Willett W.C. Intake of fat, meat, and fiber in relation to risk of colon cancer in men. Cancer Res. 1994;54:2390–2397.
50. Burkitt D.P. Epidemiology of cancer of the colon and rectum. Cancer. 1971;28:3–13.
51. Velazquez O.C. and Rombeau J.L. Butyrate. Potential role in colon cancer prevention and treatment. Adv Exp Med Biol. 1997;427:169–181.
52. Corpet D., Jacquinet C., Peiffer G., and Taché S. Insulin injections promote the growth of aberrant crypt foci in the colon of rats. Nutr Cancer. 1997;27:316–320.
53. Tran T.T., Medline A., and Bruce R. Insulin promotion of colon tumors in rats. Cancer Epidemiol Biomarkers Prev. 1996;5:1013–1015.
54. Wolever T.M.S., Jenkins D.J., Jenkins A.L., and Josse R.G. The glycemic index: methodology and clinical implications. Am J Clin Nutr. 1991;54:846–854.
55. Wolever T. The glycemic index, in *Aspects of Some Vitamins, Minerals, and Enzymes in Health and Disease*, G Bourne, Ed. 1990, Karger: Basel, Switzerland. pp. 120–185.
56. Wickramasinghe S. and Fida S. Bone marrow cells from vitamin B_{12}- and folate-deficient patients misincorporate uracil into DNA. Blood. 1994;83:1656–1661.
57. Weinberg G., Ullman B., and Martin D.W., Jr. Mutator phenotypes in mammalian cell mutants with distinct biochemical defects and abnormal deoxyribonucleoside triphosphate pools. Proc Natl Acad Sci USA. 1981;78:2447–2451.
58. Meuth M. Role of deoxynucleoside triphosphate pools in the cytotoxic and mutagenic effects of DNA alkylating agents. Somatic Cell Genet. 1981;7:89–102.

59. Sutherland G.R. The role of nucleotides in human fragile site expression. Mutat Res. 1988;200:207–213.
60. Fenech M. and Rinaldi J. The relationship between micronuclei in human lymphocytes and plasma levels of vitamin C, vitamin E, vitamin B_{12} and folic acid. Carcinogenesis. 1994;15:1405–1411.
61. Hunting D.J. and Dresler S.L. Dependence of u.v.-induced DNA excision repair on deoxyribonucleoside triphosphate concentrations in permeable human fibroblasts: a model for the inhibition of repair by hydroxyurea. Carcinogenesis (Lond). 1985;6:1525–1528.
62. James S.J., Basnakian A.G., and Miller B.J. *In vitro* folate deficiency induces deoxynucleotide pool imbalance, apoptosis, and mutagenesis in Chinese hamster ovary cells. Cancer Res. 1994;54:5075–5080.
63. Blount B.C., Mack M.M., Wehr C.M., MacGregor J.T., Hiatt R.A., Wang G., Wickramasinghe S.N., Everson R.B., and Ames B.N. Folate deficiency causes uracil misincorporation into human DNA and chromosome breakage: implications for cancer and neuronal damage. Proc Natl Acad Sci USA. 1997;94:3290–3295.
64. Holliday R. The inheritance of epigenetic defects. Science. 1987;238:163–170.
65. Cravo M., Fidalgo P., Pereira A.D., Gouveia-Oliveira A., Chavas P., Selhub J., Mason J.B., Mira F.C., and Leitao C.N. DNA methylation as an intermediate biomarker in colorectal cancer: modulation by folic acid supplementation. Eur J Cancer Prevention. 1994;3:473–479.
66. Feinberg A.P. and Vogelstein B. Hypomethylation distinguishes genes of some human cancers from their normal counterparts. Nature. 1983;301:89–92.
67. Goelz S.E., Vogelstein B., Hamilton S.R., and Feinberg A.P. Hypomethylation of DNA from benign and malignant human colon neoplasms. Science. 1985;228:187–190.
68. Feinberg A.P., Gehrke C.W., Kuo K.C., and Ehrlich M. Reduced genomic 5-methylcytosine content in human colonic neoplasia. Cancer Research. 1988;48:1159–1161.
69. Issa J.-P.J., Vertino P.M., Wu J., Sazawal S., Celano P., Nelkin B.D., Hamilton S.R., and Baylin S.B. Increased cytosine DNA-methyltransferase activity during colon cancer progression. J Natl Cancer Inst. 1993;85:1235–1240.
70. Makos M., Nelkin B.D., Lerman M.I., Latif F., Zbar B., and Baylin S.B. Distinct hypermethylation patterns occur at altered chromosome loci in human lung and colon cancer. Proc Natl Acad Sci USA. 1992;89:1929–1933.
71. Glynn S.A. and Albanes D. Folate and cancer: a review of the literature. Nutr Cancer. 1994;22:101–119.
72. Benito E., Stiggelbout A., Bosch F.X., Obrador A., Kaldor J., Mulet M., and Munoz N. Nutritional factors in colorectal cancer risk: a case-control study in Majorca. Int J Cancer. 1991;49:161–167.
73. Meyer F. and White E. Alcohol and nutrients in relation to colon cancer in middle-aged adults. Am J Epidemiol. 1993;138:225–236.
74. Ferraroni M., La Vecchia C., D'Avanzo B., Negri E., Franceschi S., and Decarli A. Selected micronutrient intake and the risk of colorectal cancer. British Journal of Cancer. 1994;70:1150–1155.
75. Freudenheim J.L., Graham S., Marshall J.R., Haughey B.P., Cholewinski S., and Wilkinson G. Folate intake and carcinogenesis of the colon and rectum. Int J Epidemiol. 1991;20:368–374.
76. Glynn S.A., Albanes D., Pietinen P., Brown C.C., Rautalahti M., Tangrea J.A., Gunter E.W., Barrett M.J., Virtamo J., and Taylor P.R. Colorectal cancer and folate status: A nested case-control study among male smokers. Cancer Epidemiol Biomarkers Prevention. 1996;5:487–494.
77. Giovannucci E., Rimm E.B., Ascherio A., Stampfer M.J., Colditz G.A., and Willett W.C. Alcohol, low-methionine-low-folate diets, and risk of colon cancer in men. J Natl Cancer Inst. 1995;87:265–273.
78. Ma J., Stampfer M.J., Giovannucci E., Artigas C., Hunter D.J., Fuchs C., Willett W.C., Selhub J., Hennekens C.H., and Rozen R. Methylenetetrahydrofolate reductase polymorphism, dietary interactions, and risk of colorectal cancer. Cancer Res. 1997;57:1098–1102.
79. Giovannucci E., Stampfer M.J., Colditz G.A., Hunter D.J., Fuchs C., Rosner B.A., Speizer F.E., and Willett W.C. Multivitamin use, folate, and colon cancer in women in the Nurses' Health Study. Ann Intern Med. 1998;129:517–524.
80. Bird C.L., Swendseid M.E., Witte J.S., Shikany J.M., Hunt I.F., Frankl H.D., Lee E.R., Longnecker M.P., and Haile R.W. Red cell and plasma folate, folate consumption, and the risk of colorectal adenomatous polyps. Cancer Epidemiol Biomarkers Prev. 1995;4:709–714.
81. Tseng M., Murray S.C., Kupper L.L., and Sandler R.S. Micronutrients and the risk of colorectal adenomas. Am J Epidemiol. 1996;144:1005–1014.
82. Giovannucci E., Stampfer M.J., Colditz G.A., Rimm E.B., Trichopolous D., Rosner B.A., Speizer F.E., and Willett W.C. Folate, methionine, and alcohol intake and risk of colorectal adenoma. J Natl Cancer Inst. 1993;85:875–884.
83. Benito E., Cabeza E., Moreno V., Obrador A., and Bosch F.X. Diet and colorectal adenomas: a case-control study in Majorca. Int J Cancer. 1993;55:213–219.

84. Paspatis G.A., Kalafatis E., Oros L., Xourgias V., Koutsioumpa P., and Karamanolis D.G. Folate status and adenomatous colonic polyps. A colonoscopically controlled study. Dis Colon Rectum. 1995; 38:64–68.
85. Steinmetz K.A. and Potter J.D. Vegetables, fruit, and cancer. II. Mechanisms. Cancer Causes Control. 1991;2:427–442.
86. Cerutti P.A. Prooxidant states and tumor promotion. Science. 1985;227:375–381.
87. Wiseman H. and Halliwell B. Damage to DNA by reactive oxygen and nitrogen species: role in inflammatory disease and progression to cancer. Biochem J. 1996;313:17–29.
88. Ames B., Shigenaga M., and Hagen T. Oxidants, antioxidants, and the degenerative diseases of aging. Proc Natl Acad Sci USA. 1993;90:7915–7922.
89. Clark L.C., Combs G.F., Jr., Turnbull B.W., Slate E.H., Chalker D.K., Chow J., Davis L.S., Glover R.A., Graham G.F., Gross E.G., Krongrad A., Lesher J.L., Jr., Park H.K., Sanders B.B., Jr., Smith C.L., and Taylor J.R. Effects of selenium supplementation for cancer prevention in patients with carcinoma of the skin. A randomized controlled trial. Nutritional Prevention of Cancer Study Group. JAMA. 1996;276:1957–1963.
90. The Alpha-Tocopherol Beta-Carotene Cancer Prevention Study Group. The effect of vitamin E and beta carotene on the incidence of lung cancer and other cancers in male smokers. N Engl J Med. 1994;330:1029–1035.
91. Wang H., Cao G., and Prior R. Total antioxidant capacity of fruits. J Agric Food Chem. 1996; 44:701–705.
92. Stahl W. and Sies H. Uptake of lycopene and its geometrical isomers is greater from heat-processed than from unprocessed tomato juice in humans. J Nutr. 1992;122:2161–2166.
93. Giovannucci E., Ascherio A., Rimm E.B., Stampfer M.J., Colditz G.A., and Willett W.C. Intake of carotenoids and retinol in relation to risk of prostate cancer. J Natl Cancer Inst. 1995;87:1767–1776.
94. Hsing A.W., Comstock G.W., Abbey H., and Polk B.F. Serologic precursors of cancer. Retinol, carotenoids, and tocopherol and risk of prostate cancer. J Natl Cancer Inst. 1990;82:941–946.
95. Parsonnet J. Bacterial infection as a cause of cancer. Environ Health Perspect. 1995;103(suppl 8):263–268.
96. World Health Organization. The World Health Report, 1997, World Health Organization, Geneva.
97. Zheng W., Blot W., Shu X., Diamond E., Gao Y., Ji B., and Fraumeni J., Jr. Risk factors for oral and pharyngeal cancer in Shanghai, with emphasis on diet. Cancer Epidemiol Biomarkers Prev. 1992; 1:441–448.
98. Winn D.M., Ziegler R.G., Pickle L.W., Gridley G., Blot W.J., and Hoover R.N. Diet in the etiology of oral and pharyngeal cancer among women from the southern United States. Cancer Res. 1984;44:1216–1222.
99. Franco E., Kowalski L., Oliveira B., Curado M., Pereira R., Silva M., Fava A., and Torloni H. Risk factors for oral cancer in Brazil: a case-control study. Int J Cancer. 1989;42:992–1000.
100. De Stefani E., Correa P., Oreggia F., Deneo-Pellegrini H., Fernandez G., Zavala D., Carzoglio J., Leiva J., Fontham E., and Rivero S. Black tobacco, wine and mate in oropharyngeal cancer. A case-control study from Uruguay. Rev Epidemiol Sante Publique. 1988;36:389–394.
101. Oreggia F., De Stefani E., Correa P., and Fierro L. Risk factors for cancer of the tongue in Uruguay. Cancer. 1991;67:180–183.
102. Martìnez I. Factors associated with cancer of the esophagus, mouth, and pharynx in Puerto Rico. J Natl Cancer Inst. 1969;42:1069–1094.
103. De Stefani E., Muñoz N., Estéve J., Vasallo A., Victora C.G., and Teuchmann S. *Mate* drinking, alcohol, tobacco, diet, and esophageal cancer in Uruguay. Cancer Res. 1990;50:426–431.
104. Vassallo A., Correa P., De Stéfani E., Cendán M., Zavala D., Chen V., Carzoglio J., and Deneo-Pellegrini H. Esophageal cancer in Uruguay: a case-control study. J Natl Cancer Inst. 1985; 75:1005–1009.
105. Victora C., Munoz N., Day N., Barcelos L., Peccin D., and Braga N. Hot beverages and oesophageal cancer in southern Brazil: a case-control study. Int J Cancer. 1987;39:710–716.
106. Yioris N., Ivankovic S., and Lehnert T. Effect of thermal injury and oral administration of N-methyl-N′-nitro-N-nitrosoguanidine on the development of esophageal tumors in Wistar rats. Oncology. 1984;41:36–38.
107. Lu S., Chui S., Yang W., Hu X., Guo L., and Li F. Relevance of N-nitrosamines to oesophageal cancer in China. IARC Sci Publ. 1991;105:11–17.
108. Sivam G., Lampe J., Ulness B., Swanzy S., and Potter J. Helicobacter pylori–in vitro susceptibility to garlic (Allium sativum) extract. Nutr Cancer. 1997;27:118–121.

109. Dorant E., van den Brandt P., Goldbohm R., and Sturmans F. Consumption of onions and a reduced risk of stomach carcinoma. Gastroenterology. 1996;110:12–20.
110. Manabe S., Tohyama K., Wada O., and Aramaki T. Detection of a carcinogen, 2-amino-1-methyl-6-phenylimidazo[4,5-b]pyridine {PhIP), in cigarette smoke condensate. Carcinogenesis. 1991; 12:1945–1947.
111. Felton J.S., Knize M.G., Shen N.H., Lewis P.R., Andresen B.D., Happe J., and Hatch F.T. The isolation and identification of a new mutagen from fried ground beef: 2-amino-1-methyl-6-phenylimidazo[4,5-b]pyridine (PhIP). Carcinogenesis. 1986;7:1081–1086.
112. Sugimura T. and Sato S. Mutagens-carcinogens in foods. Cancer Res. 1983;43:2415S–2421S.
113. Weisburger J.H. and Jones R.C. Prevention of formation of important mutagens/carcinogens in the human food chain. Basic Life Sci. 1990;52:105–118.

4

ALCOHOL AND CANCER

Silvia Franceschi

Epidemiology Unit, Aviano Cancer Center
Via Pedemontana Occ. 1
33081 Aviano (PN), Italy

1. INTRODUCTION

Epidemiological studies have demonstrated that alcohol intake is responsible, in developed countries, for about 3% of cancer deaths.[1] Such a proportion can reach 20% in men in some countries (e.g., France). Cancer sites firmly associated with alcohol consumption include oral cavity and pharynx, larynx, oesophagus, and liver. A relationship is also possible with cancer of the colon-rectum, pancreas, lung and, most notably, cancer of the breast, as reviewed in two extensive reviews by the International Agency of Research on Cancer[2] and the World Cancer Research Fund.[3]

Few agents have been studied so extensively from an epidemiological view point as alcohol, on account of two specific features of such consumption: 1) large variability between and within populations,[3] and 2) large amount and relative good quality of information available.[4] In fact, at variance with the complexity of the recall of dietary habits, alcohol derives from relatively few, well identified sources, basically wine, beer, and spirits (liquor). A good standardisation of the ethanol content of different types of alcoholic beverages is possible. Aside from a tendency to underreport alcohol intake, particularly at high consumption levels, a relatively good reproducibility and validity of self-reported alcohol consumption has been shown.

In the present review the association of alcohol consumption with cancer risk will be examined with special emphasis on cancer of the upper aero-digestive tract (where the relationship has been better quantified) and cancer of the breast, on account of the public health importance of an association with a tumour as common as the latter.

Advances in Nutrition and Cancer 2, edited by Zappia *et al.*
Kluwer Academic / Plenum Publishers, New York, 1999.

2. CANCER OF THE UPPER AERO-DIGESTIVE TRACT

2.1. The Effect of Alcohol

In the following, relative risk (RR) will often be used as a measure of the association of alcohol intake with the risk of various diseases. In brief, the RR denotes the ratio of the incidence of the disease among exposed divided by that among non-exposed individuals. Data derive from a series of case-control studies by Franceschi et al.[5] comprising 200–300 cases for each cancer site and conducted with the same study design in Italy, a country where alcohol consumption is high and socially better accepted (hence, more reliably reported) than, for instance, in the United States. Risk estimates are, however, in agreement with studies from North America and other European countries.

A highly significant trend of increasing risk with an increasing number of drinks per week emerged for each cancer of the upper aero-digestive tract. Significantly elevated risks, however, became apparent only in those who drank 35 or more alcoholic drinks per week. RR of 3.4, 3.6, 6.0, and 2.1 for cancer of the oral cavity, pharynx, oesophagus, and larynx, respectively, were seen among those persons who reported drinking 60 or more drinks per week compared to less than 20. Thus, the association with alcohol intake was highest for oesophagus and lowest for larynx. Conversely, the duration of alcohol-drinking habit did not appear to be related to risks for any of the tumours considered herein. This lack of a duration effect is typical of cancer promoters, i.e., agents which act on late-stage carcinogenesis.

2.2. The Interaction of Alcohol and Smoking

The joint effect of tobacco and alcohol intake is examined in Table 1, still based on the above mentioned Italian studies.[5,6] Cases of oral and pharyngeal cancers are considered together. Moreover, abstainers and light alcohol drinkers (<34 drinks per week) are combined since the risk appeared to be very similar. The risk of cancer of the oral cavity and pharynx for the highest levels of alcohol and smoking was increased about 80-fold relative to the lowest levels of both factors. The joint effect of smoking and drinking appeared, thus, for oral cavity and pharynx greater than multiplicative (i.e., greater than the RR in non-smoking heavy drinkers multiplied by the one in light-drinking heavy smokers). For esophageal cancer, high levels of alcohol and cigarette consumption combined increased the risk by 17.5 over the risk for the lowest levels of consumption. The pattern of combined risk for this site appears to be intermediate between additive and multiplicative. The effect of drinking in non-smokers was only slightly stronger than the effect of smoking in light drinkers (RR = 7.9 vs. 6.4). Thus, at variance with cancer of the oral cavity and pharynx, where the RRs for tobacco are appreciably higher than those of alcohol, alcohol appears to have an effect on the risk of oesophageal cancer approximately as strong as smoking. Cancer of the larynx is clearly associated more strongly to smoking than alcohol consumption. Excluding non-smokers (among whom alcohol has little impact), the RR in heavy smokers-heavy drinkers was 11.7, i.e., consistent with a multiplicative model.

Thus, data on the joint effects of smoking and alcohol in cancers of the upper aero-digestive tract in Italian as well as other populations[7] suggest that: (a) some differences in risk exist across anatomical sites for alcohol adjusted for smoking and for smoking adjusted for alcohol; and (b) the combined effects of alcohol and tobacco in

Table 1. Relative risk of cancer of the oral cavity/pharynx, oesophagus, and larynx in males according to smoking and alcohol drinking habits[5,6]

	Relative risk[a] Alcohol intake (total drinks per week)			
Smoking status[b]	<35	35–39	^{3}60	Total (Smoking adjusted)
Oral cavity/pharynx				
Nonsmokers	1[c]	1.6	2.3	1[c]
Light	3.1	5.4	10.9	3.7
Intermediate	10.9	26.6	3.4	14.1
Heavy	17.6	40.2	79.6	25.0
Total (alcohol-adjusted)	1[c]	2.3	3.4	
Oesophagus				
Nonsmokers	1[c]	0.8	7.9	1[c]
Light	1.1	7.9	9.4	2.5
Intermediate	2.7	8.8	16.7	4.0
Heavy	6.4	11.0	17.5	6.6
Total (alcohol-adjusted)	1[c]	3.1	5.7	
Larynx				
Nonsmokers	1[c]	1.6	—	1[c]
Light	0.9	5.0	5.4	1.0
Intermediate	4.5	7.1	9.5	5.4
Heavy	6.1	10.4	11.7	6.7
Total (alcohol-adjusted)	1[c]	1.4	2.8	

[a]Adjusted for age, area of residence, years of education, occupation, drinks per week, and smoking habits, as appropriate.
[b]Smoking status defined in four categories: 1) nonsmokers; 2) light, ex-smokers who quit ≥10 years before; or smokers of 1–14 cigarettes/day for <30 years; 3) intermediate, 15–24 cigarettes/day regardless of duration, 30–39 years duration regardless of amount, 1–24 cigarettes/day for ≥40 years, or ≥15 cigarettes/day for <30 years; 4) heavy, smokers of ≥25 cigarettes/day for ≥40 years.
[c]Reference category.

each of the upper aero-digestive tract sites tend to be more than additive (i.e., multiplicative, or nearly so). From a public health prospective, this is of great importance and means that alcohol drinking is associated with a far larger cancer burden in smokers than in non-smokers. Similarly, the cancer burden of tobacco smoking is larger in drinkers than in non-drinkers.

It is always difficult to make inference on biological mechanisms from the examination of results arising from statistical modelling, such as the ones in Table 1. However, anatomical gradients in risk should reflect, in a gross or crude fashion, gradients in exposure due to anatomical function; i.e., the larynx is exposed more to cigarette smoke than to alcohol, the oesophagus is exposed more to alcohol than to cigarette smoke, and the oral cavity/pharynx are exposed directly to both. It is clear that where exposure to smoking is maximal and exposure to alcohol is limited, laryngeal cancer risk is more elevated. Conversely, where alcohol intake is substantial and the level of smoking is low, the oesophageal cancer risk is higher. Increases in risk with increasing levels of smoking/drinking are substantially steeper for the oral cavity/pharynx than for either larynx or esophagus. This suggests that direct contact with alcohol can make the oral mucosa particularly susceptible to the carcinogenic insult of tobacco, possibly acting as an irritant or a solvent. Mechanical reasons can also be invoked since both alcohol and smoke particles can easily deposit and remain in contact for more prolonged periods in the oral cavity than elsewhere in the upper aero-digestive tract. In fact, the rates of carcinomas along the airway have been shown to correspond closely with the deposition pattern of smoke particles.

2.3. Attributable Risk and Conclusions

Attributable risk is perhaps the most important indicator from a public health viewpoint. About 80% of oral cavity and pharyngeal cancer could be attributed to smoking and drinking in the United States. Studies from North Italy shows attributable risks of over 80–85% for every site of the upper aero-digestive tract.[8–10] As expected, smoking showed a higher attributable risk than alcohol for cancers of the larynx while, for the oesophagus, alcohol showed a slightly higher attributable risk than smoking.

The observation that the interaction between alcohol and tobacco appears greater than multiplicative for oral and pharyngeal cancer, multiplicative for laryngeal and intermediate between additive and multiplicative for oesophageal cancer finds an interesting potential interpretation in terms of physiology and pattern of exposure to alcohol and tobacco in each of these sites. Oral cavity and pharynx, in fact, are directly exposed to both risk factors, while larynx is directly exposed to tobacco but not to alcohol and oesophagus is directly exposed to alcohol but not to tobacco.

As noted by a number of investigators,[7] the implication of a joint effect of smoking and alcohol which appears to be, on the whole, greater than additive, is a substantial reduction in the occurrence of cancers of the upper digestive tract by eliminating or moderating one or the other of these two high risk behaviours.

3. LIVER CANCER

Most studies on alcohol consumption and primary liver cancer support a positive association, particularly in low-incidence areas, such as the United States and northern Europe.[2] A dozen of prospective investigations and about the same number of case-control studies showed RR around 2-to-5 in drinkers, as compared to non- or light drinkers, depending upon the amount of alcohol drunk and the co-presence of other risk factors for liver cancer (e.g., aflatoxin exposure, hepatitis B or C virus infection, etc.). A few studies suggested that, alcohol intake being equal, the direct association with liver cancer is stronger in women than in men.

Two models of association have been proposed: 1) alcohol causes liver cirrhosis, which predisposes to cancer; and 2) alcohol is, itself, carcinogenic. The weight of the evidence is in favour of the former model. Sixty-to-90% of liver cancer occurs in cirrhotic livers. In low-incidence areas, liver cancer seems to occur as a consequence of chronic, clinical cirrhosis of alcoholic aetiology. In contrast, in high-risk areas (i.e., China and Thailand), cirrhosis is often only discovered in conjunction with a symptomatic tumour.

The question remaining is whether there is also an association of alcohol and liver cancer independent of cirrhosis. Experimental studies argue against alcohol being directly carcinogenic in the liver, even though it may be related to chromosomal damage. Alcohol may also exacerbate liver damages deriving from malnutrition and hepatitis B and C viral infection.[11]

The alcohol-virus interaction requires now better studies, since few past epidemiological studies included optimal information on hepatitis B virus, and even fewer on hepatitis C virus, whose serological markers have become available only since the early 1990s. Hepatitis C virus is a much more common cause of liver cancer than hepatitis B virus one in developed countries where alcohol has been firmly associated with liver cancer. This new knowledge points to the need of reexamining the role of alcohol in combination with accurate assessment of occurrence and outcome of viral infections.

4. BREAST CANCER

Little attention had been paid to the role of alcohol consumption in the etiology of female breast cancer prior to 1977, when the first report of an unexpected direct association emerged.[2] Breast cancer is the most common cancer in women worldwide (about 800,000 new cases per year) but its etiology and, consequently, preventive strategies remain elusive. Thus, the possibility that a common substance such as alcohol may lead to an even moderate risk increase raised great concern. An impressive bulk of data accumulated from a wide range of female populations.

A meta-analysis of 38 epidemiological studies was carried out by Longnecker in 1994[12] and showed that the RRs associated with consumption of one, two, or three drinks of alcoholic beverages per day were, respectively, 1.1, 1.2, and 1.4, all statistically significant. Although there was strong evidence of a dose-response relation, the slope of the dose-response curve was quite modest. Confounding by known risk factors for breast cancer was unlikely, since in most studies allowance for reproductive and menstrual factors and, in several instances, dietary habits and several other lifestyle characteristics was possible. Nor seemed the association restricted to any particular subgroup of women. However, studies with stronger alcohol/breast cancer associations tended to be from countries with high *per capita* alcohol intake (e.g., southern European countries).

Prospective (cohort) studies are considered especially reliable and it is, therefore, very useful to examine a combined analysis available on the six largest such studies which had information on alcohol and breast cancer.[13] The studies included over 30,000 women from Canada, the Netherlands, Sweden, and the United States and 4335 new breast cancer cases. For intakes less than 60 g/day (i.e., up to about 5 drinks), risk increased linearly with increasing intake. The pooled RR for an increment of about one drink per day was 1.1 (95% CI: 1.0–1.1). Curiously, limited data suggested that alcohol intakes of 5 or more drinks per day (i.e., levels where alcohol-related cancers such as those of the upper aero-digestive tract start increasing substantially) were not associated with further increased risk. The specific estimates for different types of alcoholic beverages (wine, beer, and spirits) were similar. The association between alcohol intake and breast cancer was not modified by factors such as menopausal status, family history of cancer, use of hormone replacement therapy and body mass.

Only recently has some understanding emerged of potential biological mechanisms by which alcohol may increase breast cancer risk. Several studies, although not all, have shown positive correlations between alcohol intake and plasma and urinary oestrogen levels. Intervention studies have found that estradiol levels increased significantly when alcohol, but not placebo, was administered to pre- and postmenopausal women who used hormone replacement therapy.[13]

In conclusion, the data consistently support a direct association between alcohol use and breast cancer. Some aspects of the observed association make, however, a causal interpretation still uncertain: 1) magnitude of the association is modest, 2) the association starts emerging at extremely low levels of alcohol consumption, where no other adverse effect has been established for any other disease, and 3) a correlation of alcohol intake with some still unknown determinant of breast cancer risk cannot be ruled out. Conversely, tobacco smoking does not increase breast cancer risk.

5. OTHER CANCER SITES

Alcohol has been related to a few other cancer sites less consistently than to the above discussed tumours. Many, albeit not all, of over 30 epidemiological studies on the topic have suggested a modest positive relation between alcohol consumption and colorectal cancer. Despite the large amount of research, including the provision of data from many thousand cases, no consensus has emerged. Confounding and bias cannot be ruled out, as explanations of positive results, which are compatible, at most, with very small increases.[2]

Alcohol use has emerged as a weak risk factor for cancer of the lung. Even if true, any impact of alcohol consumption is minor compared with that of smoking. Epidemiological data do not support a causal relationship between alcohol consumption and the risk of pancreatic cancer. An indirect association, *via* the induction of chronic (calcifying) pancreatitis, however, cannot be excluded.[2]

6. CONCLUSIONS

Second only in importance to smoking as a proven cause of cancer, alcohol consumption may be responsible, in conjunction with smoking, for 3-to-10% of cancer deaths in various developed countries. Interestingly, alcohol itself is not carcinogenic in animals.[2] The human evidence, however, is overwhelming for several cancer sites, particularly those of the upper aero-digestive tract and the liver. While in the former tumours heavy alcohol intake is implicated, even quite a small amount may cause some increase in the risk of breast cancer in women. Policies for prevention of alcohol-related cancer, however, have to take into account the global impact of alcohol consumption on human health as discussed by Doll.[14] The consumption of small and moderate amounts of alcohol reduces mortality from vascular disease by about a third. The beneficial effect, in analogy with the carcinogenic potential,[2] seems due to the content of ethanol, not to the characteristics of any particular type of drink.

ACKNOWLEDGMENTS

The work was supported by the Italian Association for Research on Cancer.

REFERENCES

1. Doll R. and Peto R. (1981). The causes of cancer: quantitative estimates of avoidable risks of cancer in United States today. J Nalt Cancer Inst 66, 1191–1308.
2. IARC (1988). Monographs on the Evaluation of the Carcinogenic Risk of Chemicals to Humans, Vol. 44, Alcohol Drinking. Lyon: International Agency for Research on Cancer.

3. World Cancer Research Fund in association with American Institute for Cancer Research (1997). Food, nutrition, and the prevention of cancer: a global perspective. Washington: World Cancer Research Fund and American Institute for Cancer Research.
4. Ferraroni M., Decarli A., Franceschi S., La Vecchia C., Enard L., Negri E., Parpinel M.T., Salvini S., and Nanni O. (1996). Validity and reproducibility of alcohol consumption in Italy. Int J Epidemiol 25, 775–782.
5. Franceschi S., Talamini R., Barra S., Barón A.E., Negri E., Bidoli E., Serraino S., and La Vecchia C. (1990). Smoking and drinking in relation to cancers of the oral cavity, pharynx, larynx, and esophagus in Northern Italy. Cancer Res 50, 6502–6507.
6. BarÒn A.E., Franceschi S., Barra S., Talamini R., and La Vecchia C. (1993). A comparison of the joint effects of alcohol and smoking on the risk of cancer across sites in the upper aerodigestive tract. Cancer Epidemiol Biomarkers Prev 2, 519–523.
7. Blot W.J., McLaughlin J.K., Devesa S.S., and Fraumeni J.F. Jr. (1996). Cancers of the oral cavity and pharynx. In: Schottenfeld D. and Fraumeni J.F. Jr. (eds) Cancer Epidemiology and Prevention. Second Edition. pg. 666–680. New York: Oxford University Press.
8. Negri E., La Vecchia C., Franceschi S., Decarli A., and Bruzzi P. (1992). Attributable risks for oesophageal cancer in Northern Italy. Eur J Cancer 28A, 1167–1171.
9. Negri E., La Vecchia C., Franceschi S., and Tavani A. (1993). Attributable risk for oral cancer in Northern Italy. Cancer Epidemiol Biomarkers Prev 2, 189–193.
10. Tavani A., Negri E., Franceschi S., Barbone F., and La Vecchia C. (1994). Attributable risk for laryngeal cancer in Northern Italy. Cancer Epidemiol Biomarkers Prev 3, 121–125.
11. Thomas London W. and McGlynn K.A. (1996). Liver cancer In: Schottenfeld D. and Fraumeni J.F. Jr. (eds) Cancer Epidemiology and Prevention. Second Edition. pg. 772–793. New York: Oxford University Press.
12. Longnecker M.P. (1994). Alcoholic beverage consumption in relation to risk of breast cancer: meta-analysis and review. Cancer Causes Control 5, 73–82.
13. Smith-Warner S.A., Spiegleman D., Yanun S.-S., van den Brandt P.A., Folsom A.R., Goldbohm R.A., Graham S., Holmberg L., Howe G.R., Marshall J.R., Miller A.B., Potter J.D., Speizer F.E., Willett W.C., Wolk A., and Hunter D.J. (1998). Alcohol and breast cancer in women. A pooled analysis of cohort studies. JAMA 279, 535–540.
14. Doll R. (1997). One for the heart. BMJ 315, 1664–1668.

5

ENERGY SOURCES AND RISK OF CANCER OF THE BREAST AND COLON-RECTUM IN ITALY

Adriano Favero,[1] Maria Parpinel,[1] and Maurizio Montella[2]

[1] Epidemiology Unit, Aviano Cancer Center
Via Pedemontana Occ. 1
33081 Aviano (PN), Italy
[2] Epidemiology Unit
National Cancer Institut Fondazione "G. Pascale"
Cappella dei Cangiani, 80131 Napoli
Italy

1. SUMMARY

Dietary habits are thought to be involved as determinant of breast and colorectal cancer. Nevertheless results of epidemiological studies on diet show several inconsistencies. This is true for the findings related to energy and its sources. Between 1991 and 1996, 2569 women with incident breast cancer (median age: 55 years) and 2588 controls (median age: 56 years), and 1953 subjects with cancer of the colon-rectum (median age: 62 years) and 4154 controls (median age: 58 years) were interviewed in the hospitals of six Italian areas. The validated food frequency questionnaire included questions on 78 foods and recipes and specific questions on individual fat intake pattern. Significant risks for breast and colorectal cancer emerged with increasing intake of energy (odds ratios in highest vs. lowest quintile were 1.32 and 1.49 respectively). Due to the high interrelations existing among the various sources of energy, the separated analysis of each macronutrient didn't achieve the independent estimates of the effects. In order to overcome this situation, we used a completely partitioned model in which all the main sources of energy were entered simultaneously as continuous variables in the regression. High intake of starch led to an increase of cancer risk (odds ratios for an addition of 100 kcal/day were 1.08 and 1.10 for breast and colorectal cancer respectively). A positive association was also found for saturated fat (odds ratios 1.16 for breast and 1.12 for colorectal cancer). High intakes of polyunsaturated fatty acids

Advances in Nutrition and Cancer 2, edited by Zappia *et al.*
Kluwer Academic / Plenum Publishers, New York, 1999.

(chiefly derived from olive and seed oils) were protective more markedly for breast cancer. A possible interpretation of the risk for starch, implies the glycemic overload and hyperinsulinemia due to the high grade of refinement of cereals (the main source of starch) eaten in Italy.

2. INTRODUCTION

Several uncertainties remain with respect to the association between diet and cancer of the breast and colon-rectum (e.g., the role of different types of fat, energy, etc.). Italy offers special research opportunities, on account of the wide variation in the dietary pattern and the relatively low awareness of diet and cancer issues in the population, which leaves probably less scope for recall bias than, for instance, in the United States.[1] We have been able to carry out and publish two large studies on diet and cancers of the breast and colon-rectum. Major study strengths, in addition to the large size and the use of a validated food frequency questionnaire[2,3] were the inclusion of six different Italian areas, including two from the South (i.e., areas still little studied from an epidemiological viewpoint).

3. MATERIALS AND METHODS

Between 1991 and 1994, 2569 women with incident histologically confirmed breast cancer and 2588 control women, in hospital with acute, non-neoplastic diseases, were interviewed in six different Italian areas (i.e., the province of Pordenone and Gorizia, Milan, Genoa, Forlì, Latina, and Naples). Similarly, between 1992 and 1996, interviews were obtained from 1225 subjects with incident histologically confirmed cancer of the colon (of whom 537 females), 728 with cancer of the rectum (of whom 291 females) and 4154 controls (of whom 2081 females). The validated food frequency questionnaire (FFQ) included questions on 78 foods and recipes grouped into six sections and specific questions on individual fat intake pattern. The same structured questionnaire and coding manual were used in each centre, and all interviewers were centrally trained and routinely supervised. The Italian Food Composition Database was used to translate foods into micronutrients.[4] Odds ratios (OR) and 95% confidence intervals (95% CI) were estimated by means of unconditional logistic regression models including terms for study centre, age, education, and alcohol intake for both sites, parity and menopausal status for breast only, and sex and physical activity at work for colon-rectum only. Tests for trend for intake quintiles of energy (the quintile cutpoints were calculated on the joined distribution of cases and controls), were based on the likelihood ratio test between the models with and without a linear term for the exposition variable.

When we turned to energy sources, the separate analysis of each macronutrient couldn't identify the peculiar effect of such interrelated variables. Particularly the notably inverse correlation between starch and saturated fat (even if considered as energy residuals it reached the values –0.5 and –0.4 for breast and colorectal set of controls respectively) didn't allow to show the independent effect of saturated fat. In order to evaluate the independent effects of the various macronutrients, we developed a model[5] in which all the main sources of energy were entered simultaneously in the logistic regression as continuous variables. For all the analysis in continuous, we set the measurement unit as 100 kcal/day.

Table 1. Odds Ratios (OR) and corresponding 95% confidence intervals (CI) of 2569 cases of breast cancer and 1953 cases of colorectal cancer according to quintile of energy intake. Italy, 1991–1996

	Breast cancer			Colorectal cancer		
Quintile of energy intake	Ca:Co	OR[a]	95% CI	Ca:Co	OR[b]	95% CI
1[c]	444:587	1		361:860	1	
2	522:509	1.28	(1.07–1.53)	387:835	1.10	(0.92–1.31)
3	534:498	1.31	(1.10–1.57)	391:831	1.18	(0.99–1.42)
4	547:485	1.40	(1.17–1.68)	384:836	1.15	(0.96–1.39)
5	522:509	1.32	(1.10–1.59)	430:792	1.49	(1.24–1.80)
Trend chi-square		9.14[d]			15.96[d]	
Continuous (100 kcal/day)		1.01	(1.00–1.02)		1.02	(1.01–1.03)

[a]Estimates from multiple logistic regression equations including terms for age, center, education, alcohol intake, parity, and menopausal status.
[b]As above minus parity and menopausal status, and plus sex and physical activity at work.
[c]Reference category.
[d]$p < 0.01$.

4. RESULTS

In Table 1, the association with energy intake is shown both according to quintile of exposition and by means of the risk in continuous for a 100 kcal/day increase in intake, separately for breast and colorectal cancer.[5,6,7] A direct association emerged from the analysis. The trend chi-square was highly significant for both sites examined, and the risk for the highest versus the lowest quintile of intake, was higher for colorectal (OR = 1.49; 95% CI: 1.24–1.80) than breast cancer (OR = 1.32; 95% CI: 1.10–1.59). This result is also reflected by the continuous risk (OR = 1.02 and OR = 1.01 respectively).

The mean daily intake among controls, and the risk for an addition of 100 kcal/day for each major source of energy, for the two sites, are shown in Table 2. Starch intake, by far the first contributor to energy accounting almost one third of total, was significantly associated with risk both for breast and colorectal cancer. Saturated fat intake

Table 2. Odds Ratios (OR) and corresponding 95% confidence intervals (CI) of 2569 cases of breast cancer and 1953 cases of colorectal cancer for a difference of 100 kcal/day in the intake of major macronutrients. Italy, 1991–1996

	Breast cancer			Colorectal cancer		
Macronutrient	Mean daily intake[a] (kc/day)	OR[b]	(95% CI)	Mean daily intake[a] (kc/day)	OR[c]	(95% CI)
Protein	348	0.93	(0.81–1.05)	370	0.86	(0.77–0.97)
Sugars	370	0.98	(0.94–1.02)	373	0.99	(0.95–1.04)
Starch	614	1.08	(1.04–1.11)	712	1.10	(1.07–1.13)
Saturated fat	243	1.16	(1.00–1.33)	264	1.12	(0.98–1.28)
Monounsaturated fat	229	0.98	(0.88–1.09)	230	1.00	(0.91–1.10)
Polyunsaturated fat	95	0.72	(0.62–0.84)	96	0.89	(0.76–1.03)
Alcohol	75	1.07	(1.01–1.12)	181	1.01	(0.98–1.03)

[a]Among controls.
[b]Estimates from multiple logistic regression equations including terms for age, center, education, alcohol intake, parity, and menopausal status and the above listed macronutrients.
[c]As above minus parity and menopausal status, and plus sex and physical activity at work.

was also directly related to cancer risk, although the association was of borderline significance for the two sites. Polyunsaturated fat was the energy source consistently associated with risk reduction, particularly for breast cancer (OR = 0.72; 95% CI: 0.62–0.84). Protein intake did not increase breast cancer risk, but moderately lowered colorectal cancer risk (OR = 0.86; 95% CI: 0.77–0.97). Finally, an increase in alcohol intake led to an elevation of breast (OR = 1.07; 95% CI: 1.01–1.12), but not colorectal cancer risk.

5. DISCUSSION

The results of these two studies contribute to the knowledge of the association between energy and its sources and breast and colorectal cancer. The association of energy intake for breast cancer is in agreement with the literature on case-control studies.[8,9] For colorectal cancer the same result is compatible with a combined analysis of several case-control studies[10] and with evidence that caloric restriction reduces cancer incidence in rodents and colorectal cell proliferation in humans.[11]

Fat has been for long time considered a likely cause of cancer of the breast and colon-rectum, as suggested by the international correlation between fat intake and cancer mortality. During our investigation, however, negative findings were emerging, especially from prospective studies,[12] while the necessity to distinguish animal fat from vegetable fat, became clear. In our study high consumption of polyunsaturated fat, whose principal sources were olive and seed oils, conferred a protection against breast cancer. Also for colorectal cancer, polyunsaturated fat intake was inversely associated to risk, particularly at the right colon.[7] The risk due to saturated fat, as seen before, is sensible to the statistical model. In fact the exclusion from the logistic regression of the other sources of energy, particularly starch, did not allow the detection of risk. The risk is compatible with two combined analysis of case-control studies on breast and colorectal cancer[8,10] that found a significant trend for the former and the risk for energy-adjusted quintiles of intake all above the unity, for the latter. With respect to starch, the risk can be seen as the consequence of the fact that starch, in our population, represents one third of total energy intake. Consequently the direct association with diseases could be influenced by the risk of energy.

Moreover, more than 80% of starch is derived from white bread, pasta, and rice.[13] High intake of refined cereal products can also be responsible for glycemic overload and a compensatory increase of blood insulin, an important growth factor of the human colonic mucosa and a mitogen of tumor cell growth in vitro.[11] The Italian population, thus, offered the rare opportunity to evaluate, in an affluent society, carbohydrates over a large intake range. One fifth of our study population derived more than 60% of their daily calories from carbohydrates, in the vast majority refined bread and pasta.

It is, however, worth noting that, in spite of the improvements in the assessment of dietary habits, all aforementioned associations were rather weak. Therefore, although dietary modifications can contribute to the prevention of cancer of the breast and colon-rectum, a large part of the aetiology of this malignancy remains elusive.

From the point of view of prevention, emphasis should be put on decrease in total energy intake, possibly coupled with an increase in energy expenditure, by means of some even moderate physical activity.[14]

ACKNOWLEDGMENTS

The work was supported by the Italian Association for Research on Cancer.

REFERENCES

1. Giovannucci E., Stampfer M.J., Colditz G.A., Manson J.E., Rosner B.A., Longnecker M., Speizer F.E., and Willett W.C. (1993). A comparison of prospective and retrospective assessments of diet in the study of breast cancer. Am J Epidemiol 137, 502–511.
2. Franceschi S., Barbone F., Negri E., Decarli A., Ferraroni M., Filiberti R., Giacosa A., Gnagnarella P., Nanni O., Salvini S., and La Vecchia C. (1995). Reproducibility of an Italian food frequency questionnaire for cancer studies. Results for specific nutrients. Ann Epidemiol 5, 69–75.
3. Decarli A., Franceschi S., Ferraroni M., Gnagnarella P., Parpinel M.T., La Vecchia C., Negri E., Salvini S., Falcini F., and Giacosa A. (1996). Validation of a food-frequency questionnaire to assess dietary intakes in cancer studies in Italy: results for specific nutrients. Ann Epidemiol 6, 110–118.
4. Salvini S., Parpinel M., Gnagnarella P., Maisonneuve P., and Turrini A. (1998). Banca Dati di Composizione degli Alimenti per Studi Epidemiologici in Italia. Ed. IEO, Milan.
5. Decarli A., Favero A., La Vecchia C., Russo A., Ferraroni M., Negri E., and Franceschi S. (1997). Macronutrients, energy intake, and breast cancer risk: Implications from different models. Epidemiology 8, 425–428.
6. Franceschi S., Favero A., Decarli A., Negri E., La Vecchia C., Ferraroni M., Russo A., Salvini S., Amadori D., Conti E., Montella M., and Giacosa A. (1996). Intake of macronutrients and risk of breast cancer. Lancet 347, 1351–1356.
7. Franceschi S., La Vecchia C., Russo A., Favero A., Negri E., Conti E., Montella M., Filiberti R., Amadori D., and Decarli A. (1998). Macronutrient intake and risk of colorectal cancer in Italy. Int J Cancer 76, 321–324.
8. Howe G.R., Hiroata T.T., Hislop T.G., Iscovich J.M., Yuan J.M., Katsouyanni K., Lubin F., Marubini E., Modan B., Rohan T., Toniolo P., and Shunzang Y. (1990). Dietary factors and risk of breast cancer: combined analysis of 12 case-control studies. J Natl Cancer Inst 82, 561–569.
9. Clavel-Chapelon F., Niravong M., and Joseph R.R. (1997). Diet and breast cancer: review of the epidemiologic literature. Cancer Detect Prev 21, 426–440.
10. Howe G.R., Aromson K.J., Benito E., Castelleto R., Cornée J., Duffy S., Gallagher R.P., Iscovich J.M., Deng-ao J., Kaaks R., Kune G.A., Kune S., Lee H.P., Lee M., Miller A.B., Peters R.K., Potter J.D., Riboli E., Slattery M.L., Trichopoulos D., Tuyns A., Tzonou A., Watson L.F., Whittemore A.S., Wu-Williams A.H., and Zheng S. (1996). The relationship between dietary fat intake and risk of colorectal cancer: evidence from the combined analysis of 13 case-control studies. Cancer Causes Control 8, 215–228.
11. Giovannucci E. (1995). Insulin and colon cancer. Cancer Causes Control 6, 164–179.
12. Hunter D.J., Spiegelman D., Adami H-O., Beeson L., van den Brandt P.A., Folsom A.R., Fraser G.E., Goldbohm R.A., Graham S., Howe G.R., Kushi L.H., Marshall J.R., McDermott A., Miller A.B., Speizer F.E., Wolk A., Yaun S.S., and Willett W. (1996). Cohort studies of fat intake and the risk of breast cancer—a pooled analysis. N Engl J Med 334, 356–361.
13. Favero A., Salvini S., Russo A., Parpinel M., Negri E., Decarli A., La Vecchia C., Giacosa A., and Franceschi S. (1997) Sources of macro- and micronutrients in Italian women: results from a food frequency questionnaire for cancer studies. Eur J Cancer Prev 6, 277–287.
14. D'Avanzo B., Nanni O., La Vecchia C., Franceschi S., Negri E., Giacosa A., Conti E., Montella M., Talamini R., and Decarli A. (1996). Physical activity and breast cancer risk. Cancer Epidemiol Biomarkers Prev 5, 155–160.

6

ORGANOCHLORINES AND BREAST CANCER

A Study on Neapolitan Women

Rossano Dello Iacovo,[1] Egidio Celentano,[1] Anna Maria Strollo,[1] Giacomo Iazzetta,[1] Immacolata Capasso,[2] and Giacomo Randazzo[3]

[1] Tumour/Tissues Central Bank; Senology Surgery
[2] Senology Surgery
National Tumour Institute, Naples
Italy
[3] Biochemistry, Science of Food Preparation
University School of Agriculture
Federico II University, Naples
Italy

INTRODUCTION

In the past two decades efforts in controlling breast cancer (BC) in developed and developing countries have faced a dramatic increase of incidence. It is currently the most frequent cancer, and one of the main causes of premature loss of life in women throughout the developed world[1] (Fig. 1). In the near 2000 the expected number of new cases of BC could be about 1.3 million.

Although the knowledge on its aetiology has advanced during the last few years, the role of diet and hormones is still a highly debated, open question.[2,3] Experimental and epidemiological studies suggest that risk is enhanced by intensity and time of the mammalian epithelium exposure to estrogens.[4] Only 5% of the BC risk in developed countries is attributable to genetic predisposition; however, the genetic susceptibility is expressed in families with BC as well as other cancers (ovarian, prostate, colon), confirming a familial non-specific tendency to genetic mutations (miscopying DNA and/or failing to repair).

All known causes explain almost 30% of BC cases. Most of these (age at menarche, first full term pregnancy, age at menopause, family medical history) are expressions of exposure to reproductive hormones. They are widely accepted as major risk factors for breast cancer (Table 1).[5]

Advances in Nutrition and Cancer 2, edited by Zappia *et al.*
Kluwer Academic / Plenum Publishers, New York, 1999.

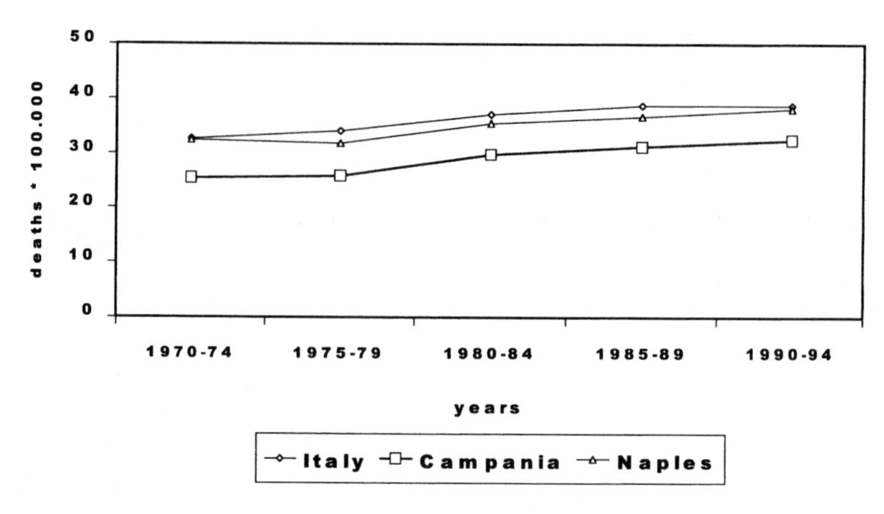

Figure 1. Trends of breast cancer mortality in Italy, Campania, Naples—1970–1994.

Environmental exposure to endocrine-disrupting chemicals may have an important role, as the international incidence variations and studies carried out on migrant populations have shown.[6–7] In spite of their overall increasing levels throughout the world, research is still needed on specific environmental chemicals that may contribute to BC risk.

Data from aetiological studies from 1976 to 1998 have suggested that BC risk may be associated with persistent organochlorine exposure.[8–19] These chemicals, highly chlorinated, characterised by strong chlorine-carbon bonds, are carcinogenic in animals,

Table 1. Risk factors for breast cancer in females

Factor	High risk	Low risk	Magnitude of risk differential
Country of birth	North America, Northern Europe	Asia, Africa	****
Socioeconomic class	Upper	Lower	***
Age	Old	Young	****
Marital status	Never married	Ever married	**
Place of residence	Urban	Rural	**
Age at first full-term pregnancy	Older than 30	Younger than 20	***
Oophorectomy	No	Yes	***
Body build, postmenopausal	Obese	Thin	***
Age at menarche	Early	Late	**
Age at menopause	Late	Early	**
Family history of premenopausal bilateral breast cancer	Yes	No	****
History of cancer in one breast	Yes	No	****
History of fibrocystic disease	Yes	No	***
First-degree relative with breast cancer	Yes	No	***
History of primary cancer in ovary or endometrium	Yes	No	***
Radiation to chest	Large doses	Minimal doses	***

Adapted from a literature review, ****, relative risk: >4.0; ***, relative risk: 2.0–4.0; **, relative risk: 1.1–1.9.

sequestered in fatty tissues, and are immune response suppressors. Their attraction to fats (vs. water), low metabolism, and chemical inertness allow them to reside for a lifetime in the body, as in the environment.[20] In addition to their fat persistence and carcinogenicity, some of these chemicals are active as reproductive hormones in laboratory tests, in experimental animals and in wildlife; among these, xenoestrogens are tumour promoters, as are natural estrogens. They have become ubiquitous in the environment. The residues, due to the high lipid solubility, are mainly stored at every level of food chain, more firmly linked to fatty meat, poultry, milk, and dairy products. The greater is their diffusion, the worse are the trends of the breast cancer incidence rates.

Organochlorines include a wide class of compounds; the most frequently investigated in humans, are: I) pesticides (α-BHC, β-BHC, γ-BHC, δ-BHC, heptachlor, aldrin, heptachlor epoxide, endosulfan I, dieldrin, pp′-DDE, endrin, endosulfan II, p p′-DDD, endrin aldehyde, endosulfan sulfate, pp′-DDT, endrin ketone, methoxychlor; II) PCB's (10 to 40 congeners classified in three groups according to estrogenic/anti-estrogenic activities and P-450 isozymes induction)[21] have been used since 1929 as dielectric or cooling fluids, but a significant association to breast cancer has never been demonstrated.

Investigations into the relationship between organochlorines exposure and breast cancer have been published during the past 22 years (Table 2); the results of these efforts are suggestive but not conclusive. Contributing to these controversial outcomes are three major causes: the studies are carried out in populations exposed to low doses over a long time period; the variability in types of hormonal disruption (mimicking oestrogen, or anti-oestrogen, etc.) due to the environmental mixed disrupter compounds exposition; the variability in host phenotype (e.g.: phase I enzyme isoforms); these add to the other differences in: year of taking the samples, study design, type of sample and drawing procedures, time span of a sample storage, evaluation of other environmental factors (e.g.: diet, pollutants) (Table 3).

It is true that science progresses through the debate among different conclusions on the same matter, but here we have too many objective and methodological differences in approaching the problem, which often lead us toward doubtful goals.

Table 2. Previous evidence of the relationship between organochlorinated compounds and breast cancer

Year	First author	Country	Cases/controls	Results
1976	Wasserman	Brazil	5/9	+
1984	Unger	Denmark	14/21	−+
1990	Mussalo-Rauhamaa	Finland	41/33	−+
1992	Falck	USA	63/59	+
1993	Wolff	USA	58/171	+
1994	Dewailly	Canada	20/17	+
1994	Krieger	USA	150/150	−+
1997	Van't Veer	North Europe	265/341	−
1997	Lopez Carrillo	Mexico	141/141	−
1997	Hunter	USA	236/236	−
1998	Moysich	USA	154/192	−+
1998	Guttes	Germany	45/20	+

Table 3. Previous evidence of the relationship between organochlorinated compounds and breast cancer

Year	First author	Samples	Unit	Design	Results
1976	Wasserman	Breast adipose tissue	p.p.m.	Clinical series	+
1984	Unger	Breast adipose tissue	p.p.m.	Clinical series	+−
1990	Mussalo-Rahumaa	Breast tissue	mg/mg fat	Clinical series	+−
1992	Falck	Breast adipose tissue	ng/g	Clinical series	+
1993	Wolff	Serum	p.p.b.–ng/ml	Cohort nested case-control	+
1994	Dewailly	Breast adipose tissue	μg/kg	Clinical series	+
1994	Krieger	Serum	p.p.b.	Cohort nested case-control	+−
1997	Van't Veer	Buttocks adipose tissue	μg/kg	Case-control	−
1997	Lopez-Carrillo	Serum	p.p.b.	Case-Control	−
1997	Hunter	Plasma	p.p.b.	Cohort nested case-control	−
1998	Moysich	Serum	ng/g	Case-control	+−
1998	Guttes	Breast adipose tissue	μg/kg	Clinical series	+

MATERIALS AND METHODS

We propose and have adopted a study plan in three steps, to be carried out on the same population sample: 1) a case control study to evaluate the association between organochlorine body burden and BC; 2) a case control study on the association of organochlorine concentrations in breast adipose tissue and BC; and 3) the study of the effect of the induced CYP-450 isoforms on BC risk.

The first step is a case control study to examine the effect on BC risk of the body burden of organochlorines able to mimic reproductive hormone actions.

A total of 170 cases and 190 community controls were examined. Cases were taken from 1997 until the first sixth months of 1998, among Neapolitan pre- and post-menopausal women undergoing their first surgical treatment for BC in the Department of Surgical Senology of the Istituto Nazionale per lo Studio e la Cura dei Tumori at Naples (ITN), they had never received other treatment. Controls were a subset of healthy women participating in a cohort study on diet and cancer, under a grant to the author by the Italian Health Ministry, being carried out since 1993 in the same institution.

The lifestyles and dietary habits data, harvested by a questionnaire, were used to identify the study subjects and recognise risk factors prevalent in the sample.[3] Cases and controls were interviewed in person by trained interviewers; the interview lasted approximately 1 hour and included assessment of reproductive history, occupational history, smoking habits, medical history, use of exogenous hormones, family history of cancer, alcohol intake, and a food frequency questionnaire that assessed usual intake in the recent years prior to the interview. Cases were interviewed after histology.

All cases and controls agreed to donate a blood sample when fasting. Cases and controls also donated breast adipose tissue. Breast cancer shares were taken in all cases. Blood and tissues were immediately processed to place into a -80°C freezer in the Tumour Tissue Bank laboratory (ITN), staffed by a trained laboratory technician, blinded to status of disease.

LABORATORY METHODS

The blood was obtained by venous puncture, collected in a Monovette® tube and kept at 37°C for 40 min. The serum was isolated following centrifugation and stored, aliquoted in 1.3 ml, in a polypropilene tube at –80°C.

The serum was thawed at room temperature and 1 ml transferred into a deactivated 10 ml glass tube with teflon lined screw cap.[22] Methanol was added, vortexed and extracted with 1 : 1 hexane:ethyl ether. After centrifugation, using a disposable Pasteur pipette previously rinsed with acetone and hexane, the organic extract was transferred into a deactivated 20 ml glass tube with teflon lined screw cap. The extraction was repeated twice.[23] The extracts were combined and reduced in solvent volume, at room temperature, to 0.5 ml under a gentle stream of prepurified nitrogen. The sample was eluted through a Florisil column with a 6% fraction ethyl ether in petroleum ether and another 15% fraction ethyl ether in petroleum ether.[12,22] The solvent was evaporated and then diluted in hexane; gas chromatography was performed, detection was by double electron capture (^{63}Ni) on double column (Rtx 5: 5% diphenyl—95% dimethylpolysiloxane, 30 m, 0.32 mm ID; Rtx 1701: 14% cyanopropyl—86% dimethylpolisiloxane, 30 m, 0.32 mm ID), in temperature program mode with splitless injection on a gas chromatograph Hewlett Packard 5890 Series II. Results are reported as nanograms per milliliter (ppb).[24] The load for each analyte was determined as the mean of background noise plus 3 standard deviations in five reagent blank samples.

The methods adopted allow the recovery of the organochlorines shown in Table 4. We prefer to process serum because of a lower organochlorines recovery in plasma (mean: –20%), added to a higher threshold noise (mean: +50%) and to an excessive need of solvents (unpublished data).

Our laboratory results were also tested in the Biochemistry Department at the University School of Agriculture, to ensure an inter-laboratory quality control.

RESULTS

The description of cases and controls are shown in Table 5.The values presented are expressed as means and standard deviations. The age is higher in cases than in controls. The lactation time is lower in controls, the prolonged breast-feeding is widely accepted as a protective factor, so this appears to be irrelevant. This is clarified by the

Table 4. Percent of extraction from contaminated serum

	%		%
α-BHC	90	p,p′-DDE	98
β-BHC	90	Endrin	106
γ-BHC	95	Endosulfan II	82
δ-BHC	90	p,p′-DDD	117
Heptachlor	96	Endrin aldehyde	97
Aldrin	92	Endosulfan sulfate	85
Heptaclor epoxide	87	p,p′-DDT	115
Endosulfan I	87	Endrin ketone	86
Dieldrin	101	Methoxychlor	108

Table 5. Description of cases and controls features

	Cases (n = 170) m (SD)	Controls (n = 195) m (SD)
Age (yrs)	54.2 (10.1)	52.0 (9.2)
Body mass index (kg/m^2)	28.7 (5.2)	27.8 (4.3)
Age at menarche (yrs)	12.8 (1.7)	12.3 (1.4)
Age at menopause (yrs)	48.8 (4.8) [109]	48.7 (5.0) [99]
Age at first pregnancy (yrs)	25.3 (5.3) [142]	25.1 (4.9) [168]
Lactation (months)	21.9 (27.2) [124]	15.2 (17.3) [150]
Serum cholesterol (mg/dl)	204.5 (39.3)	218.3 (43.3)

following data regarding the number of pregnancies. Cholesterol values in cases are lower than in controls.

The frequency distributions of some other features in cases and controls are presented in Table 6: there are less pre-menopausal women in cases, no differences in family history for breast cancer.

In our sample, parity seems to play a watershed role between cases and controls: one or two pregnancies are a protecting condition due to lactation,[25] while nulliparity or having more than two children presents risk for breast cancer. The subset having more than two children is prevalent in cases: this data is congruent with the previous, regarding a longer breast feeding period (caused in our sample by too many pregnancies, which lead to a higher risk or lower protection). Nulliparous are more frequent in cases.

A relevant factor is the high prevalence of smoking habits in Neapolitan women, particularly high in cases. There is a little difference in distribution of educational level, with a higher percentage of better educated women in controls.

The mean values and standard deviations of organochlorines detected in our sample have been adjusted for age, menopausal status, serum cholesterol, and parity. There is a significant difference only in heptachlor concentration between cases and controls (Table 7).

It is important to note that half of our sample has a homogeneous presence of an additional peak other than 4-4′-dde.

Table 6. Frequency distribution of other sample features

	Cases (n = 170) %	Controls (n = 195) %
Post-menopausal women	64.5	50.8
Positive BC family history	7.1	8.2
Parity		
Nulliparous	16.5	12.8
1–2 children	37.6	50.8
3+ children	45.9	36.4
Smoking habits		
Never	23.7	37.2
Past	13.0	14.8
Current	63.3	48.0
Education		
8 yrs	80.0	65.6
9–12 yrs	14.7	24.6
13+ yrs	5.3	9.7

Table 7. Serum concentration mean values of organochlorines detected in cases and controls

	Cases (n = 170) m (SD) n	Controls (n = 195) m (SD) n
β-BHC (ng/ml)	1.76 (1.65) 37	1.49 (0.71) 16
Heptachlor (ng/ml)	2.86 (2.17) 9	1.16 (0.54) 19
4,4′-DDE (ng/ml)	9.55 (5.42) 170	8.98 (5.17) 195
4,4′-DDT (ng/ml)	2.47 (1.08) 12	1.77 (0.85) 11
Endrin aldeide (ng/ml)	4.73 (3.64) 29	3.78 (3.44) 22
Number of peaks	1.54 (0.80) 170	1.41 (0.65) 195

The distribution of cases and controls in dde tertiles (based on the distribution of' this compound in the controls) and the trend in Odds ratio using the first tertile as reference are shown in Table 8. Odds ratios were calculated by unconditional logistic regression with 95% confidence intervals. Data were adjusted for potential confounders including age, body mass index, lactation, parity, serum lipids. As plotted, the Odds ratio confidence intervals are across one and are not significant.

No association was found between organochlorine concentrations in serum and breast cancer. The trend in Odds ratio increase starting from 4-4′-dde values in serum over 10 ng/ml should be noted (Fig. 2).

CONCLUSIONS

1) DDT and other pesticides are still produced in developed countries and widely used also in developing ones.[16] By means of agricultural trade they cross the international borders and join the servings around the world. The research on the steady rise of breast cancer occurrence in the past decades has pointed out only three factors generally agreed upon to be linked to BC: age, country of birth, and family history. These factors are not likely to change. During the last ten years clinical oncology has done a good job in treatment and diagnosis. A cure exists, but no pathways for prevention have yet been found.
2) Current studies have demonstrated that individuals in industrialised regions now carry 250 chemicals imprisoned in fat tissues. Since organochlorines mimic estrogens in experimental studies, they may serve as breast cancer promoters. The persistent organochlorines found today are a surrogate measure for previous exposure, also early in life, to a wider number of chemicals having xenoestrogen activity but a short half-life in the body and/or the traces of a remote past, the body's burden of higher doses of the same pesticides.
3) Organochlorines induce CYP-450 families.

Table 8. 4-4′-DDE serum concentrations in cases and controls

4-4′ DDE tertiles	Cases	Controls	OR (95% CI)
<6.0 ng/ml	51	66	1.0
6.0–10.2 ng/ml	49	63	0.84 (0.47–1.51)
>10.2 ng/ml	70	66	1.24 (0.70–2.20)

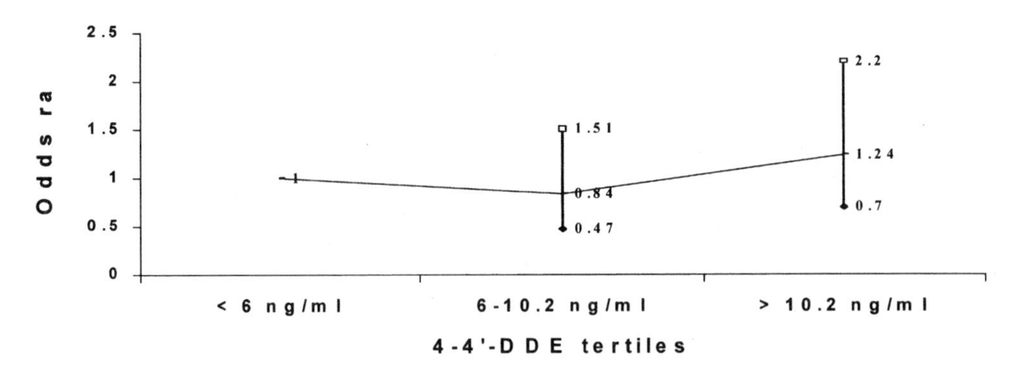

Figure 2. Plot of odds ratio confidence intervals.

a) These enzymes bio-activate many chemical carcinogens such as PAH, the prototypical animal model for chemical carcinogenesis, including breast cancer;
b) CYP-1A1, CYP-2A isoforms of CYP-450 induced by organochlorines influence the estradiol metabolism producing 16α-OH E1, the "bad" estrogen.

Therefore, the 2-OHE1/16α-OHE1 metabolite ratio is lowered, and the risk of breast cancer rises.[26,27]

Cases control studies should be carried out on the same population sample, to investigate: the serum to evaluate the organochlorines body burden, the breast adipose tissue to detect their contents in the estrogens-target organ (ER+)[13] and the CYP-450 isoforms to identify the influence of individual variability on BC risk.

The question to disentangle is of considerable interest for scientific research and public health; we cannot declare "in dubio pro reo", but to reiterate general assessment, a consensus protocol on the basic features of a study design should be achieved.

PERSPECTIVES OF PREVENTION

We should not think only to further research, to be more sure of the causal link between organochlorines and breast cancer, this would mean to play a game of chance, paying with pain and illness.

The human (and animal) variability in response to equal doses of a carcinogen is current data and shows the individual attitude to correct DNA damage(s), but also depends on the level and time of exposure to, and on the protective effects of inhibition by compounds introduced with diet and/or lifestyles derived. The same happens for xenoestrogens.

To help the spontaneous protective actions, two levels of preventive approach are possible: the individual behaviours to reduce personal exposure, and the general intervention on production, storage and distribution of foods. Promising in vitro studies on a protective effect of isoflavonoids (soy derived): biocianin a, genistein, equol, and curcumin (a turmeric derivative) could open the way to providing better answers about obtaining personal preventive behaviour through dietetic intervention and chemoprevention, since the organochlorine persistence in the environment is difficult to remove.[28,29]

Furthermore, a general intervention target is coming into sight: the transgenic agricultural production. Parasite-resistant, these products should not need to be sprayed with increasingly new pesticides that are diffused in the environment.

But this will be a task for the next millennium.

ACKNOWLEDGMENTS

This paper was prepared as part of the latest project: "Diet and cancer. . . .", funded by Italian Health Ministry to Istituto Nazionale per lo Studio e la Cura dei Tumori di Napoli, grant: 51.1/RF93.33, Head: Prof. Rossano Dello Iacovo. The author thanks: Dr. Giuseppe D'Aiuto, farseeing director of the Surgical Senology ward at Istituto Nazionale per lo Studio e la Cura dei Tumori di Napoli, and Dr. Nicola Fortunato, biologist, for his devotion to Tumor Tissue Bank management.

REFERENCES

1. Parkin D.M. et al., Estimates of the worldwide incidence of 25 major cancers in 1990. Int J Cancer 1999;80:827–841.
2. Riboli E. et al., Nutrition and cancer. Background and rationale of the European Prospective Investigation on Cancer and Nutrition (EPIC). Ann Oncol 1992;3:783–791.
3. Panico S., Dello Iacovo R. et al., Progetto ATENA, a study on the etiology on major chronic diseases in women: design, rationale, and objectives. Eur J Epidem 1992;8:601–608.
4. Colditz G.A. et al., The use of estrogens and progestins and the risk of breast cancer in postmenopausal women. N Eng J Med 1995;332:1589–93.
5. Kelsey J.L. et al., Breast cancer epidemiology. Cancer Res 1988;48:5615–5623.
6. Devra Lee Davis et al., "Recent developments on the avoidable causes of breast cancer" in Preventive Strategies for living in a chemical world: a symposium in honor of Irving J Selikoff. Annals of the New York Academy of Sciences, 837, Eula Bingham and David P. Rall, eds. (New York Academy of Sciences, New York, 1997), 514–520.
7. Regina G. Ziegler et al., Migration patterns and breast cancer risk in Asian-American women. J Natl Cancer Inst, 1993;85:1819–1827.
8. Wasserman M. et al., Organochlorine compounds in neoplastic and adjacent apparently normal breast tissue. Bull Environ Contam Toxicol 1976;15:478–484.
9. Unger M. et al., Organochlorine compounds in human breast fat from deceased with and without breast cancer and in a biopsy material from newly diagnosed patients undergoing breast surgery. Environ Res 1984;34:24–28.
10. Mussalo-Rauhamaa H. et al., Occurrence of ß-hexachlorocyclohexane in breast cancer patients. Cancer (Phila.) 1990;66:2124–2128.
11. Falck F. Jr et al., Pesticides and polychlorinated biphenyl residues in human breast lipids and their relation to breast cancer. Arch Environ Health 1992;47:143–146.
12. Wolff M.S. et al., Blood levels of organochlorine residues and risk of breast cancer. J Natl Cancer Inst 1993;85:648–652.
13. Dewailly E. et al., High organochlorine body burden in women with estrogen receptor positive breast cancer. J Natl Cancer Inst 1994;86:232–234.
14. Krieger N. et al., Breast cancer and serum organochlorines: a prospective study among white, black, and Asian women. J Natl Cancer Inst 1994;86:589–599.
15. Van't Veer P. et al., DDT (dicophane) and postmenopausal breast cancer in Europe: case-control study. BMJ 1997;315:81–85.
16. Lopez-Carrillo L. et al., Dichlorodiphenyltrichloromethane serum levels and breast cancer risk: a case-control study from Mexico. Cancer Res 1997;57:3728–3732.
17. Hunter D.J. et al., Plasma organochlorine levels and the risk of breast cancer. N Engl J Med 1997;337:1253–1258.

18. Moysich K. et al., Environmental organochlorine exposure and postmenopausal breast cancer risk. Cancer Epidemiol Biomarkers Prev 1998;7:181–188.
19. Guttes S. et al., Chlororganic pesticides and polychlorinated biphenyls in breast tissue of women with benign and malignant breast disease. Arch Environ Contam Toxicol 1998;35:140–147.
20. Dewailly E. et al., Could the rising levels of estrogen receptor in breast cancer be due to estrogenic pollutants? J Natl Cancer Inst 1997;89:888–889.
21. Wolff M.S. et al., Environmental organochlorine exposure as a potential etiologic factor in breast cancer. Environ Health Perspect 1995;103:141–145.
22. Burse V.W. et al., Partitioning of Mirex between adipose tissue and serum. J Agric Food Chem 1989;37:692–699.
23. AOAC Official Method of analysis 16th Edition 1995.
24. Long G.L. et al., Limit of detection, a closer look at the IUPAC definition. Anal Chem 1983;55:722–724.
25. Dewailly E. et al., Protective effect of breast feeding on breast cancer and body burden of carcinogenic organochlorines. J Natl Cancer Inst 1994;86:803.
26. Taioli E. et al., Ethnic differences in estrogen metabolism in healthy women. J Natl Cancer Inst 1996;88:617.
27. Stresser D.M. et al., Human cytochrome P450-catalyzed conversion of the proestrogenic pesticide methoxychlor into an estrogen. Role of cyp2c19 and cyp1a2 in o-demethylation. Drug Metab Dispos 1998;26:868–874.
28. Verma S.P. et al., Curcumin and genistein, plant natural products, show synergistic inhibitory effects on the growth of human breast cancer MCF-7 cells induced by estrgenic pesticides. Biochem Biophys Res Commun 1997;233:692–696.
29. Verma S.P. et al., The inhibition of the estrogenic effects of the pesticides and environmental chemicals by curcumin and isoflavonoids. Environ Health Perspect 1998;106:807–812.

7

OLIVE OIL CONSUMPTION AND CANCER MORTALITY IN ITALY

A Correlation Study

Amleto D'Amicis and Sara Farchi

Statistics and Food Economics Unit
Istituto Nazionale della Nutrizione
Via Ardeatina, 546, 00178 Rome
Italy

INTRODUCTION

Total dietary fat and animal fat intakes are frequently positively associated with site specific cancer mortality, the most common association being found with cancer of the breast, colon and prostate. Results from epidemiological studies are, however, controversial with respect to the relationship between total dietary fat consumption and breast cancer risk; there is more agreement that a high fat diet is associated with prostate cancer.[1]

Many epidemiological studies (ecologic, case-control, cohort) support a positive correlation between consumption of animal fat and risk of prostate cancer, but current evidence suggests that vegetable fat is not related to risk of this cancer.[2] Several studies of breast and pancreatic cancer in many Mediterranean populations have demonstrated that increased dietary intake of olive oil is associated with a slightly decreased risk or no increased risk of cancer, despite a higher proportion of overall lipid intake.[3] Other experimental animal model studies of high dietary fat and cancer also indicate that olive oil has either no effect or a protective effect on the prevention of a variety of chemically induced tumors.[4] The strongest evidence that monounsaturated fat may influence breast cancer risk comes from studies of populations living in Southern Europe, in whom intake of oleic acid sources (olive oil) appears protective.[5]

Cancer mortality rates are lower in the Mediterranean areas than in the Northern Europe.[6] The correlation between cancer and diet suggests that certain characteristics of Mediterranean diet such as relatively higher consumption of fish, olive oil,

Advances in Nutrition and Cancer 2, edited by Zappia *et al.*
Kluwer Academic / Plenum Publishers, New York, 1999.

vegetables and fruit, and lower consumption of meat and animal fat are associated with low prevalence of this disease. Several epidemiologic studies have verified that the monounsaturated fat, which is the major component in olive oil, could act as a protective factor for cancer.[5,7] In Northern Italy, where colon and breast cancer mortality rates are higher,[8] butter is the major type of fat, while in the Southern regions olive oil is most commonly consumed.[9] A case-control study evaluating dietary factors and their contribution to the geographic variation in mortality from gastric cancer within Italy,[10] suggests that the protective effects reported for consumption of fresh fruit, fresh vegetables, and olive oil may be linked to the vitamins C and E contained in these foods.

The aim of this study is to explore the existence of correlation between olive oil consumption and cancer mortality rate in Italy where the intake of seasoning fats shows relevant geographic variation.

MATERIALS AND METHODS

Mortality rates standardised for age brackets 0–74 and 75 and over, for 94 Italian USL districts were considered. These were chosen from the following administrative regions, distributed so as to be representative of the four main parts of the country: Lombardia, Trentino, Veneto, Marche, Lazio, Molise, Campania, Calabria, and Sardegna. Mortality rates referred to the years 1980–82 and causes of death were tumours in general as well as at specific sites.[8]

Dietary data were obtained from a nationwide food and eating habits survey carried out in 1980–84 by the Italian National Institute of Nutrition (INN) on a large sample of 12,000 households. Dietary data were collected by weight inventory method for seven days. To calculate the composition of the diet for subjects in the 94 USL districts, the mean values were used.[9]

Various methods are available to eliminate the effect of correlation between nutrients and calorie intake.[11,12] In this study the energy adjusted nutrient method was used, which considers the residues of regression between nutrients as dependent variables and energy, expressed in calories, as an independent variable. Linear regression models with the regression residues added to the mean intake of the various nutrients were employed, so as to obtain positive values, likely to resemble the real amounts of each item eaten.

The mortality rate for gastrointestinal tumours was calculated by adding rates for tumours of the stomach, large intestine, and pancreas. Pearson's correlation coefficients were calculated between mortality rates per tumour, the specific sites, and the energy adjusted nutrients. The coefficients of regression between mortality rates were calculated as the independent variable and energy adjusted nutrients as the dependent variable.

RESULTS

Mean intake and sources of seasoning fats (olive oil, other vegetable oils, and animal fats) are shown in Table 1; data are presented by geographical area. As we can see the olive oil consumption differs between areas: in the Centre and South the intake is more than double with respect to the Northern areas, while we observe the opposite

Table 1. Mean seasoning fat intake in 94 Italian USL (g/day/per person) by geographical area

	Northwest	Northeast	Center	South	F=
Olive oil	15.3	13.9	33.5	33.9	43.4**
Other vegetable fats	22.6	31.1	15.1	19.4	13.1**
Animal fats	17.2	16.4	7.8	7.1	35.4**

** = P < 0.0001

in the intake of other vegetable oils and animal fats, which double in the two Northern areas. All these differences are statistically significant (P < 0.0001).

Pearson's correlation coefficients between the intake of olive oil and other types of seasoning fats are negative: r = –0.52 with animal fat; r = –0.66 with other vegetable fats. This result confirms the geographic differences observed in Table 1, and stresses the fact that the habits in the consumption of seasoning fats are very deeply rooted in the several areas. The correlation coefficients between intake of seasoning fats, from different origins, and total cancer mortality rate in the 94 USL are shown in Table 2; the coefficients are reported separately for men and women.

As we can see the only statistically significant negative association with total cancer mortality is olive oil consumption. The other correlations are statistically significant but positive. No statistically significant association was found with the intake of total seasoning fats.

Correlation coefficients between olive oil consumption and intake of other nutrients, adjusted for energy, indicate the existence of a food pattern other than the simple olive oil, which represents a protective factor for cancer. The nutrients positively associated with olive oil are: oleic acid (0.83), vegetable proteins (0.64), fibre (0.61), vegetable fats (0.52), vitamin C (0.46), and iron (0.45) (maybe from vegetable origin). The nutrients inversely associated with olive oil are animal fats (–0.68), saturated fatty acids (–0.54), linoleic acid (–0.53), and animal protein (–0.44).

DISCUSSION

The main methodological problem in this type of correlation analysis concerns the data collection time. Mortality data used here refer to 1980–82 and dietary consumption were obtained from a 1980–84 food survey. These dietary data do not, therefore, reflect the past diet,[13] which is what might have given rise to risk. In order to make sense of the results of this study, the olive oil consumption should be unchanged at least during the past decade before the survey. In order to verify that diet remained

Table 2. Pearson's correlation coefficients among seasoning fat consumption and total cancer mortality in males (0–74 yrs) and females (over 74 yrs)

	Males (0–74 yrs)	Females (over 74 yrs)
Olive oil	–0.64*	–0.51*
Other vegetable fats	0.36*	0.27*
Animal fats	0.55*	0.47*
Total seasoning fats	–0.10 n.s.	–0.07 n.s.

* = P < 0.001; n.s. = not significant

Table 3. Vegetable oil consumption in Italy in 1973 (g/day/pp)

	Northwest	Northeast	Center	South
Vegetable oils	55.9	55.9	69.0	65.8

Source: ISTAT, households food basket survey[14]

unchanged over decades we compared the seasoning oil consumption in 1973, as collected by Italian National institute of Statistic (ISTAT) in the Italian households survey.[14] with seasoning oil consumption collected during the 1980–84 survey of INN. Unfortunately, ISTAT does not collect olive oil and vegetable oil data separately, so the comparison was done with the whole oil consumption. In Table 3 is shown the total vegetable oil consumption during 1973 divided by Italian areas.

As we can see the oils consumption is higher in the Centre and in the South. In Table 4 are reported the comparisons, in terms of index numbers, of oil consumption in 1973 (ISTAT) and 1980–84 (INN). The differences between the early 70's and 80's are small in the Northwest and Centre, while no differences are evident in the Northeast and the South. The differences are not relevant enough to distort the results of our analysis.

The association between seasoning fats consumption and total cancer mortality is stronger in the age class 0–74 yrs in men and over 75 yrs in women. It is possible to hypothesise that this observation could be attributable to the different age distribution among male and female cohorts and thus with the shift to older age in females the risk of tumour.

The results of this analysis support the findings of another correlation study between nutrient intake and cancer mortality in Italy.[15] In fact, the correlation between olive oil consumption and energy adjusted nutrient intake shows that the association observed between olive oil and tumour mortality could be attributable to the olive oil composition—i.e. the polyphenols: squalene, hydroxytyrosol and oleuropein for example; α-tocopherol; and monounsaturated fat acid—but also to the fact that olive oil could be a marker of a food pattern rich in vegetables, which are well represented in the Mediterranean way of eating.

Further, more statistically powerful investigation, such as case-control or follow up study, should be carried out to better understand the role of olive oil in cancer prevention, evaluating the effect of other nutrients and/or other biological variables. We can conclude that olive oil consumption, rather than other sources of fat, could be not only harmless but also protective in cancer onset. Furthermore, the associations observed between olive oil and vegetable nutrients and between olive oil and animal products suggest that the protective effects reported for this source of fat may be linked to a diet rich in vegetables and fruits.

Table 4. Comparison of index numbers of consumption of seasoning fats in 1973 (ISTAT) and 1980–84 (INN) in different Italian areas

	Northwest	Northeast	Center	South	Italy
1973	89	90	110	105	100
1980–84	76	90	97	106	100

ACKNOWLEDGMENTS

We are greatly indebted to Mrs. Antonella Pettinelli and Mr. Fabrizio Forlani for their precious help during the data process.

REFERENCES

1. Rose D.P. Dietary fatty acids and cancer. Am J. Clin Nutr, 66:4 suppl 998s–1003s, 1997.
2. Willett W. Specific fatty acids and risks of breast and prostate cancer: dietary intake. Am J Clin Nutr 66:6 suppl, 1557s–1563s, 1997.
3. La Vecchia C. Dietary fat and cancer in Italy. Eur J Clin Nutr 47:S35–8, 1993.
4. Newmark H.L. Squalene, olive oil, and cancer risk: a review and hyothesis. Cancer Epidemiol Biomarkers Prev, 6:12, 1101–3, 1997.
5. Simonsen N.R., Fernandez Crehuet Navajas J., Martin Moreno J.M., Strain J.J., Huttunen J.K., Martin B.C., Thamm M., Kardinaal A.F., vant Veer P., Kok F.J., and Kohlmeier L.
6. WHO. Healthy nutrition. Preventing nutrition-related diseases in Europe. WPT James (ed) World Health organization Regional Office for Europe, Copenhagen. WHO Regional Publications, European Series, No. 24, 1988.
7. Lipworth L., Martinez M.E., Angell J., Hsieh C.C., and Trichopoulos D. Olive oil and human cancer: an assessment of evidence. Prev med 26:2, 181–90, 1997.
8. ISTAT-ISS (Italian National Institute of Statistics)-(Italian National Health Institute); Mortalità per cause e Unità Sanitaria Locale 1980–82. Arti grafiche Vallagarina SpA, Trento 1988.
9. Saba A., Turrini A., Mistura G., Cialfa E., and Vichi M. Indagine nazionale sui consumi alimentari delle famiglie 1980–84. Alcuni principali risultati. Riv Sci Alim 4:53–65, 1990.
10. Buiatti E., Palli D., De Carli A., Amadori D., Avellini C., Bianchi S., Bisemi R., Cipriani F., Cocco P., and Giacosa A. *et al.* A case-control study of gastric cancer and diet in Italy: association with nutrients. Int J Cancer 44; 4:611–6, 1989.
11. Willett W.C. Nutritional epidemiology. Oxford University Press, Oxford, 1990.
12. Franceschi S. Assessment of fat intake in retrospective epidemiological studies. Eur J Clin Nutr 47:S39–41, 1993
13. Rosén M., Nyström L., and Wall S. Diet and mortality in the countries of Sweden. Am J Epidemiol 127:42–9, 1988.
14. ISTAT—Italian National Institute of Statistics. I consumi delle famiglie. Anno 1973. Supplemento al Bollettino Mensile di Statistica, 1975.
15. Farchi S., Saba A., Turrini A., Forlani F., Pettinelli A., and D'Amicis A. An echological study of correlation between diet and tumor mortality rates in Italy. Eur J Cancer Prev 5(2):113–120, 1996.

8

CELL DIVISION CYCLE ALTERATIONS AND HUMAN TUMORS

Fulvio Della Ragione,[1] Adriana Borriello,[1] Valentina Della Pietra,[1] Valeria Cucciolla,[1] Adriana Oliva,[1] Alfonso Barbarisi,[2] Achille Iolascon,[1] and Vincenzo Zappia[1]

[1] Institute of Biochemistry of Macromolecules
Medical School, Second University of Naples
Naples, Italy
[2] Institute of "Clinica Chirurgica Generale e Terapia Chirurgica"
General Surgery and Surgical Therapy
Medical School, Second University of Naples

INTRODUCTION

A large series of evidence has conclusively demonstrated that the development and progression of a cancer are due to the accumulation of a number of genetic alterations which finally result in a full malignant phenotype. This complex phenomenon is clearly illustrated by colorectal tumors, which often require more than a decade to be clinically evident and at least seven genetic events for completion.[1]

Several genes are structurally and/or functionally altered in human neoplasias. However, it is still difficult to predict whether such alterations are really important in the transformation process or whether they represent simply a consequence of the genomic instability peculiar to malignant cells and thus do not have any relevance in the cancerogenetic process.

Two types of genetic alterations have proven to be truly important in human cancers: some genes (the so-called cellular proto-oncogenes), when mutated, gain new functions which might promote cancerogenesis, while others after mutation lose their ability to prevent unregulated growth and are therefore called tumor suppressor genes (TSGs). Both types of alterations (gain or loss of function) occur during the evolution of a cancer, although the types of genes involved and the order of genetic alterations vary significantly in different malignancies. A third class of genes, also pivotal in human

Advances in Nutrition and Cancer 2, edited by Zappia *et al.*
Kluwer Academic / Plenum Publishers, New York, 1999.

cancerogenesis, has recently been identified which includes genes not directly involved in growth control. These genes encode protein which allow DNA repair (especially mismatch repair) which in turn reverses errors carried out by DNA polymerase.[2] Thus, the inactivation of these genes (most frequently, MSH2, MLH1, and PMS2 genes) results in the accumulation of mutations which might affect the whole genome including both proto-oncogenes and tumor suppressor genes.[3–5]

Within the last decade, substantial advances in unraveling the molecular basis of human cancerogenesis have been made as the result of several complementary experimental approaches. The first has been the genetic linkage analysis. This task allows the identification of chromosomal areas most frequently altered in a specific type of cancer (generally familial cancers) thus permitting the localization of putative tumor-related genes (proto-oncogenes or TSGs). A further clue relies on the enormously improved knowledge of the basic mechanisms responsible for cell proliferation and of their alterations in human tumors. The final clue is provided by advances in techniques in developing animal models (generally transgenic mice) with genetic abnormalities similar to those occurring in human diseases, including tumors.

One of the main conclusions emerging from such an enormous harvest of investigations is that most (if not all) tumor types show genetic aberrations modifying the proteic engine which allows cells to divide correctly. In other words, human cancers frequently have an altered cell division cycle with an unscheduled movement from G1 to S phase (see the following paragraphs for details on the division cycle process and on the types of genetic aberrations). However, the precise relationship between the molecular cell cycle alterations observed and the overall process of cancer transformation remains quite elusive. Two possible scenarios can be hypothesized to shed light on this key discovery (i.e. the interplay between an altered cell cycle and the malignant transformation), which are not mutually exclusive, but probably strongly cooperative.

The first hypothesis is that a premature S phase (where the synthesis of DNA occurs) entry could cause an ineffective genome repair. Indeed, it has been clearly demonstrated that DNA is randomly altered in human cells and that the repair of these DNA aberrations takes place during the G1 phase of the cell cycle. Thus, an impaired DNA repair (due to unregulated G1/S phases) might cause the accumulation of genetic damages, which in turn may result in cancer development and/or progression. It is noteworthy in this context, that an important tumor suppressor gene, namely the *p53* gene, is involved in the control of DNA repair and that it is often inactivated in human cancers.

Physiologically, when DNA is damaged (as, for example, after UV or other carcinogen exposure), p53 protein levels increase remarkably as the consequence of slower degradation.[6,7] Then, this molecule causes a G1 phase block which allows the cell to activate the DNA repair systems. If the DNA damage is too extensive to be repaired, p53 protein activates an alternative molecular program, i.e. the apoptotic process. In this case, the cell is destroyed by DNA digestion and protein coagulation. It is clear that if a cell presents alterations of genes regulating both the cell division cycle and the DNA repair process, it will easily develop further genetic aberrations, thus evolving towards a more aggressive phenotype.

The second scenario suggests that the lack of factors controlling cell proliferation prevents cell differentiation and/or cellular senescence. Both processes, which result in lowering the number of cycling cells, could be envisioned as important physiological mechanisms which hamper malignant progression. Thus, their inactivation might strongly contribute to tumor development and progression.

Whether or not these two hypotheses correspond to *in vivo* events, or represent only an intellectual exercise, nonetheless the role of loss of cell cycle control in the development of a full malignant phenotype still must be established. Indeed, it is probable that alterations at the level of cell division may not explain important cancer behavior, including phenomena like tissue invasion, metastases formation, and neoangiogenesis. Thus, the impairment of cell growth inhibition could be a key (but not necessarily an early) step in the dramatic genetic cascade which will give origin to an invasive tumor.

The next section will give a synthetic description of the major proteins involved in the cell division cycle; then, the most frequent tumor genetic alterations of the cell cycle will be described; and finally, the interplay beween cell division cycle and dietary habit will be discussed in the light of chemiopreventive and therapeutic strategies.

CELL DIVISION CYCLE: A BRIEF OVERVIEW

Eukaryotic cell division cycle is generally composed of four phases: the time period before DNA replication (named G1 phase), the phase of DNA duplication (S phase), the gap after DNA synthesis (G2 phase), and the phase which results in cell division (M phase)[8] (Fig. 1).

The cell cycle length is directly related to the duration of the G1 phase which can be considered as the time window during which the cell decides whether or not to

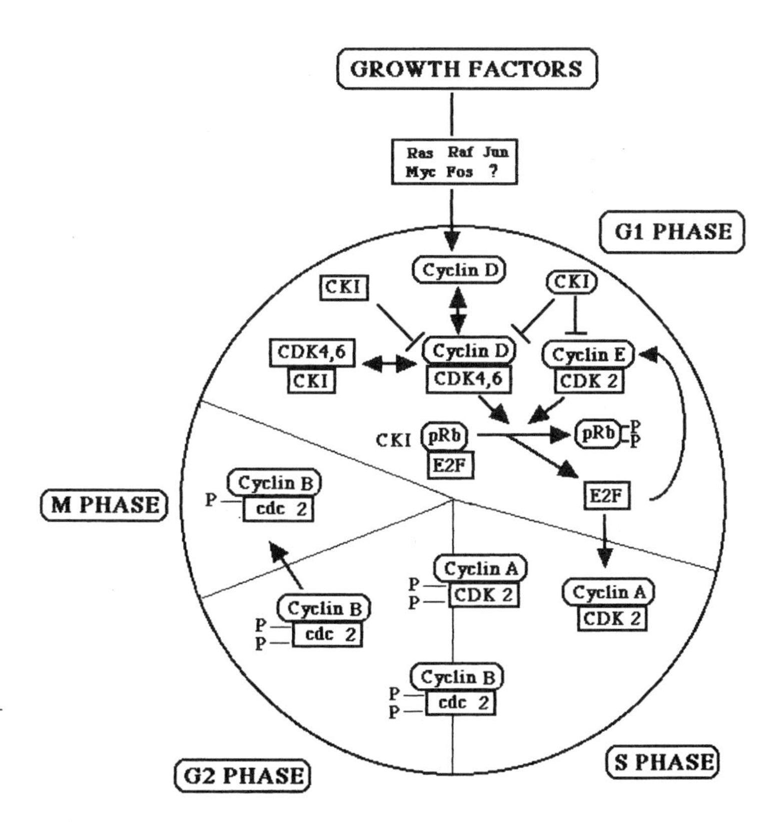

Figure 1. Schematic representation of the main steps of eukaryotic cell division cycle.

proliferate. Indeed, it is only during the G1 phase that the extracellular enviroment could influence cell cycle progression through positive and negative signals (including growth factors, cell-cell contact, extracellular matrix etc.). Conversely, once a cell overcomes the restriction point between G1 and S phases, the mitotic cycle proceeds almost totally independent on external signals. Thus the knowledge of biochemical mechanisms which allow the cell to proceed along G1 phase is essential to predict not only the rate of proliferation but also the fate of a cell.

Three families of proteins form the heart of the engine required for the cell division cycle, namely cyclins, serine-threonine cyclin-dependent kinases (CDKs) and CDK inhibitors (CKIs). Additional proteins, which do not belong to these three classes, participate on cell division process, but the description of their role is beyond the aims of this review.

The progression along the cycle corresponds to a biochemical sequence where different cyclins bind to and activate distinct CDK holoenzymes.[9,10] Then, the activated CDK/cyclin complexes regulate their protein targets by phosphorylation. Finally, these downstream effectors ultimately allow the ordered development of the cell division cycle. Thus, the regulation of CDK activities is the central event in the cell cycle progression.

Among the CDKs substrates, the product of the retinoblastoma gene, namely pRb is the most important. pRb acts as an inhibitor of cell cycle progression at G1/S boundary by virtue of its ability to sequester physically several transcription factors which are necessary for traversing the G1→S restriction point for S phase entry. Starting from mid-to-late G1 phase, pRb is more phosphorylated than during the remainder of the cell cycle by active CDK/cyclin complexes. Since it is well established that only the un- or hypophosphorylated pRb species can bind to their partner proteins, the phosphorylation of pRb results in a release of proteic factors from the pRb binding pocket and in the start of S phase.

One of the most interesting properties of cyclin molecules (and the meaning of their name) is that their levels change rapidly during the cell division cycle. These variations occur in a finely regulated fashion which is specific for each type of cyclin. Two mechanisms are responsible for the fluctuation of the cyclin levels, namely the regulation of the transcription of the relative genes and the control of the degradation rate, frequently involving the ubiquitin/proteasome pathway. Historically, the first cyclins to be characterized corresponded to those now called B cyclins (for recent reviews on cyclins see references 9 and 10), which were isolated from marine invertebrates (clam, sea urchins) and from the eggs of the amphibian Xenopus laevis. Afterwards, several additional cyclins (namely cyclins A, C, D, E, F, G, and H) were isolated and studied in detail.

Since B cyclins regulate the cell progression along the mitotic phase, they are defined as M cyclins. Conversely, cyclin A facilitates the progression through S phase in mammalian cells. The G1 cyclins, which control the earliest phases of the cell cycle and the crossing of the G1/S boundary, consist of three different D cyclins and cyclin E. These cyclins and their G1 CDK partners (CDK4, CDK6, and CDK2) are essential for the S phase entry under physiological conditions. Moreover, the cyclin D family is particularly important among other cyclins since its level is directly regulated by hormones and growth factors. In fact, one of the first events during the cell cycle is the build-up of D cyclins and the activation of their catalytic CDK partners.

Finally, cyclin E resembles cyclins A and B, in that its expression is strictly cell cycle dependent and follows that of D cyclins by extending from late G1 through the

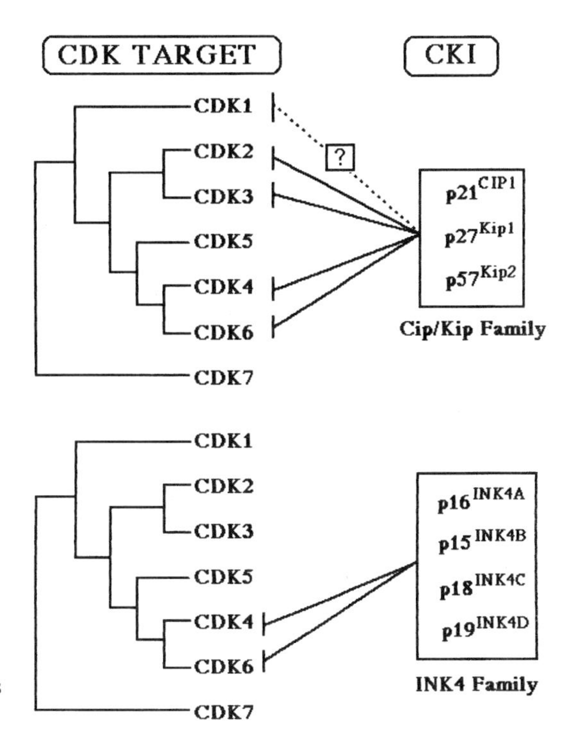

Figure 2. Cyclin-dependent kinase inhibitors (CKIs) and their target kinases (CDKs).

G1/S transition into the proper S phase. This suggests that cyclin Ds and cyclin E act by pushing cells through a critical checkpoint (the so-called restriction or R point) and thereby drive the cell division.

As mentioned in the Introduction, in addition to activation by cyclin binding, the CDK activity is regulated (inhibited) by the interaction with small proteins which have been identified within the last four years. These CKIs are of great relevance since their levels finely modulate the activity of CDK/cyclin complexes, and thus control processes like cellular differentiation, cell-cell contact inhibition, and cellular senescence. Notably, all these mechanisms are altered in human tumors.

At least seven different CDI have been identified so far and grouped into two different classes, namely CIP/Kip[11–13] and INK4[14–17] families (Fig. 2).

The first family (CIP/Kip) includes three structurally homolog members, p21^{CIP1},[11] p27^{Kip1},[12] and p57^{Kip2};[13] they contemporaneously associate with the CDK holoenzyme and its cyclin pattern to form complexes possibly involving additional proteins. Almost all the cyclin-activated CDKs are inhibited by binding with members of the CIP/Kip family, thus suggesting that these CKIs might regulate the progression through all cycle phases. Thus, the CIP/Kip family seems to play a key role in regulating the exit of cells from the division cycle, in particular towards a terminal or reversible differentiation.[18]

The INK4 gene family includes four members, namely CDKN2A (encoding p16^{INK4A} protein),[14] CDKN2B (p15^{INK4B} protein),[15] CDKN2C (p18^{INK4C} protein),[16] and CDKN2D (p19^{INK4D} protein).[17] The cloning and characterization of human CDKN2A gene was reported for the first time at the end of 1993[14] simultaneously with the identification of p21^{CIP1} gene.[11] A few months later (April 1994), two independent research groups of research identified CDKN2A gene as the putative tumor suppressor gene (TSG) mapping on the 9p21 chromosome region.[19,20] It should be underlined that a very rich body of evidence, harvested since the beginning of the eighties, strongly suggested the localization of an important TSG on the chromosome 9 short arm, and in particular at 9p21. An intense scientific debate followed the putative identification of the CDKN2A gene as the 9p21 TSG essentially arising from the discrepancy between the incidence of inactivation demonstrated in malignant cell lines and that observed in primary tumors. However, the role of this gene as an important TSG is now well established.

A major functional difference exists between the members of the CIP/Kip and CDKN2A families, in that $p21^{CIP1}$ protein (and its homologues) are up-regulated mainly as a consequence of external stimuli which inhibit cell growth and/or might cause differentiation, while $p16^{INK4A}$ protein seems to belong to an intrinsic regulatory loop of cell proliferation. Moreover, CDKN2A gene is inactivated in a large percentage of different tumors, while CIP/Kip genes (and their homologues) do not appear to be silenced in human tumors.

As previously described, an increasing body of evidence allowed the general conclusion that all the genetic abnormalities which cause a loss of cell division cycle intrinsic regulation (thus allowing the cell to escape from the environmental control) might be essential steps along the malignant transformation avenue. These genetic alterations might cause either a direct inactivation of pRb or related proteins or the activation of those cyclin-dependent kinases able to phosphorylate (and thus functionally inactivate) pRb. In particular, this second mechanism might be due to increasing levels of G1 CDKs (CDK4 and CDK6) or G1 cyclins as well as to functional inactivation of some CKI genes.

The next section will report the types, the incidence and clinical importance of alterations of the cell division cycle regulation demonstrated in human cancers.

CELL DIVISION CYCLE ALTERATIONS IN HUMAN CANCERS

1. The Retinoblastoma Gene

The retinoblastoma (*RB*) gene might be thought of as the archetype of tumor suppressor genes. It was cloned in 1986 and then was showed to include about 200 kb of human chromosome 13q14.[21] Although its initial cloning depended on the identification of large deletions, it soon became evident that the majority of inactive *RB* alleles, associated with human cancers, had lost their function through subtle alterations and particularly point mutations. These *RB* gene alterations affect the synthesis of an intact *RB* gene product (pRB), a 105–110 kDa nuclear phosphoprotein which, as described in the previous paragraph, is the preferential substrate of activated CDKs.

The association of pRB with transcription factors such as E2F/DP1 and others[22] is held responsible for pRB's ability to block the passage of cells past the G1 checkpoint. The phosphorylation of pRB renders it incapable of associating with these factors causing their release, and consequently allows the progression of cells into S phase. Loss of pRB or the functions that control its phosphorylation may result in an inability to control the passage of cells past this checkpoint, thus creating a condition that leads to tumor formation.

Phosphorylation of pRB is controlled by two different reactions: phosphorylation by protein kinases, which transform pRB into an inactive form, and dephosphorylation due to phosphatases which reconstitute the ligand-binding pRB ability. The function of phosphorylating pRB is probably exerted by enzymes of the CDK family. Fifteen consensus phosphorylation sites for these kinases are found in the primary sequence of pRB and these consensus are conserved in a wide range of animal species. Recent reports indicate that cyclin D/CDK4 (or CDK6) and cyclin E/CDK2 complexes phosphorylate pRB on different sites[23,24] and neither active CDK4 nor CDK2 is sufficient by itself for pRB full phosphorylation thus suggesting that the two kinases co-operate in regulating pRB activity. Members of the type 1 family of protein phosphatases (PP1) are held responsible for pRB dephosphorylation.[25] Since PP1 activity seems controlled by CDK-dependent phosphorylation, the role of cyclin-dependent kinases in pRB regulation appears to be further confirmed.

Inactivation of both *RB* gene alleles has been demonstrated, in addition to retinoblastoma, in several human cancers, including osteosarcomas, prostate carcinomas, a third of bladder and mammary carcinomas, and virtually all small cell lung cancers (>75% of cases). The absence of a functional pRb should result in an unscheduled activation of pRb-inhibited transcription factors and in a premature entry into S phase. The consequences of this phenomenon, which might be superimposable to those due to CDKN2A gene inactivation, will be described in more detail in the next section.

2. The CDKN2A Gene

As mentioned above, $p16^{INK4A}$ (now formally defined CDKN2A) gene was isolated at the end of 1993 in a search devoted to identify CKI proteins, but only toward the middle of the following year did its relevance in cancer become evident. This was due to the putative identification of CDKN2A as the long searched 9p21 TSG.

The structure of the human CDKN2A gene is quite complex and deserves a brief description (Fig. 3).

The gene is composed of three exons (exon 1α, 2, and 3) spanning on about 6–8 kb. Its translation results in the synthesis of a single 0.8 kb mature mRNA which is translated in a protein of 156 amino acids able to interact and inhibit CDK4 and CDK6 proteins. Interestingly, the gene contains an alternative first exon (exon 1β) which might be transcribed and, after mRNA maturation, joined to exon 2 and 3 giving a different open reading frame translated into a protein of 132 amino acids ($p14^{arf}$ protein; arf, alternative reading frame).[26] This protein has a sequence completely different from that of 156 amino acids, but importantly, it is still able to inhibit cell cycle progression by interfering with p53 degradation.[27,28] Very recent investigations have demonstrated that $p14^{arf}$ interacts with MDM2 protein and directs this last toward degradation. Since MDM2 protein is able to downregulate p53 activity by committing p53 towards degradation, the upregulation of $p14^{arf}$ levels stimulates p53 function. Additional important information has been recently obtained on the molecular events which cause an increased expression of $p14^{arf}$. In particular, it has been demonstrated that oncogenic stimuli (like oncogenic Ras) might induce $p14^{arf}$ expression, thus activating p53 upregulation.[29,30]

In conclusion, by means of two different proteins ($p16^{INK4A}$ and $p14^{arf}$) the CDKN2A gene regulates pRB and p53 pathways and consquently both the key mechanism for cell proliferation and the key checkpoint mechanism. Thus, CDKN2A inactivation, especially by homozygous deletions is particularly important in cancer development.

Recent genetic studies, based on the targeted disruption of specific genes, have demonstrated that rodents lacking both the alleles of the CDKN2A gene show a

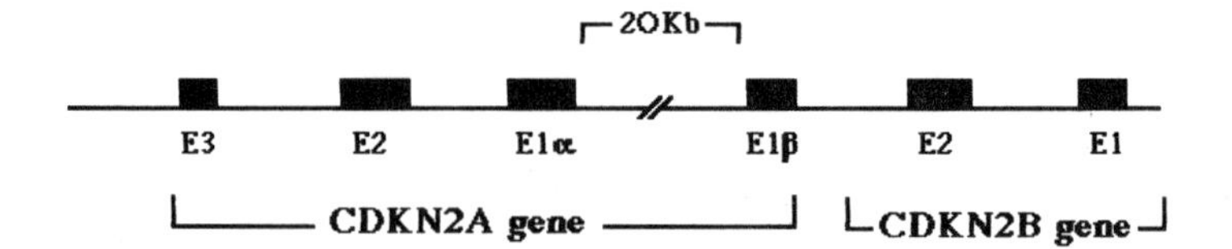

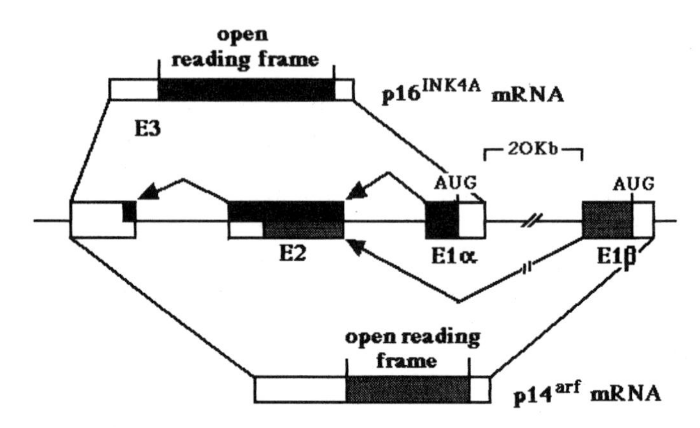

Figure 3. Structure of CDKN2A gene and maturation of its different mRNAs.

normal development, but have a remarkable sensitivity to cancerogenic agents, thus confirming the tumor suppressor function of this gene.[31] Moreover, and very interestingly, primary cultures of fibroblasts established from CDKN2A null mice do not show the senescence phenomenon *in vitro*. This is particularly important since the aging process at the cellular level (and possibly at the organism level) is now thought of as a protective mechanism against tumor development.

As discussed in the previous section, $p16^{INK4A}$ protein interacts with CDK4 and CDK6, thus hampering their kinase activity and then the pRb phosphorylation. Therefore, this CKI acts as a major factor in restraining progression towards S phase. Consequently, its absence might result in G1 phase shortening, premature S entry and possibly ineffective repair of DNA damage. Moreover, the absence of $p14^{arf}$ upregulation in the presence of oncogenic stimuli, probably abrogates a p53 response and a key cell cycle checkpoint.

Three different molecular mechanisms have been shown to cause the inactivation of $p16^{INK4A}$ gene in transformed cells, namely homozygous deletions, point mutations, and transcriptional silencing by methylation at 5′-CpG islands. Homozygous gene deletion is the most frequently observed event and seems to occur in several human cancers, while point mutations have been mainly demonstrated in familial melanomas, esophageal squamous cell carcinomas, and pancreatic adenocarcinomas. The transcriptional silencing of CDKN2A gene has been observed in cell lines, while it has scarcely been investigated in cancer specimens.

Table 1 reports the cancer types where the loss of CDKN2A gene function has been demonstrated. As shown there, an enormous variety of different tumors presents the inactivation of CDKN2A with, in several instances (childhood acute lymphoblastic leukemias of T-cell, pancreatic tumors, glioblastomas, head and neck tumors,

Table 1. CDKN2A gene inactivation in human cancers

Tumor type	CDKN2A gene homozygous deletions (%)	CDKN2A gene point mutations (%)	CDKN2A gene transcriptional silencing (%)	TOTAL (%)
Glioblastoma	50–68	—	30	80–98
Astrocytomas	52	15	—	67
Head and neck	33	7–52	30	70–100
Esophageal		52	—	52
Pancreas	37	49	—	86
Nasopharyngeal	35–67	—	—	35–67
Lung				
NSCLC*	10–80	12	35	57–100
SCLC*	—	—	—	0
Mesothelioma	72	—	—	72
Leukemias				
T-ALL**	25–83	—	—	25–83
B-ALL**	6–15	—	—	6–15
AML**	—	—	—	0
Bladder	9–19	4–10	—	13–29
Melanoma	—	15	—	15
Breast	0–15	—	—	0–15
Kidney	—	—	Observed	+
Neuroblastoma	—	—	—	0
Colon	—	—	—	0
Endometrium	—	—	—	0

*NSCLC, non-small cell lung cancer; SCLC, small cell lung cancer.
**T-ALL, T-cell acute lymphoblastic leukemia; B-ALL, B-cell acute lymphoblastic leukemia; AML, acute myeloid leukemia.
—, not observed.

mesotheliomas, and others), an extraordinary high incidence.[32,33] In summary, the CDKN2A gene must be considered as one of the most frequent genetic alterations detectable in human cancers.

3. p27^{Kip1} Protein

A third interesting alteration, related to the cell division cycle regulation, has recently been demonstrated in human cancers. In this case, however, the aberration does not involve a gene (structurally or functionally inactivated) but the level of a CKI protein, namely p27^{Kip1}. Indeed, a large amount of evidence, harvested in the past three years, has conclusively demonstrated that several cancers (listed in Table 2), showing a favorable prognosis, present a p27^{Kip1} level significantly lower than that of tumors of the same stage but with a poorer evolution.

This finding seems to be dependent on a higher p27^{Kip1} proteolytic activity occurring in aggressive tumors with a low content of the CDK inhibitor.[34] Conversely, p27^{Kip1} mRNA levels (as judged by *in situ* hybridization) appear identical in all cases independent from the tumor prognosis.[34–38] Although a definite explanation of such observation is still lacking, a probable hypothesis is the existence of a relationship between the differentiation status of malignant cells (and thus the prognosis) and the ability to remove the CDK inhibitor.

In conclusion, p27^{Kip1} cellular content seems to be an important factor during the development and progression of a neoplasia as a well as a very promising independent prognostic factor.

Table 2. Human cancers showing $p27^{Kip1}$, cyclin D1, and CDK4 alterations

Alterations	Tumors
$p27^{Kip1}$ protein level down-regulation (due to an increased degradation)	Colon cancer Breast carcinoma Lung cancer
Cyclin D1 gene (amplification, rearrangment, translocation)	**Frequent** Breast carcinoma Parathyroid adenoma and carcinoma Esophageal squamous cell carcinoma Head and neck carcinomas Centrocytic/mantle-cell lymphomas Cutaneous squamous cell carcinomas **Rare** Colonic tumors Adenomatous polyps Actinic keratoses Non-small cell lung cancer Hairy cell leukemia Prolymphocytic leukemia Non-Hodgkin's lymphomas Splenic lymphoma Multiple myeloma Ovarian carcinoma Laryngeal carcinoma Sarcomas
CDK4 gene (amplification, mutation)	Glioblastoma Melanoma Sarcomas

4. Alterations of Cyclin D1 (CCDN1) or CDK4 Genes

The CCDN1 gene on chromosome 11q13, which encodes cyclin D1, is often overexpressed via numerous alterations, including gene amplification, chromosomal rearrangements, proviral insertion, and other mechanisms. Among various types of cancers where CCDN1 gene aberrations have been reported, the data on breast cancer appear to be the most sound. First, a remarkably high percentage of breast carcinomas (and particularly invasive breast carcinoma) shows high levels of cyclin D1 protein. Second, increasing cyclin D1 mRNA can be seen in human breast neoplastic progression. Third, transgenic mice overexpressing cyclin D1 show mammary hyperplasias and carcinomas. Fourth, cyclin D1 null mice exhibit normal mammary proliferation during puberty, but not during pregnancy. Recently, a report suggests that cyclin D1 mRNA *in situ* quantitation could represent a potent tool to distinguish invasive breast carcinomas from non-malignant lesions. However, further studies are required to substantiate such a proposal.

CDK4 gene is also altered in some tumor types. Two different aberrations have been described, including gene amplification and point mutation. The last mechanism is particularly interesting, since CDK4 mutations preventing inhibitory interaction with $p16^{INK4A}$ protein were demonstrated to be involved in melanoma development. Table 2 reports the human cancers showing cyclin D1 and CDK4 gene alterations.[39,40]

DIET AND CELL DIVISION CYCLE CONTROL: A PROBABLE IMPORTANT CONNECTION

A large number of epidemiological and experimental studies have conclusively demonstrated that dietary habit plays a key role in both preventing and inducing malignant transformation.[41,42] Due to the importance of cell division cycle in human neoplasias, it is easy to predict that components of human diet could modify normal cell cycle progression by interfering with some of the steps responsible for such a process. In this paragraph, we will briefly review a few examples of naturally occurring dietary molecules which directly act on the cell division cycle engine, and in particular we will describe compounds which present anticancer activity. Conversely, the mechanism(s) by which dietary molecules indirectly (i.e. by acting as mutagens) cause genetic alterations and thus cancer will not be discussed.

The identification of specific dietary compounds responsible for a protective effect is one of the major avenues of cancer research, this both from a preventive point of view as well as for the development of new and efficacious therapeutic strategies. A major example of this positive connection is the observation that several cancers (colon, breast, prostate, and endometrial neoplasias) have a lower incidence in Asia than in Western countries.[43] Ample evidence points to plant-derived compounds as the most probable factors responsible for the chemoprotective activity. Indeed, the long-known preventive effect of vegetarian diets on tumorigenesis and other chronic diseases is well documented and is confirmed by the finding that immigrants from Asia who maintain their traditional plant-based diet do not increase the risk of developing cancer.[44]

Biochemical analyses carried out on the urine of subjects consuming a diet rich in plant products led to the identification of flavonoids as potent inhibitors of cell proliferation.[45] Moreover, the excretion of these compounds, and in particular of genistein (Fig. 4A), in urine of vegetarians is several-fold (up to 30-fold) higher than that of omnivores.[46] Thus, molecules belonging to the flavonoid family appear as potential candidates for at least part of the well-known antitumorogenic property of plant-based diets.

A number of cellular targets of these polyphenols has been identified which frequently belong to signal transduction pathways. Much less is known about the effects of these molecules on the cell division cycle engine, probably as a consequence of the scarce number of studies performed. The major findings obtained so far suggest that flavonoid and other naturally-occurring antioxidants (like vitamin E) act on the cell division cycle engine through complex and pleiotropic mechanisms. Vitamin E and genistein appear to be able to induce up-regulation of $p21^{CIP1}$ level.[47–49] The increase of the CKI levels is of relevance since it might explain the block of cell proliferation and the induction of programmed cell death observed in some cellular models after the treatment with the two molecules. In the case of vitamin E, the effect on $p21^{CIP1}$ content seems to be related to the activation of members of the C-EBP transcription factor family and in particular to modulation of the phosphorylation pattern of these proteins.[47] These studies also suggest an interesting possible correlation between the decrease of oxygen reacting species and the activation of $p21^{CIP1}$ expression.[47]

Genistein also increases $p21^{CIP1}$ level but this isoflavonoid seems to act by up-regulating both gene transcription and mRNA translation.[48,49] However, in the case of

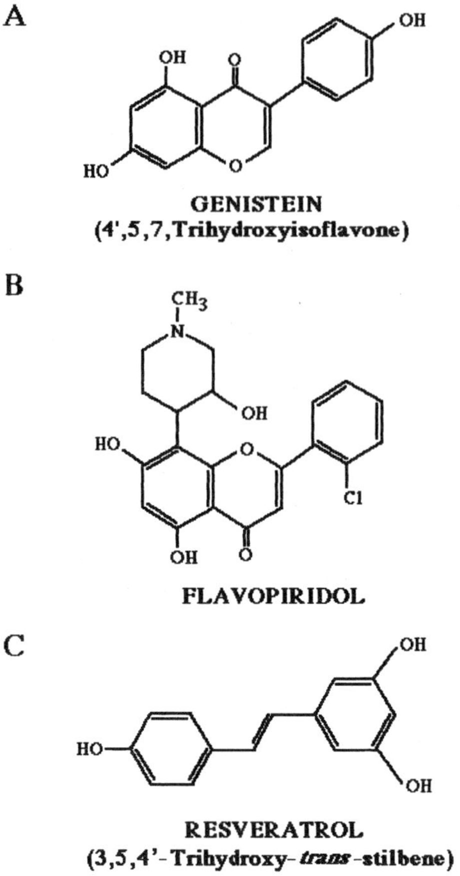

A

GENISTEIN
(4',5,7,Trihydroxyisoflavone)

B

FLAVOPIRIDOL

C

RESVERATROL
(3,5,4'-Trihydroxy-*trans*-stilbene)

Figure 4. Structure of some antioxidant molecules endowed with antiproliferative activity. A) Genistein; B) Flavopiridol; C) Resveratrol.

genistein the control of cell cycle is more complex. Indeed, this compound is a well-known inhibitor of tyrosine kinase, by acting as an ATP analog.[50] Among the tyrosine kinases inhibited by genistein is that which phosphorylates Cdk1 on Tyr15 by pushing the cyclin-dependent kinase towards an inactive form. This kinase, i.e. p56/p53lyn, is rapidly induced by treatments that trigger cell cycle checkpoints (ionizing radiation, cytosine arabinoside), suggesting that p56/p53lyn may actively delay the onset of mitosis by phosphorylating Tyr15 on Cdk1.[51] Therefore, the inhibition of this kinase might accelerate the entry in mitosis thus hampering one of the important mitotic checkpoints. This effect along with that reported on the induction of $p21^{CIP1}$, makes the evaluation of genistein's effect on the cell cycle quite complex and strictly dependent on the cell system analyzed.

In addition, since the isoflanoid inhibits several tyrosine kinases, its activity on the cell cycle might be due to interference with growth stimulating transduction pathways. This has been confirmed by recent reports which demonstrated that genistein and a related plant-derived flavonoid, quercetin, down-regulates the level of cyclin D1 (the only cyclin type which is strictly dependent on growth factors) in hepatic cells and that this effect might cause cell proliferation impairment.[52] In conclusion, genistein inhibits cell proliferation by causing an accumulation of cells in G1 phase as a consequence of up-regulation of the CKI level and decrease of the cyclin D cellular content.

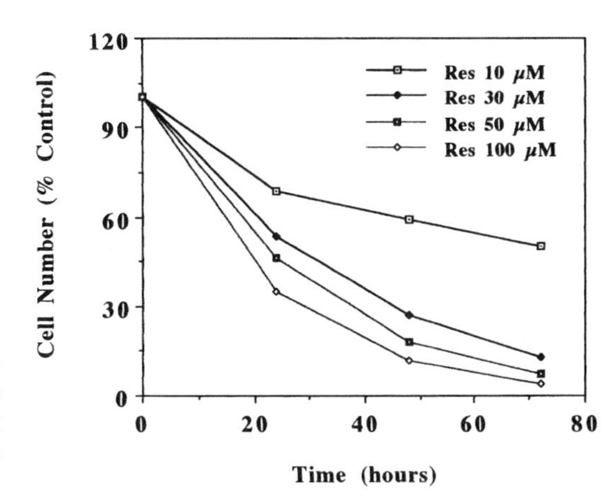

Figure 5. Effect of resveratrol on the proliferation of HL-60 cells. HL-60 cells were plated at 300,000 cells/ml and incubated with or without different resveratrol amounts. Cells were then counted daily.

In this context, it should be remembered that chemically-synthesized flavonoids are now used as specific inhibitor of CDK and have been suggested as molecules to be introduced in therapy. An example of these compounds is flavopiridol (Fig. 4B) which a selective and powerful inhibitor of CDK2, CDK4, and CDK1.[53]

A further dietary compound which has recently attracted considerable interest is a phytoalexin occurring at high levels in red wine, namely resveratrol (Fig. 4C). This molecule inhibits cell proliferation, and depending on the concentration employed and the model system used induces differentiation or apoptosis.

We have investigated its effects on cell growth of HL-60 cells, a promyelocytic cell line. As shown in Fig. 5, as low as 30 μM causes a complete block of HL-60 cells proliferatrion. Subsequently, we have studied the molecular bases of the cytostatic effect demonstrating that the compound arrests the cells in G1/S phases without inducing apoptosis.

A detailed biochemical analysis allowed us to conclude that resveratrol hampers the transition between the S to G2 phase.[54] Although the molecular basis of this effect is not yet clarified, the phytoalexin might inhibit the DNA synthesis (acting on a DNA polymerase or a topoisomerase) or alternatively can activate proteins responsible for the S→G2 checkpoint (like protein phosphatases of the CDC25 family or protein kinases activated by DNA alterations, DNA-PK and/or ATM). In conclusion, resveratrol might cause an elongation of the S phase, perhaps due to both the inhibition of DNA duplication and the activation of a differentiation (or of an apoptotic) program.

A third example of dietary molecule which could regulate the cell division cycle engine is butyric acid, a short chain fatty acid which derives from colon bacterial fiber fermentation. Although, the antiproliferative mechanism of action of this compound is discussed elsewhere in this volume (see Russo et al.), it is interesting to underline that the butyrate causes a remarkable induction of $p21^{CIP1}$ gene expression probably by modulating the level of histone acetylation.[55,56] Thus, some of the important antiproliferative effects of butyric acid are due to a cell cycle arrest at the G1→S boundary.

In conclusion, the available epidemiological and experimental data strongly suggest the possibility of discovering more potent natural dietary molecules which could regulate the cell division cycle thus slowing or hampering the development of human cancer. These molecules could represent an excellent basis for projecting the synthesis of new antiproliferative compounds. Additional studies will further highlight these important aspects in the search of new rational and efficacious chemopreventive strategies.

REFERENCES

1. Kinzler, K.W. and Vogelstein, B. Lessons from hereditary colorectal cancer. Cell, 87, 159–170, 1996.
2. Modrich, P. and Lahue, R. Mismatch repair in repliation fidelity, genomic recombination, and cancer biology. Ann. Rev. Biochem., 65, 101–133, 1996.
3. Prolla T. DNA mismatch repair and cancer. Curr. Opinion in Cell Biol., 10, 311–316, 1998.
4. Davis, T.W., Wilson-Van Patten, C., Meyers, M., Kunugi, K.A., Cuthil, S., Reznikoff, C., Garces, C., Boland, C.R., Kinsella, T.J., Fishel, R., and Boothma, D.A. Defective expression of the DNA mismatch repair protein, MLH1 alters G2-M cell cycle checkpoint arrest following ionizing radiation. Cancer Res., 58, 767–778, 1998.
5. Nicolaides, N.C., Littman, S.J.P., Kinzler, K.W., and Vogelstein, B. A naturally occurring hPMS2 mutation can confer a dominant negative mutator phenotype. Mol. Cell. Biol., 18, 1635–1641, 1998.
6. Lane, D. Awakening angels. Nature, 394, 616–617, 1998.
7. Woo, R.A., McLure, K.G., Lees-Miller, S.P., and Lee, P. DNA-dependent protein kinase acts upstream of p53 in response to DNA damage. Nature, 394, 700–704, 1998.
8. Pardee, A.B. G1 events and regulation of cell proliferation. Science, 246, 603–608, 1989.
9. Nurse P. Ordering S phase and M phase in the cell cycle. Cell, 79, 547–550, 1994.
10. Sherr, C.J. G1 phase progression: cycling on cue. Cell, 79, 551–555, 1994.
11. Xiong, Y., Hannon, G.J., Zhang, H., Casso, D., Kobayashi, R., and Beach, D. p21 is a universal inhibitor of cyclin kinases. Nature, 366, 701–704, 1993.
12. Toyoshima, H. and Hunter, T. p27, a novel inhibitor of G1 cyclin-Cdk protein kinase activity, is related to p21. Cell, 78, 67–74, 1994.
13. Lee, M.-H., Reynisdottir, I., and Massagué, J. Cloning of $p57^{KIP2}$, a cyclin-dependent kinase inhibitor with unique domain structure and tissue distribution. Genes and Dev., 9, 639–649, 1995.
14. Serrano, M., Hannon, G.J., and Beach, D. A new regulatory motif in cell cycle control causing specific inhibition of cyclin D/CDK4. Nature, 266, 122–126, 1993.
15. Hannon, G.J. and Beach, D. $p15^{INK4B}$ is a potential effector of TGF-β-induced cell cycle arrest. Nature, 371, 257–260, 1994.
16. Guan, K.-L., Jenkins, C.W., Li, Y., Nichols, M.A., Wu, X., O'Keefe, C.L., Matera, A.G., and Xiong, Y. Growth suppression by p18, a $p16^{INK4/MTS1}$ and $p14^{INK4B/MTS2}$-related CDK6 inhibitor, correlates with wild-type pRb function. Genes and Dev., 8, 2939–2952, 1994.
17. Chan, F.K.M., Zhang, J., Cheng, L., Shapiro, D.N., and Winoto, A. Identification of human and mouse p19, a novel CDK4 and CDK6 inhibitor with homology to $p16^{INK4}$. Mol. Cell. Biol., 15, 2682–2688, 1995.
18. Steiman, R.A., Hoffman, B., Iro, A., Guillouf, C., Liebermann, D.A., and El-Houssein, M.E. Induction of p21 (WAF-1/CIP1) during differentiation. Oncogene, 9, 3389–3396, 1994.
19. Nobori, T., Miura, K., Wu, A.K., Luis, K., Takabashi, K., and Carson, D.A. Deletion of the cyclin-dependent kinase 4 inhibitor gene in multiple human cancers. Nature, 368, 753–756, 1994.
20. Kamb, A., Gruis, N.A., Weaver-Feldhaus, J., Liu, Q., Harshman, K., Tavitgian, S.V., Stockert, E., Day, R.S., Johnson, B.E., and Skolnick, M.H. A cell cycle regulatory potentially involved in genesis of many tumor types. Science, 264, 436–440, 1994.
21. Wang, J.Y.J., Knudsen, E.S., and Welch, P.J. The retinoblastoma tumor suppressor protein. Adv. Cancer Res., 64, 25–85, 1994.
22. Sanchez, I. and Dynlacht, B.D. Transcriptional control of the cell cycle. Curr. Opin. Cell Biol., 8, 318–324, 1996.
23. Zarkowska, T., Harlow, E., and Mittnacht, S. Monoclonal antibodies specific for underphosphorylated retinoblastoma protein identify a cell cycle regulated phosphorylation site targeted by CDKs. Oncogene, 14, 249–254, 1997.
24. Mittnacht, S. Control of pRB phosphorylation. Curr. Opin Gen. Dev., 8, 21–27, 1998.
25. Nelson, D.A., Krucker, N.A., and Ludlow, J.W. High molecular weight protein phosphatase type 1 dephosphorylates the retinoblstoma protein. J. Biol. Chem., 272, 4528–4535, 1997.
26. Quelle, D.E., Zindy, F., Ashmun, R.A., and Sherr, C.J. Alternative reading frames of the INK4a tumor

suppressor gene encodes two unrelated proteins capable inducing cell cycle arrest. Cell, 83, 993–1000, 1995.
27. Zhang, Y., Xiong, Y., and Yarbrough, W.G. ARF promotes MDM2 degradation and stabilizes p53: ARF-INK4a locus deletionimpairs both the Rb and p53 tumor suppression pathways. Cell, 92, 725–734, 1998.
28. Pomerantz, J., Schreiber-Agus, N., Liegeois, N.J., Silverman, A., Alland, L., Chin, L., Potes, J., Chen, K., Orlow, I., Lee, H.W., Cordon-Cardo, C., and DePinho, R.A. The Ink4a tumor suppressor gene product, p19Arf, interacts with MDM2 and neutralizes MDM2's inhibition of p53. Cell, 92, 713–723, 1998.
29. Palmero, I., Pantoja, C., and Serrano, M. p19ARF links the tumour suppressor p53 to Ras. Nature, 395, 125–126, 1998.
30. Bates, S., Phillips, A.C., Clark, P.A., Stott, F., Peters, G., Ludwig, R.L., and Vousden, K.H. p14ARF links the tumour suppressors RB and p53. Nature, 395, 124–125, 1998.
31. Serrano M., Lee, H.-W., Chin, L., et al. Role of the INK4 locus in tumor suppression and cell mortality. Cell, 85, 27–38, 1996.
32. Okamoto, A., Demetrick, D.J., Spillare, E.A., Hagiwara, K., Hussain, S.P., Bennett, W.P., Forrester, K., Gerwin B., Serrano, M., Beach, D.H., and Harris C.C. Mutations and altered expression of $p16^{INK4A}$ in human cancer. Proc. Natl. Acad. Sci. USA, 91, 11045–11049, 1994.
33. Della Ragione, F., Mercurio, C., and Iolascon, A. Cell cycle regulation and human leukemias: the role of $p16^{INK4}$ gene inactivation in the development of human acute lymphoblastic leukemia. Haematologica, 80, 562–573, 1995.
34. Loda, M., Cukor, B., Tam, S.W., Lavin, P., Fiorentino, M., Draetta, G.F., Jessup, J.M., and Pagano, M. Increased proteasome-dependent degradation of the cyclin-dependent kinase inhibitor p27 in aggressive coclorectal arcinomas. Nature Medicine, 3, 231, 1997.
35. Catzavelos, C., Bhattacharya, N., Ung, Y.C., Wilson, J.A., Ronacari, L., Sandhu, C., Yeger, H., Morava-Protzner, I., Kapusta, L., Franssen, E., Pritchard, K.I., and Slingerland, J.M. Decreased levels of the cell-cycle inhibitor $p27^{Kip1}$ protein: prognostic implications in primary breast cancer. Nature Medicine, 3, 227, 1997.
36. Esposito, V., Baldi, A., De Luca, A., Groger, A.M., Loda, M., Giordano, G.G., Caputi, M., Baldi, F., Pagano, M., and Giordano, A. Prognostic role of the cyclin-dependent kinase inhibitor p27 in non-small cell lung cancer. Cancer Res, 57, 3381, 1997.
37. Porter, P.L., Malone, K.E., Heagerty, P.J., Alexander, G.M., Gatti, L.A., Firpo, E.J., Daling, J.R., and Roberts, J.M. Expression of cell-cycle regulators $p27^{Kip1}$ and cyclin E, alone and in combination, correlate with survival in young breast cancer patients. Nature Medicine, 3, 222, 1997.
38. Tan, P., Cady, B., Wanner, M., Worland, P., Cukor, B., Magi-Galluzzi, C., Lavin, P., Draetta, G., Pagano, M., and Loda, M. The cell cycle inhibitor p27 is an independent prognostic marker in small (T1a,b) invasive breast carcinomas. Cancer Res., 57, 1259, 1997.
39. Bates, S. and Peters G. Cyclin D1 as a cellular protooncogene. Sem. Cancer Biol., 6, 73–82, 1995.
40. Schimdt, E.E., Ichimura, K., Reifenberger, G., and Collins, V.P. CDKN2 (p16:MTS1) gene deletion or CDK4 amplification occurs in the majority of glioblastoma. Cancer Res., 54, 6321–6324, 1994.
41. Miller, A.B. Diet and Cancer: a review. Rev. Oncol., 3, 87–95, 1990.
42. Adlercreutz, H. Western diet and western diseases: some hormonal and biochemical mechanisms and association. Scand. J. Clin. Lab. Invest., 50, 3–23, 1990.
43. Rose, D.P., Boyar, A.P., and Wynder, E.I. International comparison of mortality rates for cancer of the breast, ovary, prostate, colon, and per capita fat consumption. Cancer (Phila), 58, 2363–2371, 1986.
44. Kolonel, L.N. Variability in diet and its relation to risk in ethnic and migrant groups. Basic Life Sci., 43, 129–135, 1988.
45. Fotsis, T., Pepper, M., Adlercreutz, H., Fleischmann, G., Hase, T., Montesano, R., and Schweigerer, L. Genistein, a dietary-derived inhibitor of in vivo angiogenesis. Proc. Natl. Acad. Sci. USA, 90, 2690–2694, 1993.
46. Adlercreutz, H., Honjo, H., Higashi, A., Fotsis, T., Hamalainen, E., Hasegawa, T., and Okada, H. Urinary excretion of lignans and isoflavonoi phytoestrogens in Japanese men and women consuming traditional Japanese diet. Am. J. Clin. Nutr. 53, 1093–1110, 1993.
47. Chinery, R., Brockman, J.A., Peeler, M.O., Shyr, Y., Beuchamp, R.D., and Offey, R.J. Antioxidants enhance the cytotoxicity of chemotherapeutic agents in colorectal cancer: A p53-independent induction of $p21^{WAF1/CIP1}$ via C/EBPβ. Nature Medicine, 3, 1233–1241, 1997.
48. Shao, Z.M., Alpaugh, M.L., Fontana, J.A., and Barsky, S.H. Genistein inhibits proliferation in estrogen receptor-positive and negative human breast carcinoma cell lines characterized by p21WAF1/CIP1 induction, G2/M arrest and apoptosis. J. Cell Biochem., 69, 44–54, 1998.

49. Lian, F., Bhuiyan, M., Li, W.Y., Wall, N., Kraut, M., and Sarkar, F.H. Genistein-induced G2-M arrest, p21WAF1 upregulation, and apoptosis in a non-small-cell lung cancer cell line. Nutr. Cancer, 31, 184–191, 1998.
50. Markovits, J., Linnassier, C., Fosse, P., Couprie, J., Pierre, J., Jacquemin-Sablon, A., Saucier, J.M., Lepecq, J.B., and Larsen, A.K. Inhibitory effects of the tyrosine kinase inhibitor genistein on mammalian topoisomerase II. Cancer Res., 49, 5111–5119, 1989.
51. Yuan, Z., Kharbanda, S., and Kufe, D. 1-β-arabinofuranosylcytosine activates tyrosine phosphorylation of p34cdc2 and its association with the Src-like p56/p53lyn kinase in human myeloid leukemia cells. Biochemistry, 34, 1058–1063, 1995.
52. Kawada, N., Seki, S., Inoue, M., and Kuroki, T. Effect of antioxidants, resveratrol, quercetin, and N-acetylcysteine, on the functions of cultured rat hepatic stellate cells and Kupffer cells. Hepatology, 27, 1265–1274, 1998.
53. Carlson, B.A., Bubay, M.M., Sausville, E.A., Brizuela, L., and Worland, P.J. Flavopiridol induces G1 arrest with inhibition of cyclin-dependet kinase (CDK) 2 and CDK4 in human breast carcinoma cells. Cancer Res., 56, 2473–2478, 1996.
54. Della Ragione, F., Cucciolla, V., Borriello, A., Della Pietra, V., Racioppi, L., Soldati, G., Manna, C., Galletti P., and Zappia, V. Resveratrol arrests the cell division cycle at S/G2 phase transition. Biochem Biophys Res Commun., 250, 53–58, 1998.
55. Slavoshian, S., Blottiere, H.M., Cherbut, C., and Galmiche, J.P. Butyrate stimulates cyclin D and p21 and inhibits cyclin-dependent kinase 2 expression in HT-29 colonic epithelial cells. Biochem. Biphys. Res. Commun., 232, 169–172, 1997.
56. Nakano, K., Mizuno, T., Sowa, Y., Orita, T., Yoshino, T., Okuyama, Y., Fujita, T., Ohtani-Fujita, N., Matsukawa, Y., Tokino, T., Yamagishi, H., Oka, T., Nomura, H., and Sakai, T. Butyrate activates the WAF-1/CIP1 gene promoter through sp1 sites in a p53-negtive colon cancercell line. J. Biol. Chem., 272, 22199–22206, 1997.
57. Candido, E.P.M., Reeves, R., and Davie, J.R. Sodium butyrate inhibits histone deacetylation in cultured cells. Cell, 14, 105–113, 1978.

9

REGULATION OF p53 FUNCTION IN NORMAL AND MALIGNANT CELLS

Vincenzo Tortora,[1] Paola Bontempo,[1] Mariantonietta Verdicchio,[1] Ignazio Armetta,[1] Ciro Abbondanza,[1] Ettore Maria Schiavone,[2] Ernesto Nola,[1] Giovanni Alfredo Puca,[1] and Anna Maria Molinari[1]

[1] Institute of General Pathology and Oncology
Second University of Naples
Larghetto Sant'Aniello a Caponapoli 2
80138 Naples, Italy
[2] Division of Hematology 2. Cardarelli Hospital
80131 Napoli, Italy

1. BACKGROUND

Tumors arise from single cells that develop into large populations sharing a series of genetic alterations. The single phenotypic feature that best characterizes a transformed cell is its ability to proliferate indefinitely when left undisturbed.[1] The recognition of specific genes that modulate proliferation, has led to studies on the cell cycle.[2] Homeostasis has a great importance in the development of tumors and is regulated by a balance among proliferation, the arrest of growth and programmed death (apoptosis).

Numerous genes and proteins participate in cell cycle checkpoints, in apoptosis or inhibiting apoptotic stimuli, favoring cell survival.[3] Among the plethora of genes and proteins which are altered in human cancer, the involvement of the gene that codes tumor suppressor protein p53 is evident.

Many tumors contain p53 mutations, such as: cancer of the uterus, breast, esophagus, skin, colon, and others. Epidemiological studies underscore the fact that p53 occupies a special role in the genesis of tumors because it is the most frequent target for abnormalities in every type of cancer. These studies have shown that there is a strong association between p53 mutation and other human cancers and that p53 has an important role in the growth control, and differentiation. It has been observed, in fact, that p53 is the key mediator in the cellular response to DNA damage, in the sense that it is responsible for the two possible results of the damage,

Advances in Nutrition and Cancer 2, edited by Zappia *et al.*
Kluwer Academic / Plenum Publishers, New York, 1999.

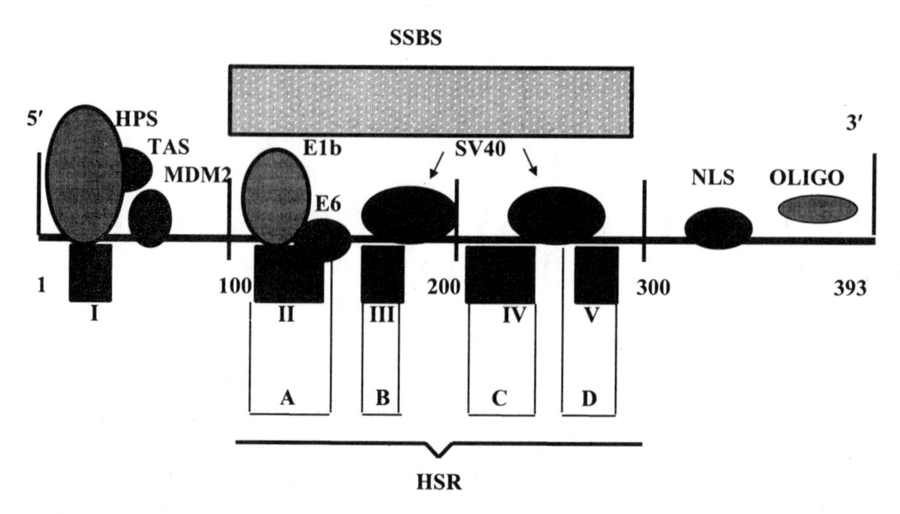

Figure 1. Structural and functional aspects of *p53* suppresor gene. The gene is approximately 20 kilobases in length, yields a 2.8 kb mRNA transcript, and encodes a 53kDa nuclear phosphoprotein composed of 393 amino acids spread through 11 exons, only the first of which is noncoding. HPS (heat shock protein); TAS (transcription activation site); SV40 large T-antigen binding sites (SV40); adenovirus E1b and papillomavirus E6 binding sites; cellular Mdm2 (mouse double minute 2) binding site; SSBS (sequence-specific DNA binding region); NSL (nuclear localization signals); OLIGO (oligomerization domain). The *p53* has five evolutionary conserved domains (HCD I–V), which correspond closely to the most frequently mutated regions in human cancer (hot spot regions: HSR, A–D).

the arrest of the cell in G1 and apoptosis. Moreover, it has been shown that in the absence of p53, cells pass on an altered chromosome which is the base of tumor transformation. In this sense, the role of the "DNA guardian" has been attributed to p53 protein.[4] In this model of tumorgenesis, p53 indirectly destroys the tumors at the bud because it reduces the number of oncogenic mutations which can arise within the cell. This model is compatible with the hypothesis that the mutation of the *p53* gene, and obviously the loss of protein function, could occur in the first stages of tumor genesis.[5]

2. STUCTURAL ASPECTS OF p53

Gene *p53* is located on the short arm of the chromosome 17 (band 13). It contains 11 exons, and is about 20kb long; it produces a mRNA transcription of 2.8kb, which gives rise to a nuclear phosphoprotein of 53kDa, composed of 393 amino acids (Fig. 1). The protein sequence is maintained in different animal species and is divided into four domains with biochemical properties.[6] The amino-terminal region (amino acids residues 1–43) corresponds to the domain that carries out transcriptional activity. This interacts with the specific areas of the DNA, and increases the activity of specific cellular genes. The region from residue 100 to the residue 300 corresponds to the domain that binds the DNA in a specific sequence. It is the most constant domain in the various species and corresponds to the domain where a great number of mutations in human tumors have been found.

The region from 320 to 360 amino acids residues corresponds to the oligomerization domain: in fact, when placed in solution, p53 forms a tetrameteric complex which binds DNA. The carboxy-terminus domain (330–393 residues) and specifically the last 26 acids, is an open domain, with nine basic amino acids residues, linked non-specifically to the nucleic acids.[7,8] These studies show that the C-terminal domain is fundamental in catalyzing the linkage of the single broken chains of nucleic amino acids in double chains, which help recognize lesions on the DNA (such as deletions of a base, breaking of a single chain, insertions, and other mutations). Together with other domains, the C-terminus favors the formation of p53 tetramers able to bind the DNA effectively in relation to distinct target genes, increasing its transcriptional activity.[9] Therefore, cells recognize DNA damage and are inclined to accumulate p53 in their nucleus.[10] For this reason, protein p53 can be considered a multifunctional transcriptional oncosuppressor factor which organizes the cellular responses after damage to the DNA and in consequence helps maintain genetic stability. The protein p53 is protected from rapid proteolysis when bound to DNA and the binding activates it as a "genome guardian" against genetic damage.[4]

This is crucial for the integrity of genetic material, and the mechanisms of activation of protein p53 are of fundamental importance. Many studies have attempted to reconstruct the events that happen within a cell for p53 activation, but doubts persist.

3. ACTIVATION MECHANISMS OF p53

It is well-known that the immediate result of DNA damage is increase of p53 activity from either prolonged half life, or to increased expression. The phenomenon is directed by a conformational transition of p53. Under the form of activated tetramer it is able to bind the damaged DNA and carry on its function as transcriptional regulator. "In vitro" studies have shown that the damaged DNA could be recognized or "acknowledged" by a protein cofactor which is able consequently, to modify p53 in an active form which normally remains within the cells in the form of an inactive tetramer. A similar mechanism has been shown[11] using short fragments of DNA oligonucleotide 5′ protruding in the attempt to mimic "in vitro" the events that occur in cells suffering from genotoxic damage (Fig. 2).

Recent studies[12,13] suggest that p53 has numerous activation mechanisms: Woo et al. have shown that p53 is made in an inactive latent state (Fig. 3A): a DNA-dependent protein kinase (DNA-PK) is needed to activate DNA binding and transcription by p53, and another factor (X) from nuclear extracts of irradiated cells is also required, which acting sequentially activate the p53 tetramers.[12] Another hypothesis admits that p53 is activated by a p53 specific phosphatase. The removal of phosphate from the serine in position 376 permits the bond of protein 14-3-3 [14] in relation to the carboxy-terminus region and actives tetramer p53 (Fig. 3B). A better example shows that p53 is in itself expressed in an active form, but constantly degraded from MDM2. The damaged DNA would stop MDM2 and other complexes which degrade p53, therefore, the cell has a great quantity of p53 at its disposal, which can be phosphorilated and function at the nuclear level (Fig. 3C). All these hypotheses may be valid, in fact the cell that expresses a normal phenotype could have numerous "pathways" to activate p53.

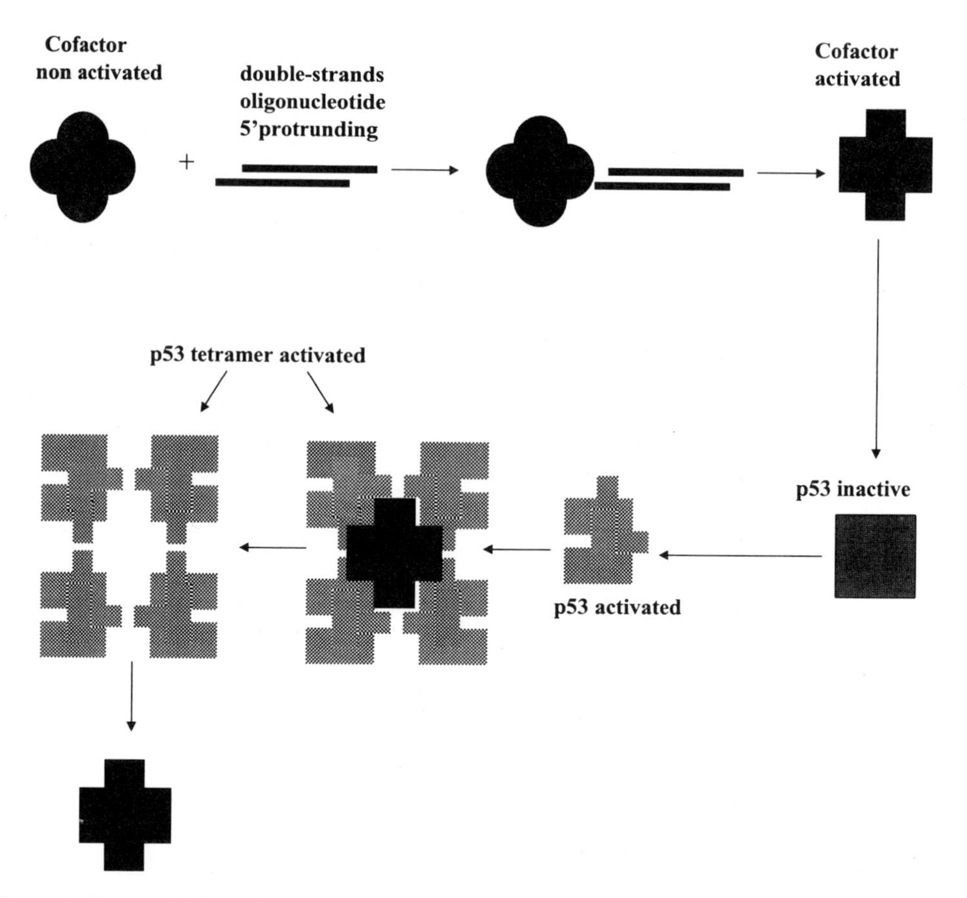

Figure 2. New model for activation of p53 tetramers by cofactor and double-strands oligonucleotide 5′ protunding.

4. PHYSIOLOGICAL FUNCTIONS OF p53

Just how protein p53 works in normal conditions is still not clear. However, it is well-known that it arrests the cell in the G1 phase of the cell cycle, actives apoptosis and promotes the differentiation of various cellular histotypes and mediates the cell response to DNA damage caused by genotoxic agents and oxidative stress.[15]

One of the major unresolved questions concerns the mechanisms that determine the specific activation pathways within the cells during embryogenesis, cellular maturing, or possible external signals which cause the activation of p53. To answer these questions, "in vitro" studies have evaluated the development and the cellular differentiation by monitoring p53 expression. The p53 protein is involved in embryogenesis and cellular differentiation; above all in the differentiation of hematopoietic cells and non-hematopoietic cells: in spermatogenesis and angiogenesis.[15]

4.1. p53 and Hematopoiesis

Several studies have shown that p53 is expressed at low levels in normal light-density human bone marrow cells and in mature cell populations identified by CD20 (B lymphocytes), CD3 (T lymphocytes), CD15 (granulocytes), and CD14

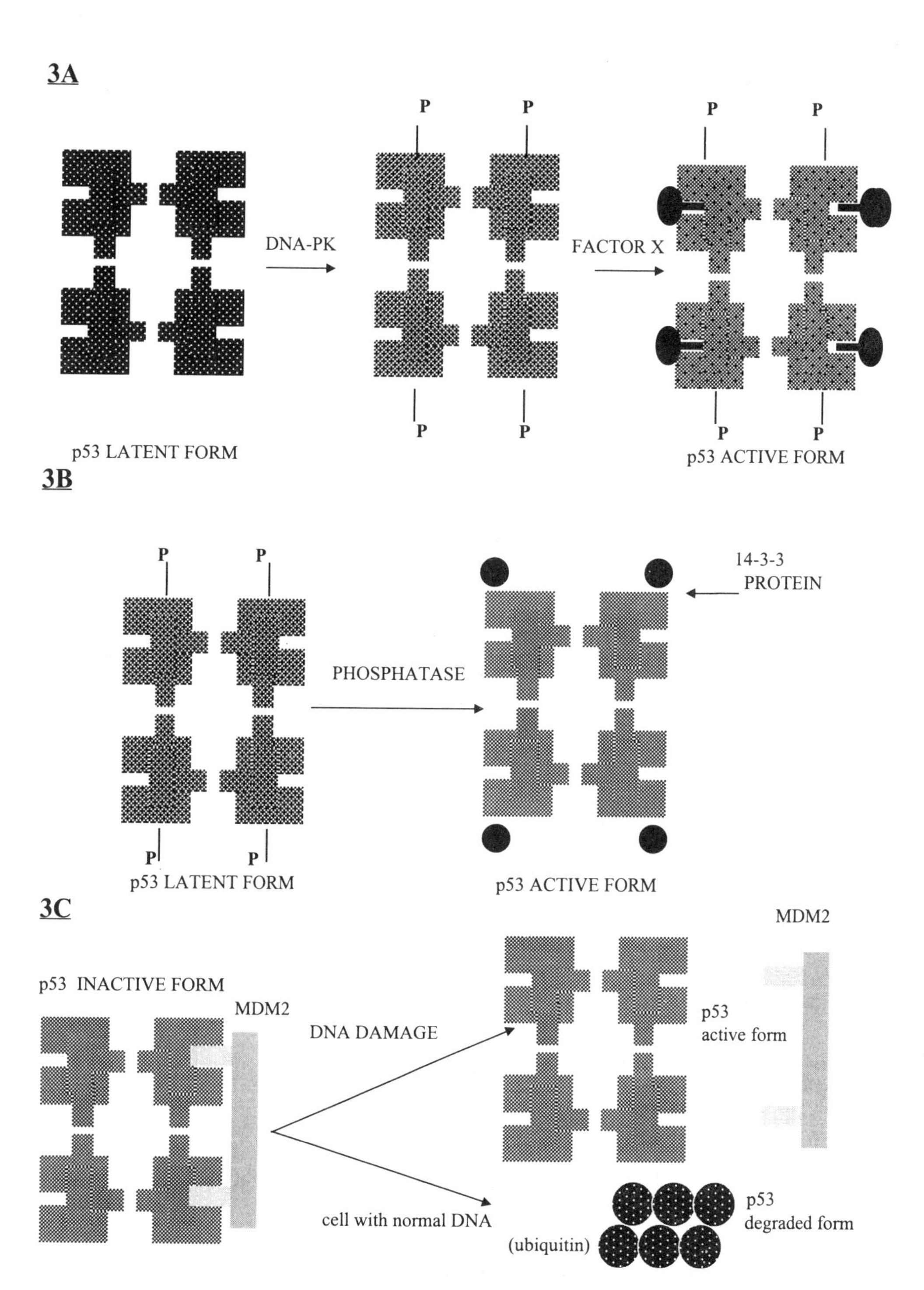

Figure 3. Possible ways to activate p53 tetramers after DNA damage.

(monocytes).[16] Moreover, it is well-known that only wt-p53 protein induces erythroid differentiation, while mutant p53 conserves the DNA binding capacity, but lacks the specific activity and is no longer able to stimulate differentiation. The expression of p53 in peripheral blood mononuclear cells implies that the *p53* gene may be transcriptionally active in these cells and that endogenous expression of wild-type p53 is increased during maturation of normal human hematopoietic cells. For example, it has been shown that p53 can transactivate the promoter and the enhancer of the light chain gene, interacting in controlling of the transactivating B cell differentiation genes. Altogether, p53 appears to be involved in lymphoid cells. In human bone marrow cells, p53 levels are very low, while in mature cells its levels are much higher and easily detected. Probably, the action of wild-type p53 protein contribute to the inhibition of proliferation that occurs during terminal differentiation and the effects of this protein are numerous and various in the normal development of the cells.

The key role played by p53, safeguarding the normal development of the differentiation and homeostasis of the hematopoietic cells, is seen in p53 mutations, which are specifically associated with the progression of disease in lymphoid and myeloid leukemias, as well as in lymphomas.[17] Moreover, in the transferring of the *p53* gene in carcinogenic blood cells it is possible to observe cellular differentiation with an increased cellular apoptosis. Hence, the restoration of expression of wt-p53 in tumor cells with a p53 mutation might eventually play a role in cancer gene therapy.

4.2. Involvement of p53 in Differentiation of Epithelial Cells

Many experimental models[15] have shown that p53 is involved in the differentiation of epithelial cells. The levels of the cellular p53 decrease during epidermal cell differentiation.[18] Furthermore, this decrease has not been observed during the growth arrest. It is likely that it blocks the cells and helps the differentiation. Similar phenomena are observable within the granulosa cells where p53 co-operates with cAMP-dependent signals in apoptosis and differentiation.[19]

Other studies have demonstrated that p53 regulates the differentiation and apoptosis of neurones and oligodendrites.[20] "In vitro" studies show p53 changes its location upon maturation and differentiation of these cells, through a well-controlled nuclear translocation mechanism.[20]

Moreover, from experimental models, p53 interacts with muscle differentiation, too. In fact, inhibiting endogenous wild type p53 stops skeletal muscle differentiation.[18]

In conclusion, this protein is associated with the differentiation and apoptosis of a variety of cells. The up-regulation of p53 is observed during differentiation of muscle cells; a decrease in epithelial cells.

4.3. p53 and Embryogenesis

p53 has an important role in embryonic development: some research suggests a correlation between embryogenesis and increased levels of p53. Schmid et al. showed that the levels of *p53* mRNA declined during the embryonic development in the mouse.[21] During the early phases of embryogenesis there is a high expression of p53 which decreases in the final phase of differentiation. The p53 protein at this time protects the genome from developmental oxidative DNA-damaging stress, preventing fetal cell mutations. During development, knock out mice present a greater rate of tumors.

However, an overexpression of p53 during embryogenesis arrests normal cell growth. This can be seen in MDM-2 knock-out mice, where the absence of down regulated p53 protein increases apoptosis. For these reasons the mechanisms that regulate the p53 activation are very important for the normal development of organs and tissues during embryogenesis.

4.4. p53 and Spermatogenesis

Several studies have indicated that p53 is induced in the meiotic process of spermatogenesis in mice.[22] It has been shown that this protein is expressed at high levels at the end of zygotene and the beginning of the pachytene stages. The pachytene stage is a particular phase where chromosome pairing, recombinations and repair of damaged DNA takes place. p53 is thus involved in DNA repair, stops the cell division to permit cells to correct and repair the damages caused to DNA. It has also been demonstrated that p53 takes part in the response of primary spermatocytes to γ-irradiation-induced DNA damage. These results show that p53 could take part in cell cycle control during spermatogenesis, where the most rapidly dividing cells mature. Some investigators believe that there are possibly two different p53 proteins in testis which may be formed by alternative splicing. The p53 protein plays a key role as "a genome guardian" during gene recombination such as spermatogenesis.

4.5. Control of Angiogenesis by p53

Angiogenesis is controlled by the local balance between factors that stimulate new vessel growth and factors that inhibit it. This balance dictates the fate of tumor development as well as the expansion of the metastatic colonies.[23] In this field p53 plays an inhibiting role on the formation of angiogenesis, confirming the function of onco-suppressive molecule. Wild-type p53 causes an endogenous down-regulated VEGF mRNA (vascular endothelial growth factor). VEGF plays a critical role in the neo-angiogenesis.[24] It is a dimeric glycoprotein and has a specific mitogenic activity in endothelial cells. Some researchers presume that VEGF down-regulation might be mediated by indirect means: wt-p53 inhibits angiogenesis through up-regulated thrombospondin-1, a powerful inhibitor of angiogenesis. In contrast to the positive role of p53 in the differentiation processes described above, in angiogenesis p53 has a negative effect. In fact, low levels of this mutated protein may enhance the angiogenic processes.

4.6. The Immune System and p53

Many studies highlight the relevant role played by p53 in the maturation of T and B lymphocytes. This is easily understandable thanks to the specific action of p53 which is activated whenever there are rearrangement processes of the DNA with the generation of double-strand breaks (DSB). This happens during the maturation of T and B cells.[25] Specific research has been carried out with the aim of showing the possible association between the DNA damage, that could occur during the maturation of lymphoid cells and p53. It is possible to observe a high incidence of T and B cell lymphomas in knock-out mice.[26] Moreover, studies led by analyzing Severe Combined Immune Deficient mice (SCID), where there is a general defect in DNA DSB repair, suggests the central role of p53 in the regulation of DNA damage in V-D-J and TCR

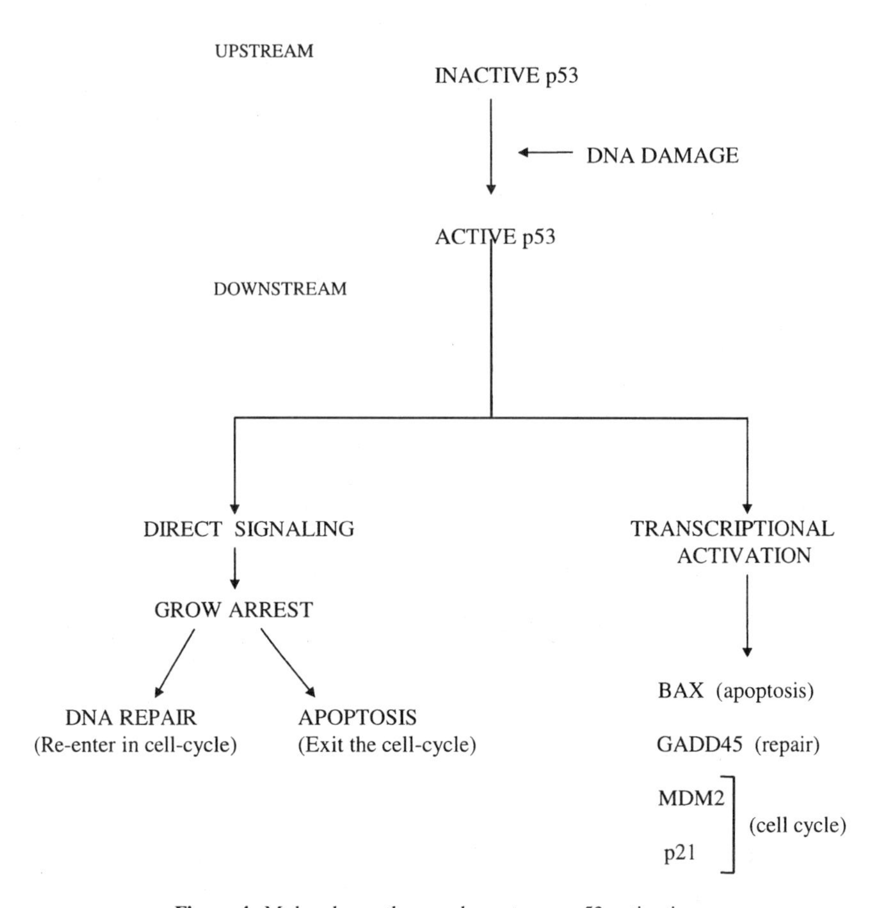

Figure 4. Molecular pathways downstream p53 activation.

rearrangements in B and T cells. In normal cells, the p53 ensures that only cells with an intact genome will proceed forward to normal differentiation and maturation.

5. MECHANISMS WHICH TAKE PLACE AFTER THE p53 ACTIVATION

The molecular basis for p53 activity is still vague. It seems, however, likely that the majority of the effects caused by the protein could be attributed to its ability to work as a powerful regulating oncosuppressor of transcription (Fig. 4).

The transcription of genes such as *MCK* (muscle creatine kinase) *MDR1* (multidrug resistance gene 1), *IL-6* (interleukin-6), *hsp-70* (heat shock protein), *c-JUN*, *c-FOS*, *GADD45*, *MDM2*, *p21/WAF1/CIP1* can be modulated by protein p53. Many studies have shown that during radiation damage to DNA, there is an increase of the transcription of some genes, such as: *GADD45*, *MDM2*, and *p21* thanks to the action of the p53 protein. In fact, in the third intron of the gene *GADD45* a concomitant sequence for the bond of p53 protein has been identified.[27] The role of *GADD45* gene in DNA repair is not clear yet. It is well known, however, that, the product of *GADD45* gene interacts directly with protein PCNA (proliferating cell nuclear antigen) a protein directly involved either in the replication of the DNA, or in the damage repair; which is associated with the product of *ERCC3* gene (transcriptional factor involved in the cut and patch of the nucleotides during damage reparation); and which interacts with the protein RPA (replication protein A) a necessary factor for replication and repair.

Even with *MDM2*, p53 protein recognizes a concomitant sequence in the promoter of the gene. In this case, however, the increase in the *MDM2* product (28) constitutes the non-activating circuit of p53 (negative feedback). The protein p21 offers the most interesting function among the induced genes of p53. The product of the *p21* gene is an inhibitor of cyclin-dependent kinase. Its activity prevents the regulating proteins of the cycle, such as those of the Rb family from being phosphorylated. Consequently, it prevents the transcriptional factors from being available to activate the genes necessary for the transition from the G1 to the S phase, blocking the cells in the G1 phase of the cycle. Moreover, some studies have shown that p21 can also directly bind to the product of *PCNA* gene, which is a subunity regulator of DNA polymerase.[29] The effect of these interactions is decreased DNA polymerase, or which makes it a compensation enzyme rather than a duplication of the DNA enzyme.

Some authors have proposed a new model for p53 activity: where the abnormal cell proliferation results in deregulated E2F-1 activity in the cell cycle. E2F-1 causes defects in the RB pathway, induces $p14^{ARF}$ that stabilizes p53. It has been shown that E2F-1 directly activates expression of the human tumor suppressor protein $p14^{ARF}$, that binds to the MDM2-p53 complex and prevents p53 degradation.[30]

Not all cells respond to the DNA damage by arresting themselves in G1.[31] Much depends on the cell type and environment. It is not clear yet what genetic products are activated by apoptosis. Some studies have shown that p53 protein can transactivate the promoter of gene *BAX*.[32] This has a concomitant sequence for p53 which makes a "gene reporter" to be transcribed with high capacity. A normal p53 protein is bound to this sequence. These results indirectly suggest that the *BAX* gene can mediate apoptosis that requires p53. In particular *BAX* seems to be activated fundamentally by p53 which is able to start a programmed death pathway. In fact, the product of *BAX* gene binds the Bcl-2 protein, preventing its principal function, allowing cell survival even when apoptotic stimuli are present.[33] For this reason, it is clear that Bcl-2 contributes to the tumor growth by helping the survival of the tumor cells, without acting on the velocity of cell proliferation. In man the *bcl-2* gene (acronym of B cell lymphoma/leukemia-2 gene because it was discovered for the first time in center follicular lymphoma with cells of the B type) codes for a 239 amino acid protein, with a M.W. of 26kDa, with specific functional domain preserved in different species.[34] The product of the *bcl-2* gene has a wide distribution in the cell. A sequence of 22 amino acids in the C-terminus is responsible for its anchorage to the membranes. Immune studies have shown its relationship with the internal membrane of mitochondria and with the endoplasmatic reticulum. This widespread localization and its relationship with the intercellular spaces, where oxygen free radicals are produced, clarify its protection against apoptosis. Protein Bcl-2 takes part in the antioxidant mechanism through a protein to protein interaction not fully understood, reducing the peroxidation of lipids and representing a negative regulator of the apoptotic process.[35] Another hypothesis is that this protein probably regulates the flux of calcium ions and the cellular oxidating state permits survival of the cells in the Go phase. Many genes code proteins that show homologous *bcl-2* regions. In particular, a family of *bcl-2* members has been identified. Among these bcl-xL, bcl-w, and Mcl-1 inhibit apoptosis while others such as bax, bik,

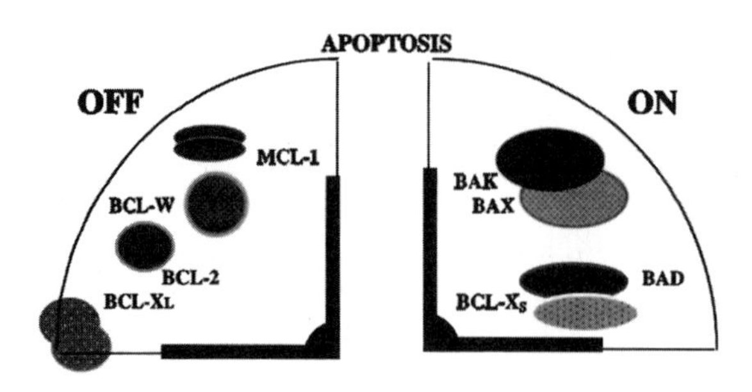

Figure 5. Action of bcl-2 family members on apoptosis.

bak, bad, and bcl-xs activate the process. All these proteins taking part in a complex network of functional balance can form the homo and heter-dimers with each other. These help or contrast the activity of the single protein monomers, quickening or inhibiting programmed cell death (Fig. 5).

6. p53 FAMILY

p53 was considered an orphan tumor suppressor gene without any related genes, until the discovery of a group of genes which are part of the *p53* gene family.[36–39] Some authors, using degenerate PCR and tissue-specific c-DNA as a template, have shown the isolation of three distinct human p53 family genes: *p53*, *p73α/73β*, and *p51A/51B*. It has been found that p73 has growth inhibitory and apoptotic promoting effects similar to p53. However, unlike p53, p73 is not up-regulated in response to DNA damage. Even p51 protein presents similar functions to p53 in knock-out p53 cells, it induces growth-suppression and apoptosis. The *p51* gene is located on chromosome 3q 28, codes two variants for splicing: p51A a 448 amino acid protein with a M.W. of 71.9 kDa; and p51B a 641 amino acid protein with a M.W. of 50.9 kDa. Other researchers have identified another member of the p53 family: a 40 kDa nuclear protein, called p53CP (p53 competing protein), that specifically binds to the consensus p53 binding sites like: *WAF-1*, *GADD 45*, *MDM2*, and *BAX*. They suggest that p53CP, sequestering wt-p53 from its DNA binding site through competitive binding, may provide a novel mechanism of p53 inactivation. The gene *p40*, like, *p51*, is mapped on humane chromosome 3q, a region where deletions are commonly associated with many species of cancer. Probably parallel multiple p53 family genes in a single cell can be a very useful defense mechanism to avoid possible mutation oncosuppressor p53. This property may be used for therapeutic applications by inducing endogenous p51 or p73 to restore p53 resembling functions in p53-deficient tumors.

7. CONCLUSIONS

Other pathways without p53 or Bcl-2 have been suggested for apoptosis that activate numerous genes, among them: *p21/WAF1/CIP1*, *GADD45*, *BAX*, and *MDM2*.[40]

Some authors have suggested a critical role for a specific co-activator of p53, known as p300.[41] In fact, they suggest that p300 is very important in orchestrating early cell cycle progression through the pathway controls mediated by E2F and p53.[42] Other studies support the idea that p300 is a negative regulator of the cell cycle progression, and can influence G_1 cellular arrest and apoptosis.

For this reason, it is clear that the complex mechanisms, which regulate the cellular homeostasis and the balance of the cellular development, are various and numerous, being influenced by many messages.[43] Further research is still necessary to clarify the organization and the interactions with the surrounding environment of normal and tumor cells in growth, differentiation, and apoptosis.

REFERENCES

1. Bishop J.M., (1987). The molecular genetics of cancer. Science, **235**:305–311.
2. Follette P.J. and O'Farrell P.H., (1997). Connecting cell behavior to pattering: lessons from the cell cycle. Cell, **88**:309–314.
3. Bukholm I.K., Nesland J.M., Karesen R., Jacobsen U., and Borresen A.L., (1997). Interaction between bcl-2 and p21 (WAF1/CIP1) in breast carcinomas with wild-type p53. Int. J. Cancer, **73**(1):38–41.
4. Lane D.P., (1992). P53, the guardian of the genome. Nature, **358**:15–16.
5. Slichenmyer W.J., Nelson W.G., Slebos R.J., and Kastan M.B., (1993). Loss of a *p53*-associated G1 Checkpoints Does Not Decrease Cell Survival Following DNA Damage. Cancer Res., **53**:4164–4168.
6. Malkin D., (1994). Germline p53 mutations and heritable cancer. Annu. Rev. Genet., **28**:443–465.
7. Prives C., (1994). How loops, β Sheets, and α Helices help us to understand p53. Cell, **78**:543–546.
8. Levine A.J., (1997). p53, the cellular gatekeeper for growth and division. Cell, **88**:323–331.
9. Lee S., Elenbaas B., Levine A., and Griffith J., (1995). p53 and its 14kDa C-terminal domain recognize primary DNA damage in the form of insertion/deletion mismatches. Cell, **81**:1013–1020.
10. Komarova E.A., Zelnick R.C., Chin D., Zeremski M., and Gudkov A.V., (1997). Intracellular localization of p53 tumor suppressor protein in γ-irradiated cells is cell cycle regulated and determined by the nucleus. Cancer Res., **57**:5217–5220.
11. Molinari A.M., Armetta I., Napolitano M., Schiavulli M., Bontempo P., and e Puca G.A. La capacità dell'antioncogene p53 di legare specifiche sequenze di DNA è modulata in vitro dalla presenza di oligonucleotidi. XXII Congresso Nazionale SIP.
12. Lane D., (1998). Awakening angels. Nature, **394**:616–617.
13. Woo R.A., McLure K.G., Lees-Miller S.P., and Lee P., (1998). DNA-dependent protein kinase acts upstream of p53 in response to DNA damage. Nature, **394**:700–704.
14. Waterman M.J., Stavridi E.S., Waterman J.L., and Halazonetis T.D., (1998). ATM-dependent activation of p53 involves dephosphorylation and association with 14-3-3 proteins. Nature Genet., **19**(2):175–178.
15. Almong N. and Rotter V., (1997). Involement of p53 in cell differentiation and development. Biochimica et Biophysica Acta, 1333 F1–F27.
16. Prokocimer M. and Rotter V., (1994). Structure and Function of p53 in Normal Cells and Their Aberrations in Cancer Cells: Projection on the Hematologic Cell Lineages. Blood, **84**:2391–2411.
17. Imamura J., Miyoshi I., and Koffler P., (1994). p53 in Hematologic Malignancies. Blood, **84**:2412–2421.
18. Soddu S., Blandino G., Scardigli R., Coen S., and Sacchi A., (1996). Interference with p53 Protein Inhibits Hematopoietic and Muscle Differentiation. The Journal of Cell Biology, **134**:193–204.
19. Amesterdam A., Keren-Tal I., and Aharoni D., (1996). Ross-talk between cAMP and p53-generated signals in induction of differentiation and apoptosis in steroidogenic granulosa cells. Steroid **61**:252–256.
20. Eizenberg O., Gottlieb E., and Schwartz M., (1996). p53 plays a Regulatory Role in Differentiation and Apoptosis of Central Nervous System-Associated Cells. Molecular and Cellular Biology, **16**:5178–5185.
21. Schmid P., Lorenz A., Hameister H., and Montenarh M., (1991). Expression of p53 during mouse embryogenesis. Development, **113**:857–865.
22. Sjoblom T. and Lahdetie J., (1996). Expression of p53 in normal and γ-irradiated rat testis suggests a role for p53 in meiotic recombination and repair. Oncogene, **12**:2499–2505.
23. Mukhopadhyay D., Tsiokas L., and Sukhatme V.P., (1995). Wild-Type p53 and v-Src Exert Opposing Influences on Human Vascular Endothelial Growth Factor Gene Expression. Cancer Research **55**:6161–6165.

24. Dameron K.M., Volpert O.V., Tainsky M.A., and Bouck N., (1994). Control of Angiogenesis in Fibroblasts by p53 Regulation of Thrombospondin-1. Science **256**:1582–1584.
25. Nelson W.G. and Kastan M.B., (1994). DNA Strand Breaks: the DNA Template Alterations That Trigger p53-Dependent DNA Damage Response Pathways. Molecular and Cellular Biology, **14**:1815–1823.
26. Aloni-Gristein R., Zan-Bar I., Alboum I., and Rotter V., (1993). Wilde type p53 functions as a control protein in the differentiation pathway of the B-cell linage. Oncogene, **8**:3297–3305.
27. Zhan Q., Chen I.T., Antinore M.J., and Fornace A.J., (1998). Tumor suppressor p53 can partecipate in transcriptional induction of the GADD45 promoter in absence of direct DNA binding. Mol. Cell Biol., May **18**(5):2768–2778.
28. Zauberman A., Barak Y., Ragimov N., Levy N., and Oren M., (1993). Sequence-specific DNA binding by p53: identification of target sites and lack of binding to p53-MDM2 complexes. The EMBO Journal, **12**(7):2799–2808.
29. Cayrol C., Knibiehler M., and Ducommum B., (1998). p21 binding to PCNA causes G_1 and G_2 cell cycle arrest in p53-deficient cells. Oncogene, **16**:311–320.
30. Bates S., Phillips C., (1998). p14 ARF links the tumor suppressor RB and p53. Nature, **395**:124–125.
31. Guillouf C., Rosselli F., Krishnaraju K., Moustacchi E., Hoffman B., and Liebermann D.A., (1995). p53 involvement in control of G_2 exit of the cell cycle: role in DNA damage-induced apoptosis. Oncogene, **10**:2263–2270.
32. Huang Y., Ray S., Reed J.C., Ibrado A.M., Tang C., Nawabi A., and Bhalla K., (1997). Estrogen increases intracellular p26Bcl-2 to p21Bax ratios and inihibits taxol-induced apoptosis of human breast cancer MCF-7 cells. Breast Cancer Res. Treat., **42**(1):73–81.
33. Miyashita T. and Reed J.C., (1995). Tumor Suppressor p53 is a direct transcriptional activator of the human bax gene. Cell, **80**:293–299.
34. Boise L.H., Gonzales-Garcia M., Postema C.E., Ding L., and Thompson C.B., (1993). bcl-x, a bcl-2 related gene that functions as a dominant regolator of apoptotic cell death. Cell, **74**:597–608.
35. Miyashita T., Harigai M., Hanada M., and Reed J.C., (1993). Identification of a p53-dependent negative response element in the bcl-2 gene. Cancer Research, **54**:3131–3135.
36. Bian J. and Sun Yi, (1997). p53CP, a putative p53 competing protein that specifically binds to the consensus p53 DNA binding sites: A third member of the p53 family?. Proc. Natl. Acad. Sci. USA, **94**:14753–14758.
37. Osada M., Ohba M., Kawahara C., and Ikawa S., (1998). Cloning and functional analysis of human *p51*, which structurally and functionally resembles *p53*. Nature Medicine vol.4 n.7:839–843.
38. Trink B., Okami K., Wu L., Sriuranpong V., Jen J., and Sidransky D., (1998). A new human p53 homologue. Nature Medicine, **4**(7):747.
39. Osada M., Ohba M., Kawahara C., and Ikawa S., (1998). Cloning and functional analysis of human *p51*, which structurally and functionally resembles *p53*. Nature Medicine **4**(7):839–843.
40. Liebermann D.A., Hoffman B., and Steinman R.A., (1995). Molecular controls of growth arrest and apoptosis: p53-dependent and independent pathways. Oncogene, **11**:199–210.
41. Avantaggiati M.L., Ogryzko V., Gardner K., Giordano A., Levine A.S., and Kelly K., (1997). Recruitment of p300/CBP in p53-Dependent Signal Pathways. Cell, **89**:1175–1184.
42. Lee Chang-Woo, Sorensen T.S., Shikama N., and La Thangue N.B., (1998). Functional interplay between p53 and E2F through co-activator p300. Oncogene, **16**:2695–2710.
43. Dickson R.B. and Lippman M.E., (1995). Growth factors in breast cancer. Endocrine Reviews, **16**(5):559–583.

THE ROLE OF MICRONUTRIENTS IN DNA SYNTHESIS AND MAINTENANCE

Robert A. Jacob

Western Human Nutrition Research Center
U.S. Department of Agriculture
Agricultural Research Service
University of California
2047 Wickson Hall
Davis, California 95616

Recent research has indicated that the micronutrients, folate, niacin, and vitamin C may be important for various aspects of DNA and chromosome integrity. Folate is an essential cofactor for biosynthesis of deoxynucleotides and DNA methylation reactions. Niacin provides ADP-ribose units for proteins which are involved in DNA replication and repair. Vitamin C, a primary intracellular antioxidant, may provide protection against oxidative DNA base damage. Evidence for the importance of these micronutrients in DNA synthesis and repair is reviewed. The evidence is substantial for some of these nutrients and merely suggestive, even controversial, for others.

1. THE ROLE OF FOLATE IN DNA SYNTHESIS AND CHROMOSOMAL INTEGRITY

Folate supplies one-carbon units required for ribonucleotide and DNA synthesis, and for numerous methylation reactions that involve S-adenosylmethionine (SAM), the activated methyl donor.[1] The importance of folate in the functioning of these pathways has been documented in a variety of studies using cell cultures, animal models, and human subjects. Studies using cell culture and rat models have shown that folate and/or methyl deficiency alters deoxynucleotide pools[2–6] and DNA methylation status.[7–9] Recent studies of blood cells from human populations have shown that folate deficiency results in DNA hypomethylation, misincorporation of uracil into DNA, and chromosome damage.[10–13]

Advances in Nutrition and Cancer 2, edited by Zappia *et al.*
Kluwer Academic / Plenum Publishers, New York, 1999.

Two forms of folate are known to be required for the biosynthesis of ribonucleotides incorporated into DNA. Methylenetetrahydrofolate serves as both one-carbon donor (via thymidylate synthase) and reducing agent for the methylation of the RNA pyrimidine deoxyuridylate (dUMP) to the DNA pyrimidine deoxythymidylate (dTMP), and 10-formyltetrahydrofolate provides the one-carbon formyl group in the synthesis of IMP, from which the purine ribonucleotides AMP and GMP are derived. The de novo synthesis of IMP requires 2 folate-dependent one carbon-transfer enzymes: glycinamide ribotide transformylase and aminoimidazolecarboxamide ribotide transformylase.[1] Studies using in vitro or animal models have shown that folate deficiency disrupts the normal cellular production of deoxyribonucleotides which are utilized for DNA synthesis and repair.

1.1. Cell and Animal Model Studies

Studies by James and co-workers using cell culture and rat models have shown that folate and methyl deficiency produces alterations in deoxynucleotide pools. Spleen cells from rats fed diets low in folate, choline, and/or methionine had decreased amounts of dTMP and dTTP, consistent with impaired folate dependent conversion of uridylate to thymidylate.[2] Mitogen stimulated rat lymphocytes cultured in low folate and/or methionine media showed alterations in nucleotide pools which were linked to retarded DNA synthesis and cell cycle progression.[4,5] Rats fed a low folate/methyl diet had increased uracil, abasic sites, and strand breaks in liver DNA, and other molecular alterations that relate to tumor promotion.[6] An increase in abasic (apyrimidinic) sites represents increased uracil base excision repair of DNA.[6]

Studies of the role of folate in DNA methylation have generally shown that lack of folate can result in hypomethylated DNA in the rat, however not all study results have been consistent. Pogribny et al.[7] showed that chronic severe folate/methyl deficiency in rats induced genome-wide and p53 gene specific hypomethylation in preneoplastic liver along with an increase in DNA strand breaks. In other studies of folate deficiency in rats, Balaghi and Wagner[8] found genomic hypomethylation of hepatic DNA after 4 weeks of folate deficiency, Kim and Christman[14] found no hypomethylation of hepatic and colonic DNA during moderate folate deficiency, and Kim et al.[9] found hypomethylation of the p53 tumor suppressor gene (but not genome-wide hypomethylation) and DNA strand breaks due to isolated folate deficiency.

1.2. Human Studies

Recent studies have illustrated the importance of folate and vitamin B12 nutriture in humans for the synthesis of DNA and maintenance of chromosomal integrity. In various populations, low plasma folate and vitamin B-12, and elevated Hcy concentrations, have been associated with misincorporation of uracil, and increased erythrocyte or lymphocyte micronuclei, measures of chromosome damage.[11,12,15–17]

Blount et al.[11] found misincorporation of uracil into DNA and increased erythrocyte micronuclei frequency in folate deficient individuals which was reversed by folate supplementation. Uracil misincorporation into DNA is believed to result in increased repair-related DNA damage and subsequent chromosomal instability.[11,12] In these studies erythrocyte micronuclei frequencies were measured in 22 healthy splenectomized individuals at baseline and in 19 of the subjects after supplementation with

5 mg folic acid for 6–8 weeks. In splenectomized individuals micronucleated erythrocytes remain in the peripheral circulation and provide an index of chromosomal damage. DNA uracil levels were also determined in three folate deficient subjects with normal splenic function.[11] The authors stated that folate-deficiency induced uracil misincorporation is a likely mechanism for the observed cytogenetic damage, but that other mechanisms, such as site specific hypomethylation, may also be involved. It is proposed that folate deficiency results in increased frequency of closely spaced uracils, and this leads to increased DNA strand breaks and chromosome damage.

Controlled folate depletion of postmenopausal women residing in a metabolic unit resulted in hypomethylation and increased dUTP/dTTP ratio in lymphocyte DNA, decreased lymphocyte NAD, and increased frequency of lymphocyte micronuclei.[10,13] The eight healthy women age 49–63 y were fed an experimental low-folate diet containing 56 μg/d of folate for 91 d.[10] Folate intake was varied by supplementing 55–460 μg/d of folic acid (pteroylglutamic acid) into the diet to provide total folate intake periods of 5 wk at 56 μg/d, 4 wk at 111 μg/d, and 3 wk at 286–516 μg/d. The Recommended Dietary Allowance of folate for U.S. women and men has recently been increased to 400 μg/d.[18] The experimental design can be seen in Fig. 1 which shows mean values for plasma folate and lymphocyte DNA methylation status. The DNA methylation assay reflects the capacity of genomic DNA to accept ^{3}H-labeled methyl groups and therefore the DNA incorporated radioactivity is inversely proportional to the level of DNA methylation.[8]

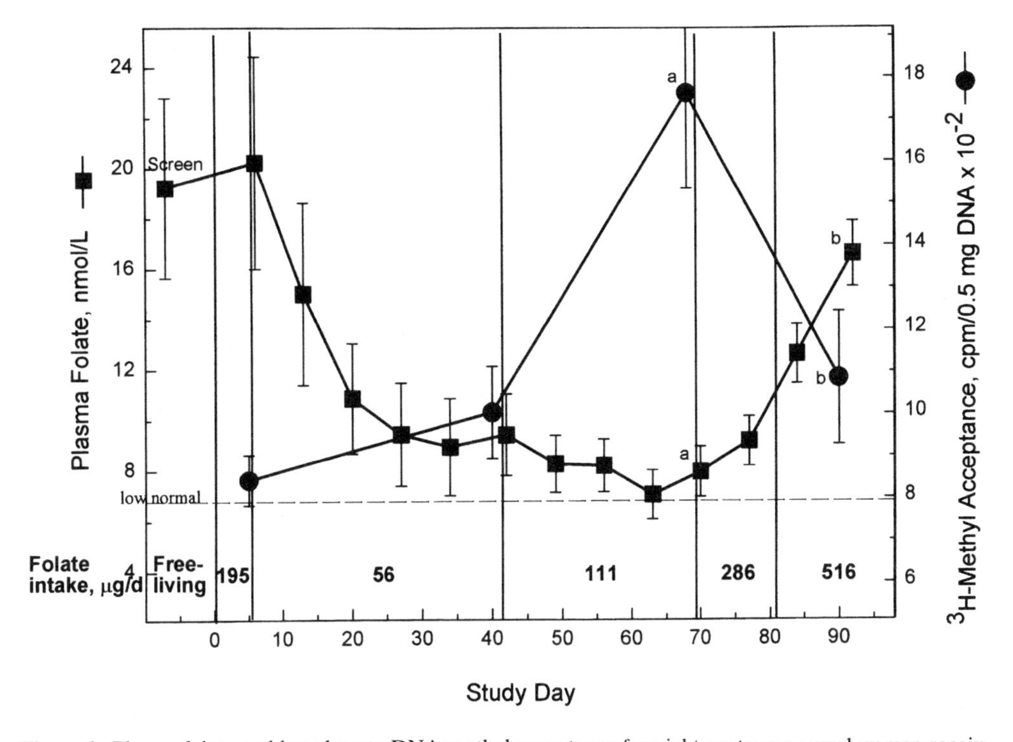

Figure 1. Plasma folate and lymphocyte DNA methyl acceptance for eight postmenopausal women receiving various dietary intakes of folate shown at bottom in μg/d. Plotted values are means ± SEM. Means at end of folate intake periods are different: [a]from d 5 or [b]from d 69, $P < 0.05$. Lower limit of normal range for plasma folate is shown at bottom as horizontal dashed line at 6.8 nmol/L. The DNA ^{3}H-methyl acceptance shown is inversely proportional to the level of DNA methylation. Data from Jacob et al.[10]

Table 1. Folate status and lymphocyte DNA indices in postmenopausal women receiving various dietary folate intakes[a]

Study day	1–5	6–69	70–91
Folate intake, μg/d	195	56–111	286–516
Plasma folate, nmol/L	19.5 ± 4.2	8.1 ± 1.2*	16.6 ± 1.6†
Plasma homocysteine, μmol/L	9.8 ± 0.4	12.6 ± 0.5*	11.8 ± 0.2*†
DNA methyl uptake, cpm/0.5 μg	795 ± 41	1738 ± 259*	1084 ± 158†
DNA strand breaks, ^{32}P cpm/μg	665 ± 242	220 ± 10*	219 ± 7*
dUTP/dTTP	2.24 ± 0.40	3.03 ± 0.40*	2.81 ± 0.57
NAD, pmol/10^6 cells	217 ± 37	126 ± 17*	120 ± 21*

[a]Means ± SEM are for end of study periods, n = 8 subjects. Means are different, $P < 0.05$: *from baseline d 5, or †from previous mean, by paired t-test or signed rank test. Data from Jacob et al.[10]

During the low folate intake periods when 56–111 μg/d folate was fed (d 6–69) plasma folate decreased significantly, but red cell folate, mean corpuscular volume, and neutrophil segmentation were unchanged. This indicates that a subclinical folate deficiency was created with low plasma folate and elevated homocysteine (Hcy), but no signs of tissue folate depletion or folate deficient hematopoiesis. Mean ± SEM values for plasma folate, homocysteine, lymphocyte NAD and DNA measures are shown in Table 1.

In addition to elevated plasma homocysteine (Hcy), the moderate folate depletion also resulted in hypomethylation of the lymphocyte DNA, an increased ratio of dUTP/dTTP in mitogen stimulated lymphocyte DNA and decreased lymphocyte NAD, changes suggesting misincorporation of uracil into DNA and increased DNA repair activity.[10] DNA methylation was assessed using the method described by Balaghi and Wagner using CpG methylase in which the uptake of ^3H-labeled methyl groups from SAM into isolated lymphocyte DNA is inversely proportional to the degree of DNA methylation.[8] The increased DNA methyl uptake seen at d 69 (Fig. 1, Table 1), a point of low plasma folate and elevated Hcy, indicates that folate depletion resulted in genome-wide DNA hypomethylation which was reversible within three weeks by folate repletion at 286–516 μg/d. No previous studies have shown that folate intake affects DNA methylation in humans. Intake levels of the dietary methyl donors methionine and choline can affect in vivo methylation capability. While the methionine intake of the women was adequate (780 mg/d), the relatively low amount of choline in the diet (147 mg/d) may have contributed to the sensitivity of the observed inverse relationship between folate intake and DNA hypomethylation.[10] The results are consistent with rat studies in which folate deficiency produced genome-wide and specific hypomethylation of the p53 tumor suppressor gene in hepatic DNA, along with increased DNA strand breaks[7,8,9] In another rat study, moderate folate deficiency produced no hypomethylation of hepatic and colonic DNA.[14] Methylation is believed to be a mechanism for regulation of gene expression, and evidence suggests that over as well as under methylation may affect tumor promotion or surveillance. For example, hypermethylation of the p16 tumor suppressor gene inactivates the expression of the related protein.[19]

Unlike findings from rat studies, DNA strand breaks were found to be decreased, not increased in the postmenopausal women with moderate folate deficiency.[10] The decrease in DNA strand breaks with folate depletion may reflect upregulation of DNA repair activity. Alternatively, reduced DNA strand breaks may reflect a cohort of surviving cells after early apoptotic elimination of DNA damaged cells. An early increase in apoptotic cell death has been previously observed in folate deficient Chinese hamster ovary cells in vitro[3] and also in liver of rats fed chronically a folate/methyl deficient diet.[20] In the latter study, increased cell apoptosis was accompanied by elevated liver poly-ADP-ribose polymerase (PARP) activity and a significant reduction in NAD concentrations.

Folate is required for the methylation of uracil to thymidine in the course of DNA synthesis and repair. The observed increase in the lymphocyte DNA dUTP/dTTP ratio and decrease in ^{3}H-deoxyuridine uptake at d 69, a point of low plasma folate (Table 1), is consistent with a block in folate-dependent de novo conversion of uracil to thymidine and misincorporation of uracil into DNA. These findings are similar those of Blount et al.[11] (discussed above) wherein elevated incorporation of uracil into DNA in folate deficient individuals was reversed by supplementation with folate. However, the dUTP/dTTP ratio in our subjects did not decrease upon folate repletion. This may be because of the much shorter and smaller folate repletion of the postmenopausal women (20 d of 286–516 μg/d) compared to the individuals in the Blount et al. study (6–8 wk of 5 mg/d).

1.3. Folate, Vitamin B12 Deficiency, and Chromosome Damage

As noted above, misincorporation of uracil into DNA due to folate deficiency is expected to stress mechanisms of DNA repair and, therefore, result in an increased level of chromosome aberrations. Study of splenectomized individuals showed that low folate status correlated with increased chromosome damage as measured by erythrocyte micronuclei.[15] Fenech et al. conducted a series of studies which showed that low folate and vitamin B12 status, and elevated plasma Hcy are associated with increased erythrocyte and lymphocyte chromosomal damage.[16,17,21] In men age 50–70 years, lymphocyte micronucleus (MN) frequency was higher in men with low blood folate and vitamin B12, and higher in men with elevated Hcy even with normal folate and vitamin B12 levels.[17] Supplementation of the diet with folic acid improved folate status but not the MN index. In a more recent study, of young adults age 18–32 years, MN frequency was related to elevated Hcy and low vitamin B12, but not to folate levels.[21] Six months dietary supplementation with folate and vitamin B12 decreased MN frequency in the half of the population with higher initial MN frequency. The MN reduction was related to increased vitamin B12 and decreased Hcy levels, but not to folate or DNA methylation status. However, none of the subjects were initially folate deficient. The two studies indicate that elevated plasma Hcy, a risk factor for vascular disease, is also associated with increased chromosome damage.[17,21]

In the folate depletion/repletion study of postmenopausal women discussed above, lymphocyte MN frequency rose and fell significantly with folate depletion and repletion, as seen in Fig. 2.[13] The strongest predictors of the MN frequency were plasma vitamin B12, folate, and baseline MN frequency. These findings are consistent with other human studies in which increased erythrocyte and lymphocyte micronuclei were associated with low plasma folate, vitamin B-12, and elevated Hcy concentrations.[15–17,21]

The recent research findings summarized above indicate that even subclinical folate and/or vitamin B12 deficiencies may result in DNA hypomethylation, uracil misincorporation, and increased chromosome damage. These effects may be involved in the reported associations of folate deficiency with cervical, bronchial, and colorectal neoplasias, and neural tube birth defects.[22–25] Recently, Fang et al.[26] reported genomic

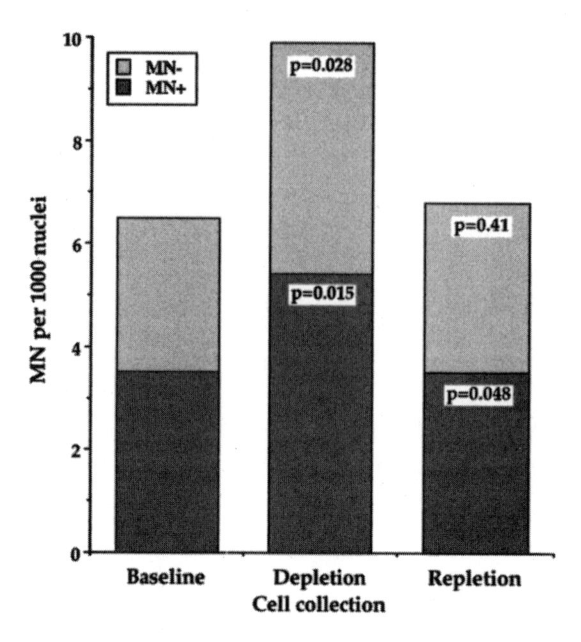

Figure 2. Kinetochore-positive (MN+) and kinetochore-negative (MN–) micronuclei in human mono- and binucleated lymphocytes during dietary folate changes in postmenopausal women. Total micronuclei were increased after folate depletion ($p = 0.014$) and decreased after repletion ($p = 0.066$) as were MN+ ($p = 0.015$ and 0.048 for depletion and repletion, respectively). MN– were elevated after depletion ($p = 0.028$), while a small decrease following repletion was not statistically significant. With permission, from Titenko-Holland et al.[13]

DNA hypomethylation of gastric cancer compared to normal tissue, and an association between hypomethylation and low serum folate concentrations in cancer patients.

2. THE ROLE OF NIACIN IN DNA REPAIR

The functional forms of niacin, NAD and NADP, are coenzyme electron acceptors or hydrogen donors for over 200 enzymes. NAD generally is involved in catabolic reactions, such as oxidation of fuel molecules, while NADP is involved in reductive biosyntheses such as fatty acids and steroids. However, recent evidence suggests a non-redox function of NAD as substrate for a class of enzymes which transer ADP-ribose units to proteins (M):

$$\text{ADPR-nicotinamide (NAD)} + \text{M} \rightarrow \text{ADPR-M} + \text{nicotinamide.}$$

These enzymes cleave the β-N-glycosylic bond of NAD to free nicotinamide and catalyze the transfer of ADP-ribose.[27] One of these enzymes, the nuclear poly-ADP-ribose polymerase (PARP) catalyzes the transfer of ADP-ribose units from NAD to an acceptor protein and also to the enzyme itself. These nuclear poly-ADP-ribose proteins seem to function in DNA replication, repair, and cell differentiation, but the molecular mechanisms are not yet known. The activity of PARP is greatly enhanced by DNA damage[28] and is strongly correlated with the life span of different species. This latter observation suggests that a higher poly-ADP-ribosylation capacity might contribute to genomic stability and thus longer life. Since the K_m of PARP is in the same range as the intracellular concentration of NAD, the niacin status may be important in the response of tissue cells to DNA damage. In niacin deficient rats, lower NAD tissue levels were correlated with an increased DNA strand break accumulation following exposure to oxidative stress.[29] Cells that are NAD deficient are more sensitive to the toxic effects of carcinogenic alkylating agents. Oxidant or DNA strand-break stressors cause rapid depletion of the cellular NAD pool in lymphocytes and murine macrophages because of excessive poly-ADP-ribose synthesis. The generation of ATP may also be impaired. There are many reports of elevated PARP activity in tumor cells accompanied by low levels of NAD. If a tissue has only a limited capacity to resynthesize NAD from the nicotinamide liberated by the polymerase, an increased PARP activity may result in tissue-specific niacin deficiencies. Decreased NAD and de novo thymidylate synthesis along with increased DNA strand breaks were found in spleen cells of rats made folate and methyl-deficient.[30] More recently, PARP activity was found to be altered and NAD concentrations significantly decreased in liver of male rats fed methyl- and folate-deficient diets with or without niacin.[31]

The decrease in lymphocyte NAD concentrations of folate deficient postmenopausal women, shown in Table 1, may be due to increased NAD utilization for DNA repair activity secondary to DNA hypomethylation and misincorporation of uracil. The decline in lymphocyte NAD occurred despite a niacin intake of 29.7 mg niacin equivalents per day, or 212% of the U.S. RDA for adult women.[18] This is the first direct evidence in humans that niacin may be an important nutrient for DNA synthesis and repair. The dietary requirement for niacin may be greater when an individual is chronically exposed to conditions that damage DNA, such as oxidant or xenobiotic stressors. Further research is needed to elucidate the mechanisms and roles of niacin in DNA processing.

3. VITAMIN C, DNA DAMAGE, AND CANCER

Two properties of ascorbic acid make it a uniquely effective physiological antioxidant: (1) it can readily donate electrons to quench a variety of reactive oxygen species, including hydroxyl, peroxyl, and superoxide radicals;[32–35] and (2) the oxidized form is easily regenerated, both enzymatically and chemically, by ubiquitous electron donors including glutathione and vitamin E.[36–38] As a major water-soluble intracellular antioxidant, vitamin C clearly has potential to provide in vivo antioxidant protection to DNA. Evidence for this is reviewed below.

3.1. Vitamin C, DNA, and Chromosome Damage

Experimental human studies which determined markers of DNA damage after controlled vitamin C intakes are summarized in Table 2. Three of the studies varied vitamin C alone while a fourth study varied vitamin C and iron. Fraga et al. reported results from a controlled vitamin C depletion/repletion study carried out on a metabolic unit.[39] When dietary ascorbic acid was reduced from baseline to 5 mg/day, the seminal plasma ascorbate decreased by half and the level of 8-oxodG, a marker of sperm DNA oxidant damage, increased by 91%, the changes reversing upon ascorbic acid repletion with 60 or 250 mg/d. The changes in ascorbic acid nutriture however did not affect lymphocyte or urine 8-oxodG, or DNA strand breaks. The authors concluded that dietary vitamin C protects human sperm from endogenous oxidative DNA damage, and that such damage could affect sperm quality and increase risk of genetic defects, particularly in smokers who have low ascorbate and increased oxidant load. Anderson et al. treated 48 healthy non-smokers with placebo, 60 or 6000 mg/d of vitamin C and found no effect on lymphocyte DNA or chromosome damage as measured by the Comet assay (with or without hydrogen peroxide challenge) and the

Table 2. Vitamin C intake and biomarkers of cellular oxidative DNA damage in humans

Reference	Subjects	Vitamin C Dose	Duration	Findings
Fraga et al. 1991[39]	8 healthy males	5 mg/d	32 days	↑ Sperm 8-oxodG (HPLC-EC)
		10 or 20 mg/d	28 days	↑ Sperm 8-oxodG (HPLC-EC)
		60 or 250 mg/d	28 days	↓ Sperm 8-oxodG (HPLC-EC) No changes in lymphocyte 8-oxodG or DNA strand breaks
Anderson et al. 1997[40]	48 nonsmokers (24 females)	60 mg/d 6000 mg/d	14 days 14 days	No change in lymphocyte DNA damage (comet assay) or chromosome breakage
Podmore et al. 1998[41]	30 subjects (16 females)	500 mg/d	6 weeks	↓ Lymphocyte 8-oxoguanine ↑ Lymphocyte 8-oxoadenine
Rehman et al. 1998[42]	10 healthy subjects	60 mg/d + 14 mg/d Fe	12 weeks	↓ Leukocyte 8-oxoguanine ↓ Leukocyte 8-oxoadenine ↑ Leukocyte 5-OH cytosine
	10 healthy subjects	260 mg/d + 14 mg/d Fe	12 weeks	↑ Leukocyte thymine glycol ↑ Total base damage at 6 wk, no change at 12 wk

chromosome aberration test without bleomycin.[40] A significant increase in bleomycin induced chromosome aberrations was found after the vitamin C supplementation. Overall, the results provided no evidence of a beneficial effect of short term vitamin C supplementation in a population of healthy individuals eating a nutritionally adequate diet.

Podmore et al. supplemented 30 healthy adults with 500 mg/d of vitamin C for six weeks and found a decrease in one oxidized DNA base from lymphocytes, 8-oxoguanine, and an increase in another, 8-oxoadenine.[41] The changes reverted back to baseline upon cessation of the supplementation. The authors concluded that the increase in 8-oxoadenine raises concern that vitamin C supplementation at 500 mg/d can result in prooxidant as well as antioxidant effects. In contrast, Rehman et al. found a decrease in both oxidized adenine and guanine bases from leukocyte DNA when healthy adults were supplemented for 12 weeks with either 60 or 260 mg/d of vitamin C along with 14 mg/d of iron.[42] However, 5OH-cytosine and thymine glycol increased due to the supplementations, again suggesting a mixed effect of vitamin C supplementation on markers of in vivo DNA damage. Because there was a transient increase in total DNA base damage after six weeks of vitamin C plus iron supplements, the authors raised concern regarding frequent use of dietary supplements containing both vitamin C and iron salts.

Results from studies which measured urinary excretion of oxidized DNA products showed no effect due to increased vitamin C intake with or without co-supplements.[39,43–45] Urinary excretion of DNA oxidant damage products represents the balance of total body DNA damage and repair and therefore is a non-specific measure for assessing changes due to micronutrient status. Other studies reported ex-vivo measurements of DNA and chromosome damage after supplementing individuals with vitamin C. Green et al. reported a decrease in lymphocyte DNA strand breaks due to vitamin C supplementation.[46] Two other studies measured chromosome breaks after treatment of lymphocytes with bleomycin, a test for genetic instability. After vitamin C supplementation of two weeks, Pohl and Reidy found decreased chromosome breaks and Anderson et al. found increased breaks.[40,47]

In summary, the results of increased vitamin C intake on biomarkers of cellular DNA damage is mixed. Some studies showed no change with increased vitamin C intake, some markers decreased, and one increased. This may be because baseline vitamin C status was already adequate in many of the healthy adult groups studied. Urinary measures of oxidized DNA products showed no change due to vitamin C intake. Two of three studies of ex-vivo DNA damage showed a benefit of vitamin C supplementation, however the relation of these results to the in vivo situation is uncertain. Therefore, results of these studies are not sufficient to conclude that supplemental vitamin C intake beyond that of a healthful diet will significantly lower DNA oxidative damage. Since cells actively concentrate vitamin C relative to the plasma, it is likely that the vitamin is more important for protecting against DNA oxidative damage in the nutritional rather than the pharmacological range of vitamin intakes.

3.2. Vitamin C and Cancer

Epidemiological studies have generally shown an inverse association between vitamin C nutriture (as assessed by dietary intake and/or plasma ascorbate levels) and some forms of cancer.[48,49] However, these types of studies cannot differentiate associative from cause and effect relationships and cannot determine the contribution of vitamin C to the reduced cancer risk among other dietary and lifestyle factors that also affect risk. Therefore, the results of observational studies provide supportive but not conclusive evidence for a protective effect of vitamin C.

Effects of vitamin C supplementation on surrogate markers of gastric and bladder cancer are summarized in Table 3. The studies show positive effects of vitamin C supplementation, i.e. decreased nitrosation and gastric mucosal DNA adduct formation, and an increase in the DNA repair enzyme, *O*-6-alkyltransferase.[50–52] Young et al. found decreased activity of beta-glucuronidase activity (linked to bladder cancer) after supplementation of healthy men with 1500 mg/d of vitamin C for one week.[53] Vitamin C treatment for one month to three years of patients with precancerous colon polyps showed either a benefit or no effect on polyp growth and cell proliferation.[54–57] While the results of the above studies of surrogate cancer markers show generally positive effects of vitamin C supplementation, the interpretation of these endpoints and relevance of the results to healthy individuals is uncertain.

The only randomized controlled intervention trial of vitamin C and cancer incidence was a cohort of the six year Linxian China micronutrient intervention trial.[58] The

Table 3. Vitamin C intake and biomarkers of gastric and bladder cancer

Reference	Subjects	Vitamin C Dose	Duration	Findings
Leaf et al. 1987[50]	7 males	2–1000 mg/d	5–12 days	↓ In-vivo nitrosation (N-nitrosoproline)
Dyke et al. 1994[51]	43 gastritis patients	1000 mg/d	28 days	↓ Gastric mucosa DNA adduct formation
Dyke et al. 1994[52]	48 gastritis patients	1000 mg/d	28 days	↑ *O*-6-alkyltransferase DNA repair enzyme
Young et al. 1990[53]	18 healthy males	1500 mg/d	1 week	↓ Urinary β-glucuronidase activity (linked to bladder cancer)

results showed no benefit to cancer incidence from a 120 mg/d vitamin C-molybdenum supplement in a Chinese population with low baseline vitamin C status and a high rate of esophageal and stomach cancer. Therefore the present limited data from randomized intervention trials are not sufficient to reach a conclusion regarding the effect of vitamin C intakes on cancer risk.

4. CONCLUSIONS

Evidence for the importance of micronutrient nutrition in DNA synthesis and repair is substantial for folate but less convincing for niacin and vitamin C. The critical roles of folate for biosynthesis of deoxynucleotides and DNA methylation reactions has been well established in both animal and human studies. Folate deficiency has been shown to result in DNA hypomethylation and strand breaks, elevated incorporation of uracil into DNA, and increased chromosome damage. These folate related disturbances in DNA and chromosome fidelity may underlie to some extent the associations of folate deficiency with precancerous epithelial lesions. Further work is needed to elucidate the possible molecular links between folate nutriture and neoplasias.

The B vitamin niacin may be an important nutrient for DNA repair because NAD appears to be required for DNA synthesis and repair as a substrate for the nuclear enzyme poly-ADP-ribose polymerase. Preliminary results suggest that niacin deficiency may predispose individuals to greater DNA damage and that stressors which stimulate DNA synthesis and repair may increase the dietary niacin requirement. Further research is needed to confirm these initial findings and hypotheses.

Vitamin C supplementation studies of oxidized DNA bases from urine or lymphocytes have shown no consistent link between vitamin C intake and measures of DNA damage. This may be because baseline vitamin C status was already adequate in many of the groups studied. Overall, the evidence does not establish that vitamin C intake beyond that of a healthful diet is important for protecting DNA against oxidative damage. More studies of the potential antioxidant protection afforded by vitamin C in the dietary rather than pharmacological intake range are needed.

While much more research is needed to elucidate the molecular roles of the above three micronutrients in DNA and cell integrity, it is clear that each nutrient's role is unique. Therefore, intake of a wide variety of micronutrients from a balanced diet promises optimal protection.

REFERENCES

1. Selhub, J. and Rosenberg, I.H., 1996, Folic acid. In: Present Knowledge in Nutrition, 7th Edition (Ziegler, E.E. and Filer Jr., L.J., eds.), pp. 206–219. International Life Sciences Institute Press, Washington, D.C.
2. James, S.J., Cross, D.R., and Miller, B.J., 1992, Alterations in nucleotide pools in rats fed diets deficient in choline, methionine, and/or folic acid. Carcinogenesis 13:2471–474.
3. James, S.J., Basnakian, A.G., and Miller, B.J., 1994, In vitro folate deficiency induces deoxynucleotide pool imbalance, apoptosis, and mutagenesis in Chinese Hamster ovary cells. Cancer Res. 54:5075–5080.
4. James, S.J., Miller, B.J., Cross, D.R., Mcgarrity, L.J., and Morris, S.M., 1993, The essentiality of folate for the maintenance of deoxynucleotide precursor pools, DNA synthesis, and cell cycle progression in PHA-stimulated lymphocytes. Environ. Health Perspectives 101:173–178.

5. James, S.J., Miller, B.J., McGarrity, L.J., and Morris, S.M., 1994, The effect of folic acid and/or methionine deficiency on deoxyribonucleotide pools and cell cycle distribution in mitogen-stimulated rat lymphocytes. Cell Proliferation 27:395–406.
6. Pogribny, I.P., Muskhelishvili, L., Miller, B.J., and James, S.J., 1997, Presence and consequence of uracil in preneoplastic DNA from folate/methyl-deficient rats. Carcinogenesis 18:2071–2076.
7. Pogribny, I.P., Basnakian, A.G., Miller, B.J., Lopatina, N.G., Poirier, L.A., and James, S.J., 1995, DNA strand breaks in genomic DNA and within the p53 gene are associated with hypomethylation in livers of folate/methyl deficient rats. Cancer Research 55:1894–1901.
8. Balaghi, M. and Wagner, C., 1993, DNA methylation in folate deficiency—use of CpG methylase. Biochem. Biophys. Res. Commun. 193:1184–1190.
9. Kim, Y.-I., Pogribny, I.P., Basnakian, A.G., Miller, J.W., Selhub, J., James, S.J., and Mason, J.B., 1997, Folate deficiency in rats induces DNA strand breaks and hypomethylation within the p53 tumor suppressor gene. Am. J. Clin. Nutr. 65:46–52.
10. Jacob, R.A., Gretz, D.M., Taylor, P.C., James, S.J., Pogribny, I.P., Miller, B.J., Henning, S.M., and Swendseid, M.E., 1998, Moderate folate depletion increases plasma homocysteine and decreases lymphocyte DNA methylation in postmenopausal women. J. Nutr. 128:1204–1212.
11. Blount, B.C., Mack, M.M., Wehr, C.M., MacGregor, J.T., Hiatt, R.A., Wang, G., Wickramasinghe, S.N., Everson, R.B., and Ames, B.N., 1997, Folate deficiency causes uracil misincorporation into human DNA and chromosome breakage: Implications for cancer and neuronal damage. Proc. Natl. Acad. Sci. USA 94:3290–3295.
12. MacGregor, J.T., Wehr, C.M., Hiatt, R.A., Peters, B., Tucker, J.D., Langlois, R.G., Jacob, R.A., Jensen, R.H., Yager, J.W., Shigenaga, M.K., Frei, B., Eynon, B.P., and Ames, B.N., 1997, Spontaneous genetic damage in man: Evaluation of interindividual variability, relationship among markers of damage, and influence of nutritional status. Mutation Research 377:125–135.
13. Titenko-Holland, N., Jacob, R.A., Shang, N., Balaraman, A., and Smith, M.T., 1998, Micronuclei in lymphocytes and exfoliated buccal cells of postmenopausal women with dietary changes in folate. Mutation Res. 417:101–114.
14. Kim, Y.-I. and Christman, J.K., 1995, Moderate folate deficiency does not cause global hypomethylation of hepatic and colonic DNA or c-myc-specific hypomethylation of colonic DNA in rats. Am. J. Clin. Nutr. 61:1083–1090.
15. Everson, R.B., Wehr, C.M., Erexson, G.L., and MacGregor, J.T., 1988, Association of marginal folate depletion with increased human chromosomal damage in vivo: demonstration by analysis of micronucleated erythrocytes. J. Natl. Cancer Inst. 80:525–529.
16. Fenech, M.F. and Rinaldi, J.R., 1994, The relationship between micronuclei in human lymphocytes and plasma level of vitamin C, vitamin E, vitamin B-12, and folic acid. Carcinogenesis 15:1405–1411.
17. Fenech, M.F., Dreosti, I.E., and Rinaldi, J.R., 1997, Folate, vitamin B12, homocysteine status and chromosome damage rate in lymphocytes of older men. Carcinogenesis 18:1329–1336.
18. Food and Nutrition Board, Institute of Medicine. 1998, Folate. In: Dietary Reference Intakes for Thiamin, Riboflavin, Niacin, Vitamin B6, Folate, Vitamin B12, Pantothenic Acid, Biotin, and Choline. National Academy Press, Washington, D.C.
19. Merlo, A., Herman, J.G., Mao, L., Lee, D.J., Gabrielson, E., Burger, P.C., Baylin, S.B., and Sidransky, D., 1995, 5′ CpG island methylation is associated with transcriptional silencing of the tumour suppresor p16/CDKN2/MTS1 in human cancers. Nature Medicine 1:686–692.
20. James, S.J., Miller, B.J., Basnakian, A.G., Pogribny, I.P., Pogribna, M., and Muskhelishvili, L., 1997, Apoptosis and proliferation under conditions of deoxynucleotide pool imbalance in liver of folate/methyl deficient rats. Carcinogenesis 18:287–293
21. Fenech, M., Aitken, C., and Rinaldi, J., 1998, Folate, vitamin B12, homocysteine status and DNA damage in young Australian adults. Carcinogenesis 19:1163–1171.
22. Butterworth, Jr., C.E., Hatch, K.D., Macaluso, M., Cole, P., Sauberlich, H.E., Soong, S.-J., Borst, M., and Baker, V.V., 1992, Folate deficiency and cervical dysplasia. J. Am. Med. Assoc. 267:528–533.
23. Glynn, S.A. and Albanes, D., 1994, Folate and cancer: a review of the literature. Nutr. Cancer 22:101–119.
24. Mason, J.B., 1994, Folate and colonic carcinogenesis: searching for a mechanistic understanding. J. Nutr. Biochem. 5:170–175.
25. Lucock, M.D., Wild, J., Schorah, C.J., Levene, M.I., and Hartley, R., 1994, The methylfolate axis in neural tube defects: in vitro characterization and clinical investigation. Biochem. Med. Met. Biol. 52:101–114.
26. Fang, J.Y., Xiao, S.D., Zhu, S.S., Yuan, J.M., Qiu, D.K., and Jiang, S.J., 1997, Relationship of plasma folic acid and status of DNA methylation in human gastric cancer. J Gastroenterol. 32:171–175.

27. Lautier, D., Lagueux, J., Thibodeau, L., •• Menard, and Poirier, G.G., 1993, Molecular and biochemical features of poly(ADP-ribose) metabolism. Mol. Cell. Biochem. 122:171–193.
28. Stierum, R.H., Vanherwijnen, M.H.M., Hageman G.J., and Kleinjans, J.C.S., 1994, Increased poly(ADP-ribose) polymerase activity during repair of (+/–)-anti-benzo[a]pyrene diolepoxide-induced DNA damage in human peripheral blood lymphocytes in vitro. Carcinogenesis 15:745–751.
29. Zhang, J.Z., Henning, S.M., and Swendseid, M.E., 1993, Poly(ADP-ribose) polymerase activity and DNA strand breaks are affected in tissues of niacin-deficient rats. J. Nutr. 123:1349–1355.
30. James, S.J. and Yin, L., 1989, Diet-induced DNA damage and altered nucleotide metabolism in lymphocytes from methyl-donor-deficient rats. Carcinogenesis 10:1209–1214.
31. Henning, S.M., Swendseid, M.E., and Coulson, W.F., 1997, Male rats fed methyl- and folate-deficient diets with or without niacin develop hepatic carcinomas associated with decreased tissue NAD concentrations and altered poly(ADP-ribose) polymerase activity. J. Nutr. 127:30–36.
32. Buettner, G.R., 1993, The pecking order of free radicals and antioxidants: lipid peroxidation, alpha-tocopherol, and ascorbate. Archives Biochem. Biophys. 300:535–543.
33. Bendich, A., Machlin, L.J., Scandurra, O., Burton, G.W., and Wayner, D.M., 1986, The antioxidant role of vitamin C, Adv. Free Rad. Biol. Med. 2:419–444.
34. Sies, H., Stahl, W., and Sundquist, A.R., 1992, Antioxidant functions of vitamins. Annals NY Acad Sci 669:7–20.
35. Frei, B., England, L., and Ames, B.N., 1989, Ascorbate is an outstanding antioxidant in human blood plasma, Proc. Natl. Acad. Sci. USA 86:6377–6381.
36. Winkler, B.S., 1992, Unequivocal evidence in support of the nonenzymatic redox coupling between glutathione/glutathione disulfide and ascorbic acid/dehydroascorbic acid, Biochim. Biophys. Acta 1117:287–290.
37. Martensson, J., Han, J., Griffith, O.W., and Meister, A., 1993, Glutathione ester delays the onset of scurvy in ascorbate-deficient guinea pigs. Proc. Natl. Acad. Sci. USA 90:317–321.
38. Jacob, R.A., 1995, The integrated antioxidant system. Nutr. Research 15:755–766.
39. Fraga, C.G., Motchnik, P.A., Shigenaga, M.K., Helbock, H.J., Jacob, R.A., and Ames, B.N., 1991, Ascorbic acid protects against endogenous oxidative DNA damage in human sperm. Proc. Natl. Acad. Sci. USA 88:11003–11006.
40. Anderson, D., Phillips, B.J., Yu, T., Edwards, A.J., Ayesh, R., and Butterworth, K.R., 1997, The effects of vitamin C supplementation on biomarkers of oxygen radical generated damage in human volunteers with low or high cholesterol levels. Environ. Mol. Mutagen 30:161–174.
41. Podmore, I.D., Griffiths, H.R., Herbert, K.E., Mistry, N., Mistry, P., and Lunec, J., 1998, Vitamin C exhibits pro-oxidant properties. Nature 392:559.
42. Rehman, A., Collis, C.S., Yang, M., Kelly, M., Diplock, A.T., Halliwell, B., and Rice-Evans, C., 1998, The effects of iron and vitamin C co-supplementation on oxidative damage to DNA in healthy volunteers. Biochem. Biophys. Res. Commun. 246:293–298.
43. Witt, E.H., Reznick, Z., Viguie, C.A., Starke-Reed, P., and Packer, L., 1992, Exercise, oxidative damage and effects of antioxidant manipulation. Am. Inst. Nutr. 122:766–773.
44. Loft, S., Vistisen, K., Ewertz, M., Tjonneland, A., Overvad, K., and Poulsen, H.E., 1992, Oxidative DNA damage estimated by 8-hydroxydeoxyguanosine excretion in humans: influence of smoking, gender and body mass index. Carcinogenesis 13:2241–2247.
45. Prieme, H., Loft, S., Nyyssonen, K., Salonen, J.T., and Poulsen, H.E., 1997, No effect of supplementation with vitamin E, ascorbic acid, or coenzyme Q10 on oxidative DNA damage estimated by 8-oxo-7,8-dihydro-2′-deoxyguanosine excretion in smokers. Am. J. Clin. Nutr. 65:503–507.
46. Green, M.H.L., Lowe, J.E., Waugh, A.P.W., Aldridge, K.E., Cole, J., and Arlett, C.F., 1994, Effect of diet and vitamin C on DNA strand breakage in freshly-isolated human white blood cells. Mutation Res. 316:91–102.
47. Pohl, H. and Reidy, J.A., 1989, Vitamin C intake influences the bleomycin-induced chromosome damage assay: implications for detection of cancer susceptibility and chromosome breakage syndromes, Mutation Res. 224:247–252.
48. Block G., 1991, Vitamin C and cancer prevention: the epidemiologic evidence. Am. J. Clin. Nutr. 53:270S-282S.
49. Fontham, E.T.H., 1994. Vitamin C, vitamin C-rich foods and cancer: Epidemiologic studies. In: Frei B., ed. Natural antioxidants in health and disease. San Diego: Academic Press 157–197.
50. Leaf, C.D., Vecchio, A.J., Roe, D.A., and Hotchkiss, J.H., 1987, Influence of ascorbic acid dose on N-nitrosoproline formation in humans. Carcinogenesis 8:791–795.
51. Dyke, G.W., Craven, J.L., Hall, R., and Garner, R.C., 1994, Effect of vitamin C supplementation on gastric mucosal DNA damage. Carcinogenesis 15:291–295.

52. Dyke, G.W., Craven, J.L., Hall, R., and Garner, R.C., 1994, Effect of vitamin C upon gastric mucosal *O*-6-alkyltransferase activity and on gastric vitamin C levels. Cancer Letters 86:159–165.
53. Young, J.C., Kenyon, E.M., and Calabrese, E.J., 1990, Inhibition of beta-glucuronidase in human urine by ascorbic acid. Human & Experimental Toxicology 9:165–170.
54. Busey, H.J.R., DeCosse, J.J., Deschiner, E.E., Eyers, A.A., Lesser, M.L., Morson, B.C., Ritchie, S.M., Thomson, J.P.S., Wadsworth, J., 1982, A randomized trial of ascorbic acid in polyposis coli. Cancer (Philadelphia) 50:1434–1439.
55. McKeown-Eyssen, G., Holloway, C., Jazmaji, V., Bright-See, E., Dion, P., and Bruce, W.R., 1988, A randomized trial of vitamins C and E in the prevention of recurrence of colorectal polyps. Cancer Res. 48:4701–4705.
56. Cahill, R.J., Osullivan, K.R., Mathias, P.M., Beattie, S., Hamilton, H., and Omorain, C., 1993, Effects of vitamin antioxidant supplementation on cell kinetics of patients with adenomatous polyps. Clin. Med. 34:963–967.
57. Hofstad, B., Almendingen, K., Vatn, M., Andersen, S.N., Owen, R.W., Larsen, S., and Osnes, M., 1998, Growth and recurrence of colorectal polyps: A double-blind 3-year intervention with calcium and antioxidants. Digestion 59:148–156.
58. Blot, W.J., Li, J.Y., Taylor, P.R., Guo, W., Dawsey, S., Wang, G.Q., Yang, C.S., Zheng, S.F., Gail, M., Li, G.Y., Yu, Y., Liu, B.Q., Tangrea, J., Sun, Y.H., Liu, F., Fraumeni Jr., J.F., Zhang, Y.H., and Li, B., 1993, Nutrition intervention trials in Linxian, China: supplementation with specific vitamin/mineral combinations, cancer incidence, and disease-specific mortality in the general population. J. Nat. Cancer Inst. 85:1483–1492.

11

BIOLOGICAL EFFECTS OF HYDROXYTYROSOL, A POLYPHENOL FROM OLIVE OIL ENDOWED WITH ANTIOXIDANT ACTIVITY

Caterina Manna, Fulvio Della Ragione, Valeria Cucciolla, Adriana Borriello, Stefania D'Angelo, Patrizia Galletti, and Vincenzo Zappia

Institute of Biochemistry of Macromolecules
Medical School, Second University of Naples
Via Costantinopoli, 16, 80138 Naples
Italy

A number of epidemiological studies indicate that dietary factors may influence the development of some types of cancer and degenerative pathologies, including cardiovascular diseases and cataract. In this respect, it is well documented that daily consumption of fruits and vegetables is associated with a lowered risk of these diseases.[1] Polyphenols are bioactive substances that are widely distributed in the vegetable kingdom[2,3] and therefore are present in high concentrations in typical components of the Mediterranean diet, such as fruit, vegetables, red wine, and olive oil. The aim of this article is to overview the most recent data on the nutritional value of the phenolic fraction of virgin olive oil in the ongoing studies on its beneficial effects on human health.

1. OLIVE OIL COMPOSITION

Like all vegetable oils, olive oil is composed of two major fractions,[4] the saponifiable one, made of triglycerides, accounting for 98–99% of the total, and the non-glyceride fraction, containing several liposoluble molecules listed in Table 1. As far as the saponifiable fraction is concerned, olive oil has a unique fatty acid composition, with the monounsaturated oleic acid (18:1 ω9) ranging from 55% to 83%, while the polyunsaturated linoleic acid (18:2 ω6) represents 3–12% of the total.

Although present in low percentage, the non-glyceride fraction plays a major role in the nutritional and organoleptic properties of virgin olive oil. Among these minor

Advances in Nutrition and Cancer 2, edited by Zappia *et al.*
Kluwer Academic / Plenum Publishers, New York, 1999.

Table 1. Nonglyceride components of olive oil

Hydrocarbons
Nonglyceride esters
Tocopherols
Alkanols
Anthocyanins
Terpenic acids
Sterols
Phenolic compounds
• Simple phenolic fraction:
hydroxytyrosol
tyrosol
caffeic acid
p-coumaric acid
sinapic acid
vanillic acid
protocatechuic acid
gentisic acid
p-hydroxybenzoic acid
syringic acid
• Hydrolyzable phenolic fraction:
oleuropein aglycone
ligstroside aglycone

components, polyphenols are particularly relevant from a sensory point of view, being responsible for the bitter and pungent aroma.[4] Their concentration differs greatly among olive oils, depending on various conditions, such as the cultivar, the soil composition, the climate, the degree of ripeness of the fruit, and the oil extraction procedures.[5] In this respect, virgin olive oils can be classified on the basis of their polyphenol content,[6,7] ranging from 100 mg up to 1000 mg/Kg. Table 1 reports the phenolic compounds identified and characterized in virgin olive oil, including (3,4-dihydroxyphenyl)ethanol (hydroxytyrosol; DPE). This hydrosoluble and liposoluble molecule, recalling the structure of cathecol, is present either as simple phenol or esterified with elenolic acid to form oleuropein aglycone[8] (Fig. 1). The term oleuropein, referring to the complex phenol typical of olives, derives from *Olea europaea* which is the taxonomic name of the most common olive species cultivated in the Mediterranean countries.

DPE is an efficient scavenger of peroxyl radicals,[9,10] which prevents the autooxidation of polyunsaturated fatty acids[11,12] and therefore contributes to the determination of the shelf-life of the oil. The free radical scavenging activity of DPE is higher than synthetic and natural antioxidants, as reported in Table 2.

2. OLIVE OIL AND HUMAN HEALTH

Olive oil represents the main lipidic source of the Mediterranean diet[13] (29% of total energy intake), a dietary habit which has been associated with a low incidence of several pathologies, including coronary heart diseases (CHD).

The advantages of olive oil intake with respect to CHD have been confirmed in a large number of studies, starting with the original observation by Ancel Keys,[14] who defined the Mediterranean Diet as a way "To eat well and stay well". Since then, a large

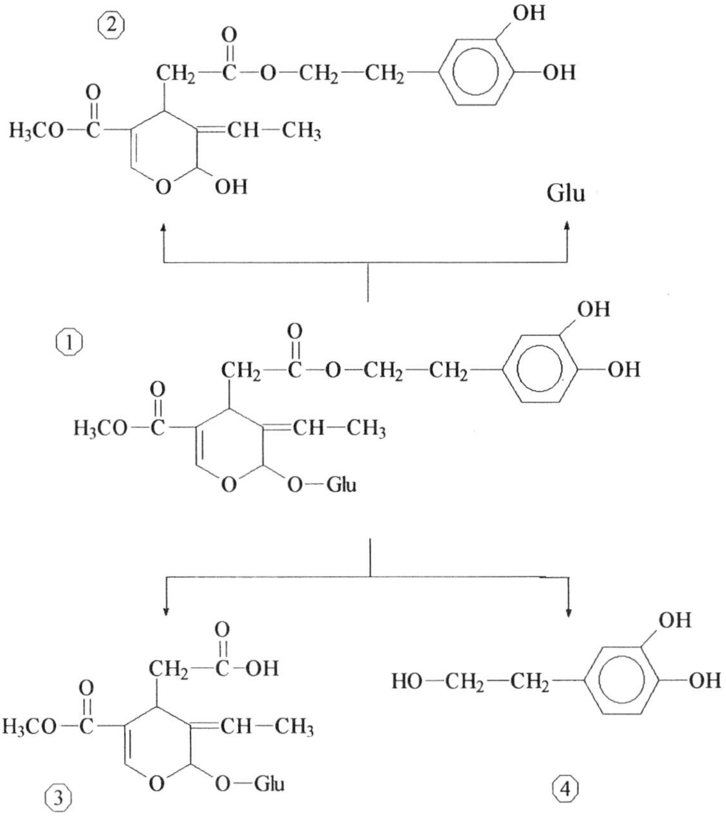

Figure 1. Chemical structure of oleuropein and its derivatives. 1 = oleuropein; 2 = oleuropein aglycone; 3 = elenolic acid; 4 = hydroxytyrosol.

body of evidence has been accumulated showing the beneficial health effects of dietary monounsaturated fatty acids (MSFA), compared to both saturated (SFA) and polyunsaturated fatty acids (PUFA). Indeed, the analysis of the serum lipidic profile of patients fed diets enriched in MSFA shows a lower level of total cholesterol, compared to that observed for people fed SFA-rich diets, without lowering HDL-cholesterol, as observed for PUFA.[15,16] Furthermore, according to the "oxidation hypothesis" of

Table 2. Free radical scavenging activity of DPE compared to other synthetic and natural antioxidants

Compound	EC_{50}
Vitamin C	1.31×10^{-5}
Vitamin E	5.04×10^{-6}
BHT	1.05×10^{-4}
DPE	2.60×10^{-7}
Oleuropein	3.63×10^{-5}

Free radical scavenging activity was measured by the 2,2-diphenyl-1-picrylhydrazyl (DPPH) quenching test. The test is based on the decrease of DPPH absorbance at 517 nm upon reduction of this stable radical by the antioxidant compounds.
BHT = butylated hydroxytoluene.
Reproduced with permission from ref. 10.

atherosclerosis,[17] which implies that oxidized LDL are more atherogenic than native forms, diets enriched in olive oil are believed to slow down the progression of the atheromatous plaque also because oleic acid-rich LDL has proven to be highly resistant to oxidative modifications.[18] Finally, more recently, several investigators have stressed that olive oil's beneficial effects should also be linked to the whole composition of olive oil, which contains several antioxidants, inhibitors of platelet aggregation, and anti-thrombotic molecules.

Moreover, numerous epidemiological studies (both case-control and cohort investigations) have shown that the Mediterranean diet is associated with a statistically significant reduction of breast and colon cancer risk.[19–22] There is clear evidence that olive oil consumption, together with vegetable and fruit intake, contributes to the reduction of malignant transformation rate.[19–22] Even in this case, the protective effect of olive oil can be attributed both to its specific fatty acid composition and to its high content of vitamin and non-vitamin antioxidants.

3. ROS AND HUMAN DISEASES: ANTIOXIDANT PROPERTIES OF OLIVE OIL POLYPHENOLS

The hypothesis that the minor components of olive oil greatly contribute to its benefits on human health, prompted several research groups to investigate the biological effects of olive oil phenolic compounds, including DPE, in different *in vitro* experimental models, mainly focusing on their antioxidant properties. It is well known that reactive oxygen species (ROS) induce a number of molecular alterations on cellular components, leading to changes in cell morphology and viability.[23–24] ROS are generated during both normal and xenobiotic metabolism;[25–26] therefore, to counteract these potentially injurious oxidizing agents, cells are naturally provided with an extensive array of protective enzymatic and non-enzymatic antioxidants. However, this multifunctional protective system is not 100% efficient; the balance between antioxidants and prooxidants in the human body, indeed, is always slightly tipped in favor of the latter. Moreover, when ROS are overproduced or when the endogenous defenses are impaired, the situation referred to as oxidative stress will result. In scheme I are reported the mechanisms proposed as responsible for the ROS-mediated tissue damage.[27] It should be underlined, in this respect, that even though a causative and direct role of ROS has been demonstrated in only a few human diseases, such as retinopathy in premature babies and reperfusion injury, in the last few years extensive data has demonstrated the major role of oxidative damage in a large number of human degenerative pathologies, listed in Table 3, including cancer.

The correlation between oxidative stress and cancer has been envisioned in the past primarily as a consequence of DNA damage.[23,26] Recent data[28–31] reporting that ROS themselves can act as pivotal molecules in signal transduction and in modulating gene expression, more closely correlate oxidants, antioxidants, and cancer initiation, promotion, and progression. As a matter of fact, the activated oncogenic form of *ras* generates high levels of superoxide, which mediates the mitogenic signal; moreover, the ability of ROS to regulate the transcription of specific genes has been frequently reported and seems mostly related to the activation or inhibition of transcription factors, including p53. Therefore, it is conceivable that the complex endogenous defense system, including natural scavengers present in diet, may exert an important chemoprevention role.

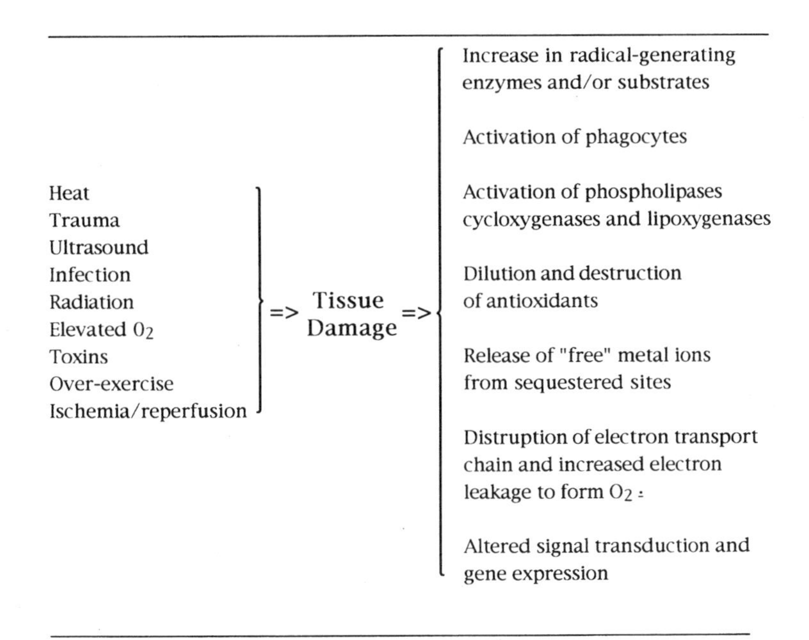

Scheme I. Mechanisms of oxidative tissue damage. Reproduced with permission from ref. 27.

The *in vivo* antioxidant activity of the olive oil phenolic fraction was first proposed in the early 90s by Scaccini and coworkers,[32] from data obtained on animal models. Lipoproteins purified from rats fed with olive oil showed, in fact, a lower content of lipid peroxidation products compared to controls receiving a synthetic diet based on the same fatty acid composition and vitamin E. More recently, similar data on the *in vitro* peroxidation of LDL purified from New Zealand white rabbits were reported.[33]

The first direct evidence of antioxidant activity of the olive oil phenolic fraction on biological models comes from the studies of C. Galli's group[34–36] in Milan (see also Table 5). These authors tested different olive oil-derived phenols for their effect on lipoprotein peroxidation and reported a protective effect played by DPE, either free

Table 3. ROS-mediated human diseases

Malnutrition
Intoxication
carbon tetrachloride
paraquat
cigarette smoking
Excess radiation exposure
Cataract and eye injury
Reperfusion injury
Arthritis and rheumatic disorders
Ulcerative colitis
Atherosclerosis
Cancer
Neurological disorders

or esterified in oleuropein derivatives, at a concentration as little as 10 μM. LDL were oxidized *in vitro*, either by chemical or enzymatic treatment, and several oxidative markers were evaluated, including vitamin E content, formation of thiobarbituric acid-reacting substances (TBARS), lipid peroxides and conjugated diene, level of PUFA, and protein modifications.

More recent data from our laboratory, discussed in the next two paragraphs, clearly demonstrate that DPE at μmolar concentrations counteracts ROS-induced cytotoxicity in human cellular systems, such as an intestinal cell line[37] (Caco-2 cells) and erythrocytes.[38]

3.1. DPE Protective Effect against ROS-Induced Cytotoxicity in Caco-2 Cells

Caco-2 cells, a cell line originally derived from a human colon carcinoma, when grown in culture on polycarbonate filters, undergo enterocytic differentiation, and form a polarized monolayer closely resembling, both morphologically and functionally, the human small intestine epithelium.[39] The cells were grown in a Transwell cell culture chamber on a collagen-treated polycarbonate membrane, separating two different compartments, the basal and the luminal one, where the apical part of the polarized epithelium is oriented.

The oxidative stress was induced in differentiated Caco-2 cells by adding, in the apical compartment of the Transwell, increasing amounts of H_2O_2 or the enzyme xanthine oxidase (XO) and its substrate xanthine. This enzymatic system was chosen because the production of ROS from xanthine oxidase is considered a primary mechanism of cellular damage in several pathologies of the gastrointestinal tract.[40] The cellular injury was checked by means of: i) the neutral red uptake assay, one of the most reliable methods of assessing cell viability; ii) the evaluation of intracellular malondialdehyde (MDA) concentration, that directly results from membrane lipid peroxidation; iii) the paracellular transport of inulin, a fructo-polysaccharide that physiologically does not cross the epithelium, as a marker of the monolayer integrity.

The results indicate that Caco-2 cells are sensitive to oxidative stress, either when induced by millimolar concentrations of H_20_2 or when enzymatically generated by the activity of XO. As a matter of fact, both treatments result in a significant decrease in cell viability.[37] This marker was then utilized to verify if DPE could prevent ROS-induced oxidative damages on Caco-2 cells. The possible antioxidant activity of the analogue 2-(4-hydroxyphenyl)ethanol (tyrosol; PE), which lacks the o-diphenolic structure, was also investigated. When cells are pretreated with DPE before being challenged with peroxide, no decrease in cell viability is observable, indicating that the phenol suppresses the ROS-induced cytotoxicity (Fig. 2). The molecule is active in a micromolar concentration range, providing complete protection against oxidative stress at about 250 μM. In contrast, no protection is exerted by PE up to a concentration of 500 μM. Similar results were obtained when oxidative stress was induced by the xantine/XO enzymatic system. When the oxidative stress is induced by the XO activity, a complete protection from the oxidative stress is observable at a concentration as low as 100 μM DPE.[37]

Among the cytotoxic effects of ROS on mammalian cells, particularly severe is the damage induced to membrane phospholipids. It is known that PUFA are susceptible to free radical attack, which starts the chain of lipid peroxidation, leading to the

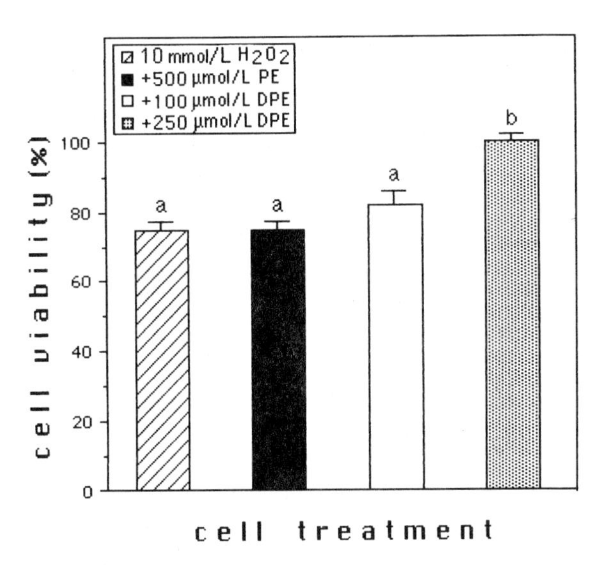

Figure 2. Effect of PE and DPE on H_2O_2-induced cytotoxicity of Caco-2 cells. The cells were preincubated for 4h at 37°C with increasing amounts of selected phenols and then treated for 20h with 10mM H_2O_2. At the end of incubation, cell viability was measured by the neutral red uptake assay as described in ref. 37. Values are means ± SD. The data were analyzed by Newman-Keuls test; values that do not share a letter are significantly different ($P < 0.01$). Reproduced with permission from ref. 37.

formation of hydroperoxides, which in turn degrade to MDA. As shown in Table 4, H_2O_2-treated Caco-2 cells show a significant increase in MDA level, indicating a severe peroxidative degradation of the enterocytic cellular membranes. Preincubation of cells with 250µM DPE, however, completely prevents MDA formation, while 500µM PE is inactive.[37]

Finally, the effect of the oxidative stress, as well as the protective role of DPE, on the integrity of the intestinal monolayer were evaluated by measuring the paracellular transport of labeled inulin.[37] In physiological conditions, indeed, the epithelium is impermeable to this fructo-polysaccharide, and only trace amounts of it cross the membrane via an extracellular route. As shown in Table 4, the oxidative stress induces a six-fold increase in transepithelial inulin flux, indicating that the oxidative treatment severely affects the barrier function of the intestinal cells. This effect is prevented by 250µM DPE; also in this case, PE did not show any significant effect, up to a concentration of 500µM. The lack of protection against the oxidative stress by PE confirms literature data on the influence of chemical structure on the biological antioxidant activity of polyphenols.[41]

As far as the molecular mechanisms of DPE antioxidant effect are concerned, this phenol, endowed with peroxyl radical scavenging activity, likely prevents ROS-induced cytotoxicity by acting as a chain-breaking inhibitor of lipid peroxidation.

Table 4. Effect of PE and DPE on MDA levels and ^{14}C-inulin paracellular transport in H_2O_2-treated Caco-2 cells

Sample and treatment	MDA (nmol/10^6 cells)	^{14}C-inulin paracellular transport (%)
control	0.99 ± 0.13^a	0.6 ± 0.05^a
+10mmol/L H_2O_2	2.10 ± 0.42^b	3.5 ± 0.12^b
+250µmol/L DPE + 10mmol/L H_2O_2	0.87 ± 0.40^a	0.6 ± 0.04^a
+500µmol/L PE + 10mmol/L H_2O_2	2.28 ± 0.27^b	3.4 ± 0.15^b

Modified from ref. 37. Data were analyzed by Newman-Keuls test; in each line, values that do not share a superscript letter are significantly different ($P < 0.01$).

DPE might also contribute to the regeneration of vitamin E by reacting with its α-tocopheroxyl radical.[35] In addition, the ability of DPE to prevent oxidative damage could also be related to the ability of the dihydroxyphenols to chelate iron ions. This metal, indeed, is known to initiate and propagate lipid peroxidation via a series of reactions involving ROS formation. Finally, the possibility that the antioxidant activity of DPE might be mediated by the induction of newly synthesized antioxidant proteins should also be taken into consideration. In fact, Khan et al.[42] reported that oral feeding of the polyphenolic fraction of green tea to female SKH-1 hairless mice induces an increase of antioxidant enzymes, including glutathione peroxidase and catalase.

3.2. DPE Protective Effect against ROS-Induced Oxidative Damages in Human Erythrocytes

To further elucidate the antioxidant properties of DPE in human cells and to identify the molecular mechanisms responsible for its cytoprotective effect, human erythrocytes (RBC) were selected. RBC are particularly exposed to oxidative hazard, because of their specific role as oxygen carriers. In fact, the spontaneous autoxidation of hemoglobin continuously produces anion superoxide which dismutates to H_2O_2.[43,44] In the presence of reduced metal ions, particularly iron, both compounds in turn promote the formation of the highly reactive hydroxyl radical. Under physiological conditions, ROS are rapidly removed by the endogenous defense system, in RBC as also occurs in other mammalian cells. However, if ROS are overproduced or if the endogenous defenses are impaired, severe oxidative damage to both plasma membrane and cytosolic components can occur, eventually leading to hemolysis.[45] In this respect, considerable evidence supports the hypothesis that chronic oxidative stress is an important factor in the etiology of some hereditary anemias, characterized either by an impaired antioxidant defense system (glucose-6-P dehydrogenase deficiency)[46] or by the increased production of ROS which overwhelms the endogenous defenses (β-thalassemia[47] and sickle cell anemia[48]). Based on these findings, the use of antioxidants has been proposed as therapeutic agents to counteract oxidative alterations in such pathologies.

In our experimental system, intact human RBC were pretreated with micromolar amounts of DPE and then exposed to H_2O_2 over different time intervals to test the ability of the phenolic compound to prevent hemolysis and membrane lipid peroxidation.[38] The capability of DPE to protect RBC from ROS-induced cytotoxicity was tested in conditions of moderate and extensive oxidative lysis, induced by 200 μM and 300 μM H_2O_2, respectively. The incubation of RBC over different time intervals in the presence of 300 μM H_2O_2 results in extensive hemolysis, in a time-dependent manner, reaching up to 40% after 2 h treatment. Moreover, a reduced lysis is observable in cells pretreated with μmolar concentrations of DPE, before they are challenged with H_2O_2, indicating that this phenol effectively protects RBC against ROS-induced cytotoxicity[38] (Fig. 3). In the experimental conditions (200 μM H_2O_2) inducing a moderate hemolysis (about 20%), a DPE concentration as little as 50 μM shows a significant effect, while 100 μM completely prevents the oxidative damage.[38]

In order to verify the effect of DPE on the specific molecular alterations which ultimately result in oxidative hemolysis, lipid peroxidation product MDA was measured. When RBC are treated with 200 μM H_2O_2, a dramatic increase in MDA

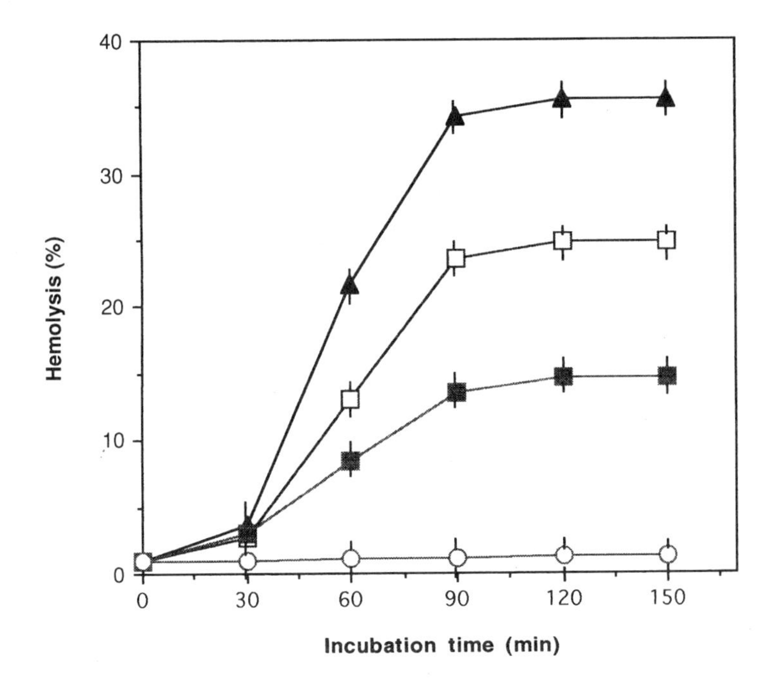

Figure 3. Time course of the effect of DPE on H_2O_2-induced hemolysis of RBC. The cells were pretreated for 15 min at 37°C with 50 μM (□) and 100 μM (■) DPE and then incubated over different time intervals in the presence of 300 μM H_2O_2; parallel sets of samples received only H_2O_2 treatment (▲) or no treatment at all (○). At the end of incubation, hemolysis was measured as described in ref. 38. Values are means ± SD. The data were analyzed by Duncan's test. Reproduced with permission from ref. 38.

concentration is observable, indicating a severe peroxidative damage of the RBC membrane. Pretreatment of cells with increasing μmolar DPE concentrations, however, prevents the phospholipid oxidative alteration in a concentration-dependent manner.[38]

The effect of oxidative stress on RBC membrane transport systems, as well as the protective role of DPE were also investigated in conditions of non-hemolytic mild H_2O_2 treatment. Under these experimental conditions, a marked decrease in the energy-dependent methionine and leucine transport is observable[38] (Fig. 4). This finding is consistent with previous observations by Rohn et al.[49] on functional alterations of RBC isolated membrane, following incubation *in vitro* in the presence of ferrous sulfate and EDTA. The authors report that ion transport ATPases are a major target for oxidative injury, suggesting that *in vivo* ion gradients are disrupted by conditions that promote ROS formation. The ROS-induced decrease in amino acid transport can be prevented by DPE pretreatment.

Our data demonstrate that DPE prevents membrane oxidative alterations, even in a metabolically simplified model system, such as RBC. In fact, in these cells, which lack transcriptional and translational machinery, the defense mechanisms mediated by the induction of protein synthesis are not operative. Therefore, in this experimental system, a direct protective effect of DPE against peroxide-induced cytotoxicity can be proposed.

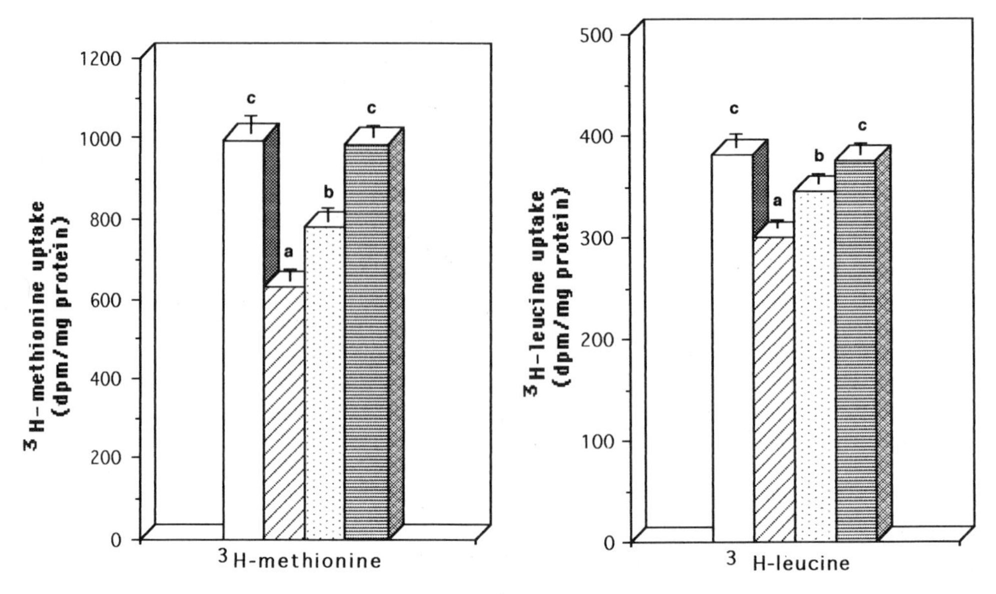

Figure 4. Effect of DPE on H_2O_2-induced decrease in ^{3}H-methionine and ^{3}H-leucine transport in RBC. The cells were pretreated for 15 min at 37°C with either 50 μM (▨) or 100 μM (▤) DPE and then incubated for 2 h with 150 μM H_2O_2; parallel sets of samples received only H_2O_2 treatment (▨) or no treatment at all (□). At the end of incubation, amino acid transport was measured as described in ref. 38. Values are means ± SD of five experiments. The data were analyzed by Newman-Keuls test. Values that do not share a letter are significantly different ($P < 0.05$). Reproduced with permission from ref. 38.

4. ROLE OF DPE ON THE PROLIFERATION AND DIFFERENTIATION OF HUMAN MALIGNANT CELLS

Several lines of evidence suggest that oxidative stress is involved in the cellular events associated with tumor initiation, promotion, and progression. Indeed, as discussed previously, it has been postulated that a major step in malignant transformation is the induction of ROS-dependent DNA alteration, which is followed by an increased mutation rate and the acquisition of a cell-transformed phenotype.[26] Moreover, it has been found that cells with an activated oncogenic form of *ras* generate high levels of superoxide, and that the mitogenic signal generated by mutated *ras* depends on its superoxide production.[30] Thus antioxidants could prove to be important in reducing the risk of developing a cancer and in down-regulating oncogene-dependent growth. In addition, the cellular redox state appears to modulate, by means of different ROS, the major processes required for cell proliferation, including signal transduction pathways and transcription factor activities.[28–29]

Since virgin olive oil is rich with antioxidant molecules, it is highly probable that these components might be responsible, at least in part, for its chemopreventive activity. As amply discussed above, there is clear epidemiological evidence that olive oil intake has been associated with a diminished risk of cancer at various anatomical sites.[19–22,50]

Although the olive oil antioxidant DPE could have a pivotal role, so far no investigations have been carried out to highlight the effects of this phenol on the proliferation and differentiation of human cells. These studies might be extremely important in the light of a newer dimension in the management of neoplasia, which refers to the dietary intake of nutrients able to prevent the initiational and promotional events associated with carcinogenesis, thus reducing mortality and morbidity.[51]

On the basis of these findings, a third line of investigation carried out by our research group deals with the effects of DPE and PE on the proliferation rate and differentiation of HL-60, a cell line established from promyelocitic acute leukemia.

Figure 5A reports the effect of increasing amounts of DPE on the proliferation rate of HL-60. The molecule exerts a significant inhibitory effect, with more than 80% inhibition at 50 μM concentration after a four day period. As shown in Fig. 5B, the incubation of cells with DPE did not cause a definite commitment towards a non-proliferating state. In fact a normal proliferation rate was recovered when DPE was washed out from HL-60 cells preincubated for 48 hours with 50 μM phenol. PE, failed to exert any inhibitory effect, indicating that the two ortho-hydroxyl groups are critical for both antioxidant and antiproliferative effects.

Subsequently, we investigated the effect of the two phenols on cell differentiation, employing vitamin D_3 as a positive control. DPE caused a significant and

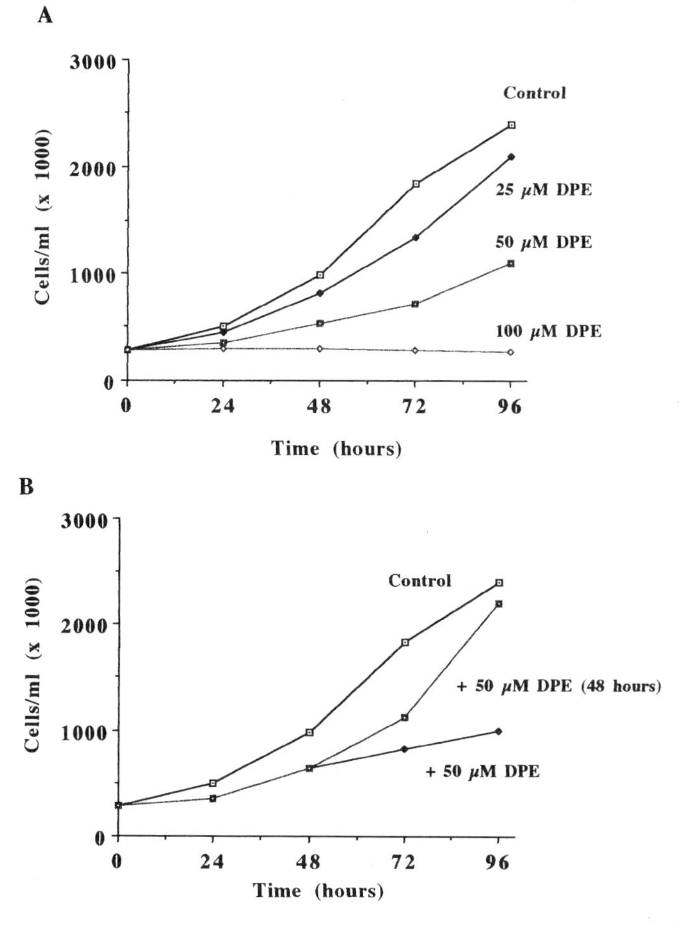

Figure 5. Effect of DPE on HL-60 cell proliferation. A) HL-60 cells were incubated with or without increasing DPE amounts. Cells were counted daily. B) HL-60 were grown in the presence of 50 μM DPE. After 48 hrs, DPE was removed and the cells resuspended in fresh medium. Cells were counted daily.

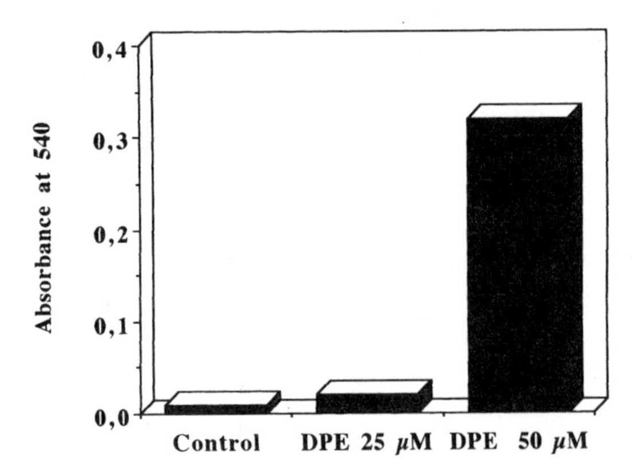

Figure 6. Effect of DPE on HL-60 differentiation. HL-60 cells were grown with or without DPE at the indicated concentrations. After 48 hrs, cells were collected and assayed for nitroblue tetrazolium reduction activity, a marker of myelo-monocytic differentiation. The enzymatic activity was determined colorimetrically as absorbance (Abs) at 540 nm.

dose-dependent differentiation of HL60 cell towards the myelo-monocytic lineage as monitored by the increase in the nitroblue tetrazolium reduction, while PE did not induce any specific cell lineage (Fig. 6).

Finally in order to elucidate a potential apoptotic effect of DPE, HL6O cells were incubated for 48 hours in the presence of increasing concentrations of the phenol and the apoptotic cells were evaluated by flow cytometry analysis. The results indicated a noticeable percentage of cells undergoing programmed cell death at DPE concentrations higher than 50 μM (data not shown). Presently, studies are underway to investigate the molecular bases of DPE apoptotic effect by analyzing the various biochemical pathways involved in this process. In this context, it must be underlined that HL-60 is a p53-negative cell line,[52] and thus the programmed cell death must be activated through p53-independent pathways.

In conclusion, our data demonstrate for the first time that DPE directly causes, at concentrations attainable under *in vivo* conditions, a remarkable growth restraint along with an induction of cell differentiation. Moreover, the activation of an apoptotic program appears particularly intriguing to explain the observed chemopreventive activity of olive oil, since a large majority of therapeutic strategies are based on the apoptotic activities of anticancer drugs.

5. CONCLUSIONS AND FUTURE PERSPECTIVES

Taken together the reported findings allow us to speculate that extra virgin olive oil, characterized by a high DPE content, could exert a protective effect against those pathologies, including cancer, whose etiology has been related to ROS-mediated injuries. In this respect, it should be stressed that, even though DPE concentrations achievable *in vivo* by dietary intake of olive oil have not yet been evaluated, in oils with high total phenol content, DPE is present in concentrations of up to 3 mM, far exceeding those required for its biological effects.

In Table 5 are summarized all the biological effects exerted by DPE so far reported in the Literature, that are probably all related to its antioxidant activity.[53–57] The lack of biological effects of PE in all the model systems investigated, indeed, clearly indicates that the two ortho-hydroxyl groups of the phenolic ring are essential with respect to both the protection towards oxidative stress and its cytostatic and differentiating effects.

Table 5. Biological effects of DPE

	Effective concentrations (μM)	References
In vitro inhibition of LDL oxidation	5–10	34–36
Inhibition of platelet aggregation and eicosanoid production	100–400	53
Protection against ROS-mediated cytotoxicity of human intestinal cells and human erythrocytes	50–250	37–38
Inhibition of platelet and leukocytes arachidonate lipoxygenases	10–100	54–55
Inhibition of PMA-induced respiratory burst in human neutrophils	10–100	10
Increase of NO production in mouse macrophages*	10–100	56
Inhibition of peroxynitrite-dependent tyrosine nitration and DNA damage	50–500	57
Antiproliferative and differentiating effects. Induction of apoptosis results	50–100	unpublished

*Effect exerted by oleuropein

The antiproliferative, differentiating, and apoptotic activities of DPE indicate that the molecule probably affects the pivotal cell programs. The efficacy of DPE in inhibiting cell proliferation and in inducing apoptosis deserves further investigation. A major question to be addressed is to determine just which molecular mechanism(s) is (are) responsible for this action. As is well known, cell death program is an active process requiring a cascade of zymogen proteases activation (procaspases-> active caspases), which finally results in DNA degradation and protein coagulation. Different signals might activate the apoptotic program, including DNA damage (*via* p53-dependent pathways), extracellular stimuli (*via* cell death membrane receptors) and intracellular signals (*via* autocrine loops or cytochrome C release). Our findings, based on experiments employing HL-60 cells, suggest that DPE-dependent apoptosis occurs *via* a p53-independent pathway. A further point which deserves more extensive investigation is related to the importance of the cell phenotype in the induction of apoptosis. This can be achieved by testing DPE effects either on several cell types or on a specific cell lineage, grown under different conditions.

The apoptotic effect of DPE, and therefore of olive oil, may also be relevant in the development of new therapeutic strategies. In this context, it should be underlined that DPE is able to efficaciously inhibit 5- and 11-lipoxygenase activities.[54–55] Since these enzymes, which are involved in leukotriene synthesis, seem to play a role in the process of metastases formation, it can be hypothesized that, at least in part, DPE could also exert a positive effect on the cancer progression process.

ACKNOWLEDGMENTS

This work was partially supported by grants from the "International Olive Oil Council."

REFERENCES

1. Ferro-Luzzi, A. and Ghiselli, A. (1993) Protective aspects of the Mediterranean Diet. In *Advances in Nutrition and Cancer* (Zappia, V.; Salvatore, M. and Della Ragione, F., eds.) pp. 137–144. Plenum Publishing Corporation, New York.

2. Decker, E.A. (1995) The role of phenolics, conjugated linoleic acid, carnosine, and pyrroloquinoline quinone as nonessential dietary antioxidants. *Nutr. Rev.*, **53**, 49–58.
3. Bravo, L. (1998) Polyphenols: chemistry, dietary sources, metabolism, and nutritional significance. *Nutr. Rev.*, **56**, 317–333.
4. Boskou, D. (ed.) (1996) Olive Oil. Chemistry and Technology, AOCS Press, Champain, Illinois.
5. Montedoro, G. and Servili, M. (1992) Olive oil quality parameters in relationship to agronomic and technological aspects. *La rivista italiana delle sostanze grasse.* **LXIX**, 563–573.
6. Montedoro, G., Servili, M., Baldioli, M., and Miniati, E. (1992) Simple and hydrolyzable phenolic compounds in virgin olive oil. 1. Their extraction, separation, and quantitative and semiquantitative evaluation by HPLC. *J. Agric. Food Chem.*, **40**, 1571–1576.
7. Montedoro, G., Servili, M., Baldioli, M., and Miniati, E. (1992) Simple and hydrolyzable phenolic compounds in virgin olive oil. 2. Initial characterization of the hydrolyzable fraction. *J. Agric. Food Chem.*, **40**, 1577–1580.
8. Capasso, R., Evidente, A., Visca, C., Gianfreda, L., Maremonti, M., and Greco, G. Jr. (1996) Production of glucose and bioactive aglycone by chemical and enzymatic hydrolysis of purified oleuropein from Olea europea. *Appl. Biochem. Biotech.*, **60**, 365–377.
9. Aeschbach, R., Loliger, J., Scott, B.C., Murcia, A., Butler, J., Halliwell, B., and Aruoma, O.I. (1994) Antioxidant actions of thymol, carvacrol, 6-gingerol, zingerone, and hydroxytyrosol. *Food Chem. Toxic.*, **32**, 31–36.
10. Visioli, F. Bellomo and Galli, C. (1998) Free radical-scavenging properties of olive oil polyphenols. *Biochem. Biophys Res. Commun.*, **247**, 60–64.
11. Tsimidou, M., Papadopoulos, G., and Boskou, D. (1992) Phenolic compounds and stability of virgin olive oil. *Food Chem.*, **45**, 141–144.
12. Papadopoulos, G. and Boskou, D. (1991) Antioxidant effect of natural phenols on olive oil. *J. Am. Oil Chem. Soc.*, **68**, 669–671.
13. Ferro-Luzzi, A. and Sette, S. (1989) The Mediterranean diet: an attempt to define its present and past composition. *Eur. J. Clin. Nutr.*, **43**, 13–29.
14. Keys, A. and Keys, M. (1975) "How to Eat Well and Stay Well, The Mediterranean Way", Doubleday & Co. Inc., New York.
15. Mattson, F.H. and Grundy, S.M. (1985) Comparison of dietary saturated, monounsaturated, and polyunsaturated fatty acids on plasma lipids and lipoproteins in man, *J. Lipid Res.*, 26:194–202.
16. Riccardi, G. and Rivellese, A. (1993) An update on monounsaturated fatty acids. *Current Opinion in Lipidology*, **4**:13–16.
17. Witztum, J.L. (1994) The oxidation hypothesis of atherosclerosis. *Lancet*, **344**, 793–795.
18. Parthasarathy, S., Khoo, J.C., Miller, E., Witztum, J.L., and Steinberg, D. (1990) Low density lipoprotein rich in oleic acid is protected against oxidative modification: implication for dietary prevention of atherosclerosis. *Proc. Natl. Acad. Sci. USA*, **87**, 3894–3898.
19. Martin-Moreno, J.M., Willett, W.C., Gorgojo, L., Banegas, J.R., Rodriguez-Artalejo, F., Fernandez-Rodriguez, J.C., Maisonneuve, P., and Boyle, P. (1994) Dietary fat, olive oil intake, and breast cancer risk. *Int. J. Canc.*, **58**, 774–780.
20. Trichopoulou, A., Katsouyanni, K., Stuver, S., Tzala, L., Gnardellis, C., Rimm, E., and Trichopoulos, D. (1995) Consumption of olive oil and specific food groups in relation to breast cancer risk in Greece. *J. Natl. Canc. Inst.*, **87**, 110–116.
21. Willett, W.C. (1997) Specific fatty acids and risks of breast and prostate cancer: dietary intake. *Am. J. Clin. Invest.*, **66**, 1557S–1576S.
22. Braga, C., La Vecchia, C., Franceschi, S., Negri, E., Parpinel, M., Decarli, A., Giacosa, A., and Trichopoulos, D. (1998) Olive oil, other seasoning fats, and the risk of colorectal carcinoma. *Cancer*, **82**, 448–453.
23. Sies, H. (ed.) (1991) Oxidative stress, oxidants, and antioxidants. London and New York: Academic Press.
24. Scott. Gerald (ed.) (1997) Antioxidants in science, technology, medicine, and nutrition, Albion Publishing Chichester.
25. Ames, B.N., Shigenaga, M.K., and Hagen, T.M. (1993) Oxidants, antioxidants, and the degenerative diseases of aging. *Proc. Natl. Acad. Sci. USA*, **90**, 7915–7922.
26. Frei, B. (ed.) (1994) Natural antioxidant in human health and disease, Academic Press.
27. Halliwell, B., Murcia, M.A., Chirico, S., and Aruoma, O.I. (1995) Free radicals and antioxidants in food and in vivo: what they do and how they work, *Critical Rev in Food Sci and Nutr.*, **35**, 7–20.
28. Montero, H.P. and Stren, A. (1996) Redox modulation of tyrosine phosphorylation-dependent signal transduction pathways. *Free Rad. Biol. Med.*, **3**, 323–333.

29. Palmer, H.J. and Pauling, K.P. (1997) Reactive oxygen species and antioxidants in signal transduction and gene expression. *Nutr. Rev.*, **55**, 353–361.
30. Irani, K., Xia, Y., Zweier, J.L., Sollot, S.J., Der, C.J., Fearo, E.R., Sundaresan, M., Finkel, T., and Goldschmidt-Clemont, P.J. (1997) Mitogenic signaling mediated by oxidants in ras-transformed fibroblasts. *Science*, **275**, 296–299.
31. Cimmino, F., Esposito, F., Ammendola, R., and Russo, T. (1997) Gene regulation by reactive oxygen species, Current topics in cellular regulation, **35**, 123–147.
32. Scaccini, C., Nardini, M., D'Aquino, M., Gentili, V., Di Felice, M., and Tomassi, G. (1992) Effect of dietary oils on lipid peroxidation and on antioxidant parameters of rat plasma and lipoprotein fractions. *J. Lipid Res.*, **33**, 627–633.
33. Wiseman, S.A., Mathot, J.N., de Fouw, N.J., and Tijburg, L.B. (1996) Dietary non-tocopherol antioxidants present in extra virgin olive oil increase the resistance of low density lipoproteins to oxidation in rabbits. *Atherosclerosis.*, **120**, 15–23.
34. Salami, M., Galli, C., De Angelis, L., and Visioli, F. (1995) Formation of F_2-isoprostanes in oxidized low density lipoprotein: inhibitory effect of hydroxytyrosol. *Pharmacol. Res.*, **31**, 275–279.
35. Visioli, F., Bellomo, G., Montedoro, G.F., and Galli, C. (1995) Low density lipoprotein oxidation is inhibited in vitro by olive oil constituents. *Atherosclerosis.*, **117**, 25–32.
36. Visioli, F. and Galli, C. (1998) The effect of minor constituents of olive oil on cardiovascular disease: new findings. *Nutr. Rev.*, **56**, 142–147.
37. Manna, C., Galletti, P., Cucciolla, V., Moltedo, O., Leone, A., and Zappia, V. (1997) The protective effect of the olive oil polyphenol (3,4-dihydroxyphenyl)ethanol counteracts reactive oxygen metabolite-induced cytotoxicity in Caco-2 cells. *J. Nutr.*, **127**, 286–292.
38. Manna, C., Galletti, P., Cucciolla, V., Montedoro G., and Zappia, V. (1999) Olive oil hydroxytyrosol protects human erythrocytes against oxidative damages. *J. Nutr. Biochem.*, **10**, 159–165.
39. Hidalgo, I.J., Raub, T.J., and Borchardt, R.T. (1989) Characterization of the human colon carcinoma cell line (Caco-2) as a model system for intestinal epithelial permeability. *Gastroenterology*, **96**, 736–749.
40. Granger D.N., McCord, J.M., Parks, D.A., and Hollwarth, M.E. (1986) Xanthine oxidase inhibitors attenuate ischemia-induced vascular permeability changes in the cat intestine. *Gastroenterology*, **90**, 80–84.
41. Nakayama, T. (1994) Suppression of hydroperoxide-induced cytotoxicity by polyphenols. *Cancer Res.*, **54**, 1991s–1993s.
42. Khan, S.G., Katiyar, S.K., Agarwal, R., and Mukhtar, H. (1992) Enhancement of antioxidant and phase II enzymes by oral feeding of green tea polyphenols in drinking water to SKH-1 hairless mice: possible role in cancer chemoprevention. *Cancer Res.*, **52**, 4050–4052.
43. Misra, H.P. and Fridovich, I. (1972) The generation of superoxide radical during the autoxidation of hemoglobin. *J. Biol. Chem.*, **247**, 6960–6962.
44. Van Dyke, B.R. and Saltman, P. (1996) Hemoglobin: a mechanism for the generation of hydroxyl radicals. *Free Rad. Biol. Med.*, **20**, 985–989.
45. Snyder, L.M., Fortier, N.L., Trainor, J., Jacobs, J., Leb, L., Lubin, B., Chiu, D., Shohet, S., and Mohandas, N. (1985) Effect of hydrogen peroxide exposure on normal human erythrocyte deformability, morphology, surface characteristics, and spectrin-hemoglobin cross-linking. *J. Clin. Invest.*, **76**, 1971–1977.
46. Johnson, G.J., Allen, D.W., Cadman, S., Fairbanks, V.F., White, J. G., Lampkin, B.C., and Kaplan, M.E. (1979) Red-cell-membrane polypeptide aggregates in glucose-6-phosphate dehydrogenase mutants with chronic hemolytic disease. *N. Engl. J. Med.*, **301**, 522–527.
47. Shinar, E. and Rachmilewitz, E.A. (1990) Oxidative denaturation of red blood cells in thalassemia. *Semin. Hematol.*, **27**, 70–82.
48. Hebbel, R.P., Eaton, J.W., Balasingam, M., and Steinberg, M.H. (1982) Spontaneous oxygen radical generation by sickle erythrocytes. *J. Clin. Invest.*, **70**, 1253–1259.
49. Rohn, T.T., Hinds, T.R., and Vincenzi, F.F. (1993) Ion transport ATPases as targets for free radical damage. Protection by an aminosteroid of the Ca^{2+} pump ATPase and Na^+/K^+ pump ATPase of human red blood cell membranes. *Biochem. Pharmacol.*, **46**, 525–534.
50. Boyle, P., Zaridze, D.G., and Smans, M. (1985) Descriptive epidemiology of colorectal cancer. *Int. J. Cancer*, **36**, 9–18.
51. Greenwald, P., Kelloff, G.J., Burch-Whitman, C., and Kramer, B.S. (1995) Chemoprevention. *CA Cancer J. Clin.*, **45**, 31–49.
52. Wolf, D. and Rotter, V. (1985) Major deletions in the gene encoding the p53 tumor antigen cause lack of p53 expression in HL-60 cells. *Proc. Natl. Acad. Sci. USA*, **82**, 790–794.

53. Petroni, A., Blasevich, M., Salami, M., Papini, N., Montedoro, G.F., and Galli, C. (1995) Inhibition of platelet aggregation and eicosanoid production by phenolic components of olive oil. *Thromb. Res.*, **78**, 151–160.
54. Kohyama, N., Nagata, T., Fujimoto S., and Sekiya, K. (1997) Inhibition of arachidonate lipoxygenase activities by 2-(3,4-dihydro xyphenyl)ethanol, a phenolic compound from olives. *Biosci. Biotech. Biochem.*, **61**, 347–350.
55. de la Puerta, R., Ruiz Gutierrez, V., and Hoult, JR. (1999) Inhibition of leukocyte 5-lipoxygenase by phenolics from virgin olive oil. *Biochem Pharmacol.*, **57**, 445–449.
56. Visioli, F., Bellosta, S., and Galli, C. (1998) Oleuropein, the bitter principle of olives, enhances nitric oxide production by mouse macrophages. *Life Sci.*, **62**, 541–546.
57. Deiana, M., Arouma, O.I., Bianchi, M. de L.P., Spencer, J.P.E., Kaur, H., Halliwell, B., Aeschbach, R., Banni, S., Dessi, M.A., and Corongiu, F.P. (1999) Inhibition of peroxynitrite dependent DNA base modification and tyrosine nitration by the extra virgin olive oil-derived antioxidant hydroxytyrosol. *Free Rad. Biol. Med.*, **26**, 762–769.

12

PROTECTIVE EFFECTS OF BUTYRIC ACID IN COLON CANCER

Gian Luigi Russo,[1] Valentina Della Pietra,[2] Ciro Mercurio,[2] Rosanna Palumbo,[1] Giuseppe Iacomino,[1] Maria Russo,[1] Mariarosaria Tosto,[3] and Vincenzo Zappia[2]

[1] Institute of Food Science and Technology
National Research Council
83100 Avellino, Italy
[2] Institute of Biochemistry of Macromolecules
II University of Naples
80138 Naples, Italy
[3] Zoological Station "Anton Dohrn"
80121 Naples, Italy

1. INTRODUCTION

A great deal of evidence indicates that cancer is the result of a reciprocal interaction between genetic susceptibility and environmental factors. Apparently, only 5% of all cancers is linked to genetic, inheritable alterations, while the remainder is associated with environmental conditions that act in concert with individual susceptibility.[1] In this context, dietary habit covers a fundamental role. Following the recent publication from the American Institute for Cancer Research[2] (AICR), a "correct" diet could decrease the cancer rate by as much as 20%. For some specific cancers, the effect is even more dramatic: from 33 to 50% of breast cancer can be prevented through diet, and as much as 33% of lung cancer and 75% of colorectal cancers are prevented by correct diet choices. Colorectal cancer represents one of the well studied cancers at the molecular level: here, the steps of tumor progression show *in vivo* an accumulation of morphological changes that result *in vivo* in mutations in specific genes. Loss of function of the *APC* tumor suppressor gene causes hyperproliferation of normal epithelium and early adenoma; the subsequent activation of *k-ras* oncogene and inactivation of *DCC* and p53 tumor suppressor genes are associated with late adenoma and carcinoma.[3,4] Colorectal cancers are very common causing over 50,000 deaths per year in the United States, 11% of total deaths from cancer. They result from the combined

Advances in Nutrition and Cancer 2, edited by Zappia *et al.*
Kluwer Academic / Plenum Publishers, New York, 1999.

effect of inherited factors and stochastic genetic mutations, partially related to environmental conditions such as diet composition.[5] Considerable epidemiological data support the hypothesis that a diet rich in fiber is associated with a decreased risk of intestinal cancers.[6,7] Although this views appears simplistic and remains controversial for many aspects, it is currently generally accepted. In fact, one of the fifteen recommendations of the AICR advises to "eat a variety of vegetables and fruits all year round".[2] Several mechanisms have been considered in order to explain the protective effect of dietary fiber (DF). They include dilution of luminal carcinogen concentration, fiber-associated changes in colonic transit that decrease colonic enterocyte exposure to luminal carcinogens, interactions with intestinal contents, production of tissue factors by stimulation of colonic mucosa, role of bile acids and direct antineoplastic activity by DF components.[8–13] A different mechanism by which fiber may modulate carcinogenesis is related to the production of butyric acid (BA), a component of the short-chain fatty acids (SFCA) obtained by degradation of poorly fermented fiber by colonic microflora. This review will focus on the molecular aspects that affect the activity of BA on *in vivo* and *in vivo* models.

1.1. Structure and Composition of Dietary Fiber

The term dietary fiber was first coined more than 40 years ago to indicate the component of plant cell wall present in the human diet. Since then, the parallel efforts of carbohydrate chemists and physiologists clarified the composition and nutritional function of DF. It was immediately clear that DF included a portion of carbohydrates that, unlike starch, was "not available" to release glucose. In parallel, physiological studies showed that dietary fiber is not digested in the small intestine, but passes through the alimentary tract and is fermented by microflora present in the colon, producing several components that provide energy for the colonocytes. Since the study of DF was approached from different points of view, its definition is probably still incomplete. Currently, two definitions, that are not mutually exclusive, are given to indicate this component of food. Following the 'structural' definition, DF consists of non-starch polysaccharides of plant cell walls; the 'physiological' view, instead, defines DF as polysaccharides and lignin that are undigested by the enzymes of the small intestine.[15] In fact, the amylases secreted into the small intestine hydrolyze the α-1,4 bonds present in starch releasing smaller units of glucose, but not β-1,4 bond present in the so-called DF.

DF can be classified accordingly to its solubility. Soluble DF includes: pectins, beta-glucans, mucilages, plant exudate gums, legume seed gums, seaweed polysaccharides, and bacterial polysaccharides. Insoluble DF is cellulose and hemicelluloses. Lignin (a polyphenylpropane unit) is generally included among insoluble DF although it is not a polysaccharide. However, lignin is associated with plant cell wall and behaves as a fiber in terms of gastrointestinal physiology. Resistant starch requires special mention. This definition includes the type of starch that crystallizes in indigestible structures following heating-cooling processes (*e.g.* during food processing). Finally, DF includes several minor components, as well as phytic acid, tannins and cutins.[9,15]

1.2. Dietary Fiber in Cancer Prevention

The original hypothesis of Burkitt on the role of DF in cancer prevention[6,7] was based on four major points (Table 1, A–D). Further evidence indicates that other

Table 1. Possible causes of colorectal cancer prevention by dietary fiber

Causes	Effects	Evidence in Human[8]
A. Stool bulking	Dilution of potential carcinogens present in the diet	Yes
B. Increase transit thought the colon	Reduction time for bacteria to produce metabolites	Yes/No
C. Changes in bacterial flora	Elimination of 'dangerous' bacteria	No
D. Changes in physicochemical condition of the colon	Less favorable condition for bacteria to produce carcinogens	No
E. Production of SCFA	Production of butyrate has been associated with anticancer properties	Yes

possible causes of DF protection are the influence of bile acid metabolism, and the production of SCFA.[8,16,17] Unfortunately, these studies are largely inconclusive, since the mechanisms by which fiber protects against colorectal carcinogenesis are still far from being determined. Several epidemiological data support the 'Burkitt hypothesis'. Case-control studies indicate a strong, inverse association between colon cancer and DF intake. This conclusion comes from two independent studies performed analyzing either 2,000 cases of colon cancer in the USA,[18] or elaborating case-control data from different countries.[19] On the contrary, cohort studies are less consistent. In one work, the inverse association between colon cancer and fiber intake was weak and related only to fruit fiber intake;[20] in other studies, no significant relationship was reported between colon cancer and fiber.[21,22] Finally, a large, recent prospective study from Willett denies any association in women between colon cancer and adenoma and high intake of DF.[23] In a recent review, Negri and La Vecchia[24] attempted to explain the controversial data obtained from epidemiological studies suggesting that stool bulk, transit time, and fermentation are important factors to be considered, but their extent depends on the type and origin of the fiber. They analyzed the protective role of fiber paying great attention to the separate effect of various types of fiber. In a recent work, these Authors following this approach, demonstrated a protective and independent effect of fiber on colorectal cancer in a large case-control study conducted in Italy.[25,26]

In summary, analyzing the data reported in Table 1, points A and B have found experimental evidence in humans, while C and D are currently less supported.[8] In the next paragraphs, we will focus our attention on the effect of butyric acid production as a potential mechanism in colon cancer prevention.

1.3. Production of SCFA

According to Miller and Wolin,[27] the main products of bacterial fermentation of DF are CH_4, CO_2, H_2O, and SCFA (acetic acid, propionic acid, butyric acid) at a concentration in the human large intestine estimated to be approximately 75 mM acetic acid (AA), 30 mM propionic acid (PA), and 20 mM BA.[28] The blood concentration of SCFA in humans is 375 ± 70 mM,[29] with the main contribution given by AA since BA and PA are taken up by the liver. The rate of different SCFA intake depends on the physiology of the intestinal tract and by different mechanisms of transport across the colon. For instance, at the level of the proximal colon occurs the maximal intake of acetate, while butyrate is rapidly absorbed in the left colon.[30] The precise mechanism for the absorption process of SCFA is unclear; several models have been proposed describing the dependence of SCFA absorption from rates on luminal pH and P_{CO2}, as well as on fluxes of water, protons, and inorganic ions[31–33] through the colonic mucosa. A recent work concluded that the rat colon non-ionic diffusion is the most important if not the only mechanism of SCFA absorption at least for PA and BA.[34] The fate of SCFA is to be β-oxidized in the cells via the mitochondrial fatty acyl-CoA β-oxidizing system. For each molecule of BA two acetyl-CoA are obtained.[9] It is generally accepted that BA represents the main fuel for colonocytes providing more than 70% of the energy required for the cell to survive (see Conclusions). It is clear that SCFA exert a trophic effect on colonic epithelium: lack of intraluminal SCFA may be associated with several pathological status alterations. Probably multifactorial mechanisms can explain the trophic activity of SCFA on colonic mucosa: they include mucine production, effect on barrier function, and effect on colonocytes proliferation. A review on these arguments has recently been published.[30]

2. BUTYRIC ACID AND COLON CANCER

Among the short chain fatty acids, BA has most attracted the attention of scientists as a potential candidate to explain the supposed protective effects of DF against colon cancer. Strong evidence from *in vivo* and *in vivo* studies support the antimitotic effects of BA acting as a differentiating agent and pro-apoptotic factor (see paragraph 2.2.2). In this section, we will examine the most recent data on BA activity in carcinogenesis, referring to studies performed on *in vivo* and *in vivo* models.

2.1. *In Vivo* Effects of Butyric Acid

The relative production of different SCFA depends on the type of DF considered. In rats fed with high fiber diet the ratio AA:PA:BA moves from 69:21:10 to 92:7:1 compared to animals fed with fiber-free diet.[35] More interestingly, the amount of BA compared to AA and PA increased significantly in rats and pigs fed with insoluble fiber (oat bran) compared to a diet rich in soluble fiber (pectins).[36,37] These studies and others, both in animals and in humans[38–40,10] confirmed a higher production of BA from a diet rich in insoluble fiber, compared to lower production of BA from fermentation of soluble fiber. In other words, although the total amount of SCFA produced by fermentation of soluble fiber is higher compared to a diet rich in soluble fiber, the relative production of BA compared to AA and PA is higher following a diet rich in insoluble fiber. This consideration is most important considering that in animal models the protective effect of DF against experimentally induced colon cancer is mainly associated with insoluble fiber, while no consensus exists for water-soluble fibers, gums, and pectins.[41–43] In carcinogens-treated rats, Young observed that a diet rich in insoluble fiber (wheat bran, 25% soluble), was significantly more effective in preventing colorectal cancer than a diet with guar gum (85% soluble).[44,10] The same author also verified that an increased concentration of BA in the distal large bowel was associated with a protective effect. On the other hand, the production of BA is lowered compared to total SCFA in patients suffering from polyp-colon cancer,[45] colon cancer, and adenoma.[46]

The protective effect of BA against colon cancer has been actively debated, and, in our opinion, only recently has evidence been accumulated that more strongly support the "Burkitt hypothesis". In 1992, Klurfeld concluded his review stating that no beneficial effects could be associated to SCFA and BA in colon cancer prevention.[12] However, in rat models, the production of BA from DF protects against large bowel cancer,[47] while in experimental colitis butyrate reduces colon cancer risk.[48] More recently, Medina et al.[49] reported that direct application of butyrate significantly decreased the total number of tumors and the incidence of malignancies and carcinoma in rat colon. Similarly, Caderni proposed an anticancer, pro-apoptotic activity of BA in rats treated with azoxymethane.[50]

All these *in vivo* studies on animal models show a general consensus on the effect of BA in carcinogenesis. More controversial are the results obtained in humans. Several studies published considered as an experimental model those patients suffering from ulcerative colitis in order to assay the protective effect of BA. The rationale for this choice is that increased cell proliferation in colonic epithelial cells has been found in familial polyposis, sporadic adenomas, and ulcerative colitis, and it has been indicated as a marker of increased cancer risk.[51] Moreover, considering BA a 'cancer preventing agent' than an anticancer drug, the occurrence of a pre-cancer pathology is justified to test butyrate activity. In 1993, Bradburn[52] suggested that a decreased production of BA by colonic carbohydrate fermentation may predispose a patient to colon cancer, since he measured a significant decrease of BA in patients with familial adenomatous polyposis. Two years later, Chapman[53] reported that butyrate oxidation was impaired in colonic mucosa of patients with quiescent ulcerative colitis. More cautious were the conclusions of a recent work on similar subjects.[54] These Authors showed a more marked effect of BA on crypt cell proliferation than on parameters of inflammation in patients with active ulcerative colitis.[55] More recently, a Japanese group associated the increased production of BA from a diet enriched in barley foodstuff with a positive effect on ulcerative colitis regression.[56] Accordingly, *Plantago ovada* seed-derived butyrate production was effective in maintaining remission in ulcerative colitis.[57] Despite this large amount of positive evidence, a Canadian group demonstrated no effect of butyrate treatment on ulcerative colitis remission.[58]

2.2. Effect of Butyric Acid on Cell Division Cycle Regulation

The great popularity of BA as a potential anticancer molecule certainly comes from its effects on cell cultures. Exposure of several colon cancer-derived cell lines to butyrate induces block of cell growth, differentiation followed by apoptosis in several days (Fig. 1). These changes are also common to other tumor cell lines with a different origin compared to colorectal adenocarcinoma. Table 2 summarizes the major changes observed on cell lines treated with 2–5 mM concentration of BA. The number of papers published every year on the *in vivo* effect of butyrate has increased exponentially in the last five years making it almost impossible to cite all the work done. Several reviews have been published in recent years representing an excellent source of information

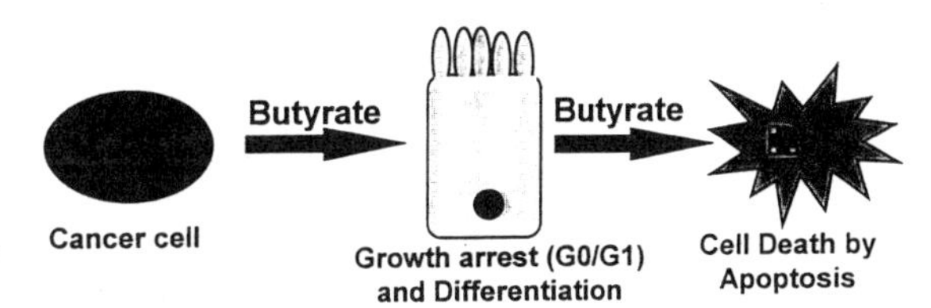

Figure 1. Scheme representing the activity of BA on cancer cell lines.

Table 2. Effects of butyric acid on cultured cells

Alterations described	References
Increased doubling times	10
Reduced colony-forming ability in soft agar	10
Morphological alteration	10
Terminal differentiation	9, 65–70
Increases in alkaline phosphatase and other enzymes	10, 71
Induction of carcinoembryonic antigen	10
Modification in proto-oncogene expression:	
c-*myc*	14, 72–74
c-*myb*	128
c-*jun*	72
c-*fos*	75
Ha-*ras*	76
Src	77
Alterations in glycoprotein synthesis	10
Increased expression of cytochrome c oxidase	78
Increased expression of the basolateral Na^+/H^+ exchanger	10, 79
Altered activity at the level of gene expression:	
Histone acetylation	9, 80–82
Phosphorylation of histone H1 and H2A	9
Chromatin structure	9
Poly(ADP)-ribosylation	9, 83
Hormone-dependent gene expression	9
Effect on cell cycle regulation	89–105, 129
Induction of apoptosis	123–125, 127, 150

(see Table 2, and[9–11,59–64]). In this section, we will focus our attention mainly on the effect of BA on cell cycle regulation and apoptosis.

2.2.1. Cell Cycle Arrest. As shown in the scheme in Figure 1, cancer cell lines of different origins treated with BA stop growth and arrest their cell cycle in G0/G1 phase.[84,85,76] In HT-29 cells, derived from a colon adenocarcinoma,[86] growth arrest is observed after 24 h of induction with 2 mM BA, in parallel with the start of the differentiation program, as shown by the increase of alkaline phosphatase activity.[85,87] In addition, BA differentiates HT-29 into a layer of polarized epithelial cells exhibiting tighter junctions and well defined basolateral and brush border membranes with a phenotype that closely resembles that of a normal colonocyte[88] (Fig. 1). According to a modern view of cell cycle regulatory machinery (reviewed in[95–99]), we expect to observe an induction of genes specific for G0/G1 transition, and a repression of the S-G2/M genes. However, before analyzing the direct effect of BA on cell cycle regulation, we will briefly review the main protagonists, at the molecular level, that regulate phase transition in mammalian cells.

Cell division cycle is regulated by three families of molecules (Table 3). Cdks kinases (Cdk1, Cdk2, Cdk4, Cdk6, and others) are associated with two classes of regulatory subunits: their activators, cyclins (B, A, E, D, G, and others), that allow Cdks to act as kinases, or their inhibitory subunits CKI (mainly, p16, p15, p18, p19, p21, p27, p57). These three groups of molecules are associated differently to the specific phase of the cell cycle. For example, in order to have an active complex in G1, Cdk4, or Cdk6, must bind to one of the D-type cyclins. In S phase, the main active complex is

Table 3. Molecular complexes that regulate the cell cycle

Cell cycle phase	Cell Division Kinases	Cyclins	Cell division Kinase Inhibitor
G2/M	Cdk1	A, B	p21
S	Cdk2	A, E	p21, p27, p57
G1	Cdk4/6	D	p21, p15, p16, p18, p19
	Cdk3	E	?
G2/M	Cdk7	H	?
G2/M	Cdk8	C	?

Cdk2/Cyclin E responsible for driving the cell cycle in S phase and initiating DNA replication. Finally, in mitosis, Cdk1/Cyclin B forms the active Cdk complex. The CKI inhibitors work at different levels in the cell cycle blocking specific Cdk-Cyclin complexes. As an example, the p16^{INK4A} family (including the other members, p15, p18, p19) inhibits the kinase activity of Cdk4(6)/Cyclin Ds, while p21$^{WAF1/Cip1}$ seems to be a more specific inhibitor of G1/S phase transition blocking Cdk2/Cyclin E complex. However, the final regulatory pathway is more complex since all Cdks mentioned are regulated not only by physical binding with their partners (cyclins and CKI), but also by phosphorylation/dephosphorylation events, that are cell cycle specific.

Two key molecules link the cell cycle to the gene expression: p53 and Rb, the product of the retinoblastoma gene. The first induces the transcription of p21$^{WAF1/Cip1}$, the CKI that arrests cells at G1/S transition. Rb, in its dephosphorylated state, binds E2F and inhibits its transcriptional activity including expression of several molecules responsible for the passage in G2/M as well as cyclin E, cyclin A, and Cdk2. The phosphorylation of Rb by Cdk4/Cyclin D frees E2F from RB binding and enables the passage into S phase.

In this scenario, BA acts at several levels as a molecule that inhibits cell cycle progression and blocks cells in G1. Accordingly, BA stimulates the expression of cyclin D and p16^{INK4A} in HT-29 cell line, and inhibits expression of Cdk2,[89] although in a different system of normal fibroblasts, BA exerts the opposite effect down-regulating cyclin D1 expression.[90] The activity on cell cycle arrest of BA is confirmed by its ability to inhibit phosphorylation of Rb,[91,92] and by the observation that BA analogs inhibit the enzymatic activity of Cdks that phosphorylate and inhibit Rb.[93,94,129] However, the most evident effect of BA on cell cycle arrest is the induction of the CKI, p21$^{WAF1/Cip1}$. This molecule is a potent inhibitor of G1/S transition by suppressing the phosphorylation of Rb via inhibition of G1 Cdks.[100] It is well accepted that the transcription of the p21$^{WAF1/Cip1}$ is activated by wild-type p53,[101] a key factor in apoptosis induction (see below). Therefore, in several systems, the p53-dependent growth arrest is mediated by p53-dependent transcription of p21$^{WAF1/Cip1}$. On the other hand, several reports agree with the demonstration that BA induces cell cycle arrest by activating p21$^{WAF1/Cip1}$ transcription in a p53-independent fashion.[102–105,126] Using mutants of p21$^{WAF1/Cip1}$ promoter, it has been shown that the butyrate-responsive elements are two Sp1 sites at –82 and –69 relative to the transcription start site.[103] A more recent study indicated that activation of p21$^{WAF1/Cip1}$ by BA is mediated by histone hyperacetylation providing strong evidence for the first time that the activity of BA on histone acetylation reflects the transcription of BA-specific gene responsible for BA effects.[105] The same Authors pointed out the critical importance of p21$^{WAF1/Cip1}$ in BA-mediated growth arrest since stable overexpression of p21$^{WAF1/Cip1}$ cause growth arrest, while a p21$^{WAF1/Cip1}$-deleted cell line (–/–) is insensitive to BA activity.[105]

2.2.2. Butyric Acid and Apoptosis. Apoptosis, or programmed cell death, is a process that activates and synthesizes gene products in order to direct cellular self-destruction. The difference between apoptosis and cell necrosis is mainly based on different morphological changes, and by the involvement of an inflammatory process characteristic of cell necrosis that is observed only slightly in apoptosis.[106] In evolutionist terms, apoptosis represents one solution that organisms selected to eliminate damaged cells and prevent injury in remaining parts of the tissues. In fact, apoptosis is a common, physiological mechanism occurring during development of multicellular organisms,[107] as well as in regulating thymocyte homeostasis during the immunological response.[108] Original studies on the nematode *Caenorhabditis elegans* allowed the identification and cloning of essential genes of the cell death machinery. Homologs of these genes have been cloned in human and other organisms. Briefly, a class of compounds, named cell death receptors, transmit the 'death' signal from the extracellular environment to the nucleus with the activation of the apoptotic program. Using specific adapters, the receptors transmit the message to a class of proteolytic enzymes known as caspase that start a cascade reaction resulting in disassembly of the cells. Other essential actors in the apoptotic pathways are the mitochondria and the activity of the Bcl-2 family of apoptotic factors (for a recent and exhaustive review on apoptosis, see[109–113,147]). Apoptosis has also been studied in terms of potent defense mechanisms against cancer development: when DNA repair mechanisms are not able to eliminate mutations that predispose cancer, cells answer by suppressing growth, or triggering their suicide.[109] The tumor suppressor gene p53 plays an important role in this process. In response to DNA damage, the cell induces cell repair mechanisms; if they fail, p53 activation occurs with two possible consequences: cell arrest via transcription of $p21^{WAF1/Cip1}$, or cell death.[109] BA induces apoptosis in tumor cell lines in a p53-independent manner.[9,69,114–119] Several mechanisms have been proposed to characterize the effect of BA to induce apoptosis in tumor cell lines. They involve a pivotal role of mitochondria;[120] the over-expression of heat-shock protein;[68,121] the upregulation of histone deacetylase genes;[80] or the deregulation of Bcl-2.[122] Actually, the scenario of BA-induced apoptosis is even more complex, and the role of p53 more intriguing. In fact, recent works indicate that BA down-regulates the expression of wild type and mutated p53 in cultured colonic adenoma cells and colonic epithelial cells.[123] In addition, BA may act as co-mutagen since it attenuates p53-dependent DNA damage response in non-tumor cell line.[124] Finally, BA seems to be essential in blocking natural apoptosis in cultured colonocytes via a mechanism that requires down-regulation of Bax.[125]

2.3. Protein Kinase CKII: A Link between Cell Growth Arrest and Apoptosis in Butyric Acid Activity

From the data reported above, it appears that BA exerts a role on cell cycle regulation and apoptosis, although a link between these two BA-mediated events, if any, has not been clearly shown. We postulate that protein kinase CKII (formerly known as casein kinase II) may represent one of the missing elements that mediates BA activity at least in tumor cell lines. In fact CKII is involved in the regulation of both p53 and $p21^{WAF1/Cip1}$. CKII activity has been detected in all tissues and cell lines investigated so far (reviewed in[131–135]). In vertebrate cells, it is composed of two catalytic subunits, α and α', and a regulatory subunit β that is phosphorylated *in vivo* and *in vitro* by Cdk1 kinase,[136,137] and by autophosphorylation.[133,136] A number of circumstantial observations led us to the hypothesis that CKII could be a potential candidate for BA activity. Firstly, butyrate induces cell cycle arrest in early G1 and CKII is involved in cell cycle regulation in HeLa cells[138–140] and in yeast.[141] Secondly, butyrate is a potent differentiating agent, and CKII activity is reported to be regulated by several exogenous factors that induce post-translational modifications of its subunits.[132,134,142] Thirdly, the expression of several oncogenes and tumor suppressor genes is modified following treatment of cell lines with BA; the same molecules are also regulated by CKII phosphorylation.[131–135] Finally, CKII β subunit binds and inhibits p53[143–145] and $p21^{WAF1/Cip1}$.[146] We recently demonstrated that CKII is a target of BA activity.[85] In the human adenocarcinoma cell line HT29, CKII activity decreases 50% at 24 and 48 hours after drug addition. The enzyme down-regulation is not due to changes in protein amount since the levels of the different CKII subunits remain constant during butyrate treatment. Our data provided the first evidence that CKII down-regulation is involved in the signal transduction pathway started by BA.[85] Further studies indicate that the activity of BA on CKII regulation is mediated by CKII-β autophosphorylation. BA treatment of HT-29 cell line indicates a strong increase *in vivo* of CKII-β autophosphorylation without affecting expression of the regulatory subunit (Russo unpublished). This result may suggest a role for CKII-β in targeting different substrates associated to BA activity, and a function for CKII-β phosphorylation in mediating the binding activity of CKII-β. Alternatively, it may be that the different level of CKII-β auto-phosphorylation might 'drive' the catalytic subunit, α or α′, towards specific substrates in the differentiating and pro-apoptotic process triggered by BA in tumor cells.

3. CONCLUSION: THE "PARADOX BUTYRATE"

The potential use of BA as an anti-cancer drug in human trials has been considered as early as 1983. BA was first employed against acute myelogenous leukemia with partially encouraging results.[9] The positive trend of using BA as a drug has been improved by the synthesis of several analogs characterized to be more stable, with an increased half-life *in vivo*, and free of the unpleasant BA odor. At the moment, phase I clinical trials are in progress for tributyrin and other BA derivatives.[9] In addition, a new interesting immune therapeutic approach to cancer cure consists of combining administration of interleukin-2 and BA.[64] Several reports on the use of BA as anti-cancer drug have been recently published.[9,64,70,148–149,151]

The potential pharmacological use of BA raises the problem of its controversial activity on tumor versus normal cells. In fact, as reported above and in other studies, BA provided 70% of energy for colonic mucosa in order to grow and proliferate, while the same molecule inhibits growth, induces differentiation and apoptosis in colorectal derived cell lines. In addition, to render this view even more controversial, two recent papers reported a possible "negative" effect of BA on normal cells. Hass demonstrated that BA withdrawal from normal colonic mucosa isolated from upper intestinal crypts, results in a very rapid and massive apoptosis and increased levels of Bax protein.[125] In other words, BA stimulates proliferation, although this effect is not necessarily comparable with tumor promoting. Janson, using a non-tumor cell line expressing wild-type p53 and low passage fibroblasts from p53 knockout mice, verified the effects of BA on p53-dependent DNA-damage checkpoint.[124] Both cell types reacted similarly in terms of G1 arrest, induction of apoptosis, and expression of $p21^{WAF1/Cip1}$. Preincubation of $p53^{+/+}$ cells with BA resulted in attenuated p53 accumulation and apoptosis in mitomycin C treated cells; since mitomycin C is a well known DNA-damage agent, BA acted as a co-mitogen decreasing the ability of these cells to balance the cellular insult induced by mitomycin C in a p53-dependent apoptotic mechanism. This conclusion is even more important considering that colon cancer "prefers" the left part of the colon that is more exposed and dependent, in terms of energy source, on BA presence.

In the attempt to draw a coherent view that takes advantage of all data available on BA activity, certainly we can assume that BA acts with different mechanisms in normal versus transformed cells, suggesting that the molecule triggers pathways that sustain cell growth in normal colonocytes, or induce differentiation and apoptosis in tumor cells (Fig. 2). One attractive explanation regards the levels of BA in tumor cells compared to normal colonocytes. It appears that colorectal tumor cells present a higher concentration of BA because of their reduced ability to oxidize it, and the more anaerobic metabolism of colon carcinoma.[150] Both these conditions determine an increase of intracellular levels of BA that might explain its activity on cell growth and differentiation: normal colonocytes rapidly utilize BA as the major energetic fuel reducing its intracellular concentration to values unable to inhibit cell growth and induce differentiation. In transformed cells, when BA concentration reaches a critical threshold, the pathways responsible for cell cycle arrest, differentiation, and apoptosis are activated.[150] Accordingly, BA stimulates the expression of cyclin D in tumor cell lines, while in normal fibroblasts, it exerts the opposite effect down-regulating cyclin D1 expression.[89,90] This model cannot explain all the published data on BA effects, but it is a good starting point to address further studies.

At the molecular level, it is clear that the targets of BA activity in normal versus tumor cells differ. Apoptosis is induced in both cell types; however, in normal cells, it is considered a physiological process, necessary to maintain the correct homeostasis in the intestinal crypts by balancing the synthesis of new cells from the bottom of the crypt.[130] BA-induced apoptosis in normal cells is a rapid process independent from cell differentiation, and probably involving down-regulation of p53 gene,[123] induction of p53-independent p21$^{WAF1/Cip1}$,[124] and up-regulation of Bax.[125] On the contrary, in tumor cells, BA triggers apoptosis with a slow process, occurring after several days of treatment.[9,115] This event represents the end-point of a cell differentiation process induced by BA, and dependent from p21$^{WAF1/Cip1}$ expression, but certainly p53-independent. A clear role of p53 in mediating BA activity is currently missing, since its described down-regulation following BA treatment has been observed on both cell types harboring

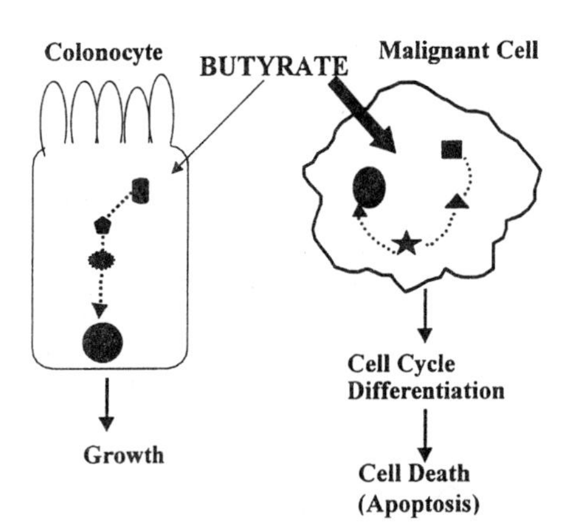

Figure 2. Scheme representing a summary of BA activity on normal and transformed cells according to Hague et al.[150] Large and narrow arrows correspond to higher and lower intracellular concentration of BA, respectively. Symbols represent different targets in the BA-depending pathways leading to the activation of specific genes in the nucleus (N).

wild-type[123] or mutated[73] p53. This leads to BA-induced changes in p53 expression as possible epi-phenomena present in all transformed cell lines.

The controversial epidemiological data on the effects of DF in cancer prevention do not allow us to draw any final conclusion on the role of BA *in vivo* as cancer preventing agent. More conclusive are the data supporting a pharmacological use of BA as anticancer drug. In this sense, a better understanding of its mechanism of action will strongly enhance its use. It is critical to clarify how BA acts at molecular level on normal colonocytes comparing to malignant cells. We must to keep in mind that comparing a pure tumor cell population growing in monolayer, with a normal mucosa that is a three dimensional structure is a condition very far from the *in vivo* situation. A strong effort is currently being made to develop stable colonic epithelial cells to better investigate the effects of BA on normal cells.[150]

Abbreviations: AICR, American Institute for Cancer Research; DF, dietary fiber; BA, butyric acid; SFCA, short chain fatty acids; AA, acetic acid; PA, propionic acid.

REFERENCES

1. Perera, F.P. 1996. Molecular epidemiology: insights into cancer susceptibility, risk assssment, and prevention. J. Natl. Cancer Institute. 88:496–509.
2. American Institute for Cancer Research. 1997. Food, nutrition, and the prevention of cancer: a global perspective.
3. Aaltonen, L.A., P. Peltomaki, F.S. Leach, P. Sistonen, L. Pylkkanen, J.P. Mecklin, H. Jarvinen, S.M. Powell, J. Jen, S.R. Hamilton, et al. 1993. Clues to the pathogenesis of familial colorectal cancer. Science. 260:812–816.
4. Fearon, E.R. and B. Vogelstein. 1990. A genetic model for colorectal tumorigenesis. Cell. 61: 759–767.
5. Rustgi, A.K. and D.K. Podolsky. 1992. The molecular basis of colon cancer. Ann. Rev. Med. 43:61–68.
6. Burkitt, D.P. 1971. Epidemiology of cancer of the colon and rectum. Cancer. 28:3–13.
7. Burkitt, D.P. 1971. Some neglected leads to cancer causation. J. Natl. Cancer Inst. 47:913–919.
8. Hill, M.J. 1998. Mechanisms proposed for the protective action of fiber against colorectal cancer. Gastroenterology Int. 11 (suppl. 1):58–61.
9. Smith, J.G., W.H. Yokoyama, and B. German. 1998. Butyric acid from the diet: actions at the level of gene expression. Critical Rev. Food Sci. 38:259–297.
10. Bugaut, M. and M. Bentéjac. 1993. Biological effects of short-chain fatty acids in nonruminant mammals. Annu. Rev. Nutr. 13:217–241.
11. Hassing, C.A., J.K. Tong, and S.L. Schreiber. 1997. Fiber-derived butyrate and the prevention of colon cancer. Chem. Biol. 4:783–789.
12. Klurfeld, D.M. 1992. Dietary fiber-mediated mechanisms in carcinogenesis. Cancer Res. (suppl.). 52:2055s–2059s.
13. Eastwood, M.A. 1992. The physiological effect of dietary fiber: an update. Ann. Rev. Nutr. 12:19–35.
14. Buckley, A.R., M.A. Leff, D.J. Buckley, N.S. Magnuson, G. de Jong, and P.W. Gout. 1996. Alteration in pim-1 and c-myc expression associated with sodium butyrate-induced growth factor dependency in autonomous rat Nb2 lymphoma cells. Cell Growth Differ. 7:1713–1721.
15. Gurr, M.I. and N.-G. Asp. 1994. Dietary Fibre. ILSI Press, Brussels.
16. Hill, M.J. and F. Fernandez. 1990. Bacterial metabolism, fiber, and colorectal cancer, pp. 417–429. *In* D. Kritchevsky, C. Bonfield, and J.W. Anderson (ed.), Dietary Fiber. Plenum, New York.
17. Giacosa, A. 1998. RPC consensus cereals, fiber, and colorectal cancer. Gastroenterology Int. 11 (suppl. 1):66–68.
18. Slattery, M.L., J.D. Potter, A. Coates, et al. 1997. Plant foods and colorectal cancer: an assessment of specific foods and their related nutrients (United States). Cancer Causes Control. 8:575–590.
19. Howe, G.R., E. Benito, R. Castelleto, J. Cornee, J. Esteve, R.P. Gallagher, J.M. Iscovich, J. Deng-ao, R. Kaaks, G.A. Kune, et al. 1992. Dietary intake of fiber and decreased risk of cancers of the colon and rectum: evidence from the combined analysis of 13 case-control studies. J. Natl. Cancer Inst. 84:1887–1896.

20. Willet, W.C., M.J. Stampfer, G.A. Colditz, et al. 1990. Relation of meat, fat, and fiber intake to the risk of colon cancer in a prospective study among women. N. Engl. J. Med. 323:1664–1672.
21. Giovannucci, E., E.B. Rimm, M.J. Stampfer, et al. 1994. Intake of fat, meat, and fiber in relation to risk of colon cancer in men. Cancer Res. 54:2390–2397.
22. Steinmetz, K.A., L.H. Kushi, R.M. Bostick, et al. 1994. Vegetables, fruit, and colon cancer in the Iowa women's health study. Am. J. Epidemiol. 139:1–15.
23. Fuchs, C.S., E.L. Giovannucci, G.A. Colditz, D.J. Hunter, M.J. Stampfer, B. Rosner, F.E. Speizer, and W.C. Willett. 1999. Dietary fiber and the risk of colorectal cancer and adenoma in women. N. Engl. J. Med. 340:169–176.
24. Negri, E. and C. La Vecchia. 1998. Dietary fiber and colorectal cancer prevention. Gastroenterology Int. 11 (suppl. 1):56–57.
25. Negri, E., S. Franceschi, M. Parpinel, and C. La Vecchia. 1998. Fiber intake and risk of colorectal cancer. Cancer Epidemiol. Biomarkers Prev. 7:667–671.
26. Levi, F., C. La Vecchia, F. Lucchini, and E. Negri. 1995. Cancer mortality in Europe, 1990–92. Eur. J. Cancer Prev. 4:389–417.
27. Miller, T.L. and M.J. Wolin. 1979. Fermentations by saccharolytic intestinal bacteria. Am. J. Clin. Nutr. 32:164–172.
28. Rombeau, J.L., S.A. Kripke, and R.G. Settle. 1990. Short chain fatty acids. Production, absorption, metabolism, and intestinal effects, pp. 317–323. *In* D. Kritchevsky, C. Bonfield, and J.W. Anderson (ed.), Dietary Fiber. Plenum Press, New York.
29. Cummings, J.H., E.W. Pomare, W.J. Branch, C.P.E. Naylor, and G.T. Macfarlane. 1987. Short chain fatty acids in human large intestine, portal, hepatic, and venous blood. Gut. 28:1221–1227.
30. Latella, G. 1998. Effects of SCFA on human colonocytes. Gastroenterology Int. 11 (suppl. 1):76–79.
31. Bugaut, M. 1987. Occurrence, absorption and metabolism of short chain fatti acids in the digestive tract of mammals. Comp. Biochem. Physiol. B. 86:439–472.
32. Rechkemmer, G., K. Ronnau, and W. von Engelhardt. 1988. Fermentation of polysaccharides and absorption of short chain fatty acids in the mammalian hindgut. Comp. Biochem. Physiol. A. 90:563–568.
33. Titus, E. and G.A. Ahearn. 1992. Vertebrate gastrointestinal fermentation: transport mechanisms for volatile fatty acids. Am. J. Physiol. 262:R547–553.
34. Charney, A.N., L. Micic, and R. Egnor. 1998. Nonionic diffusion of short-chain fatty acids across rat colon. Am. J. Physiol. 274:G518–G524.
35. Maczulak, A.E., M.J. Wolin, and T.L. Miller. 1993. Amounts of viable anaerobes, methanogens, and bacterial fermentation products in feces of rats fed high-fiber or fiber-free diets. Appl. Environ. Microbiol. 59:657–662.
36. Knudsen, K.R., B.B. Jensen, and I. Hansen. 1993. Oat bran but not a beta-glucan-enriched oat fraction enhances butyrate production in the large intestine of pigs. J. Nutr. 123:1235–1247.
37. Lupton, J.R. and P.P. Kurtz. 1993. Relationship of colonic luminal short-chain fatty acids and pH to in vivo cell proliferation in rats. J. Nutr. 123:1522–1530.
38. Boffa, L.C., J.R. Lupton, M.R. Mariani, M. Ceppi, H.L. Newmark, A. Scalmati, and M. Lipkin. 1992. Modulation of colonic epithelial cell proliferation, histone acetylation, and luminal short chain fatty acids by variation of dietary fiber (wheat bran) in rats. Cancer Res. 52:5906–5912.
39. Nordgaard, I., H. Hove, M.R. Clausen, and P.B. Mortensen. 1996. Colonic production of butyrate in patients with previous colonic cancer during long-term treatment with dietary fibre (Plantago ovata seeds). Scand. J. Gastroenterol. 31:1011–1020.
40. McIntyre, A., G.P. Young, T. Taranto, P.R. Gibson, and P.B. Ward. 1991. Different fibers have different regional effects on luminal contents of rat colon. Gastroenterology. 101:1274–1281.
41. Jacobs, L.R. 1986. Modification of experimental colon carcinogenesis by dietary fiber. Adv. Exp. Med. Biol. 206:105–118.
42. Jacobs, L.R. 1986. Relationship between dietary fiber and cancer: metabolic, physiologic, and cellular mechanisms. Proc. Soc. Exp. Biol. Med. 183:299–310.
43. Jacobs, L.R. and J.R. Lupton. 1986. Relationship between colonic luminal pH, cell proliferation, and colon carcinogenesis in 1,2-dimethylhydrazine treated rats fed high fiber diets. Cancer Res. 46:1727–1734.
44. Young, G.P., A. McIntyre, T. Taranto, P. Ward, and P.R. Gibson. 1991. Butyrate production from dietary fiber protects against large bowel cancer in a rat model. Gastroenterology. 100:A411.
45. Weaver, G.A., J.A. Krause, T.L. Miller, and M.J. Wolin. 1988. Short chain fatty acid distributions of enema samples from a sigmoidoscopy population: an association of high acetate and low butyrate ratios with adenomatous polyps and colon cancer. Gut. 29:1539–1543.

46. Clausen, M.R., H. Bonnen, and P.B. Mortensen. 1991. Colonic fermentation of dietary fibre to short chain fatty acids in patients with adenomatous polyps and colonic cancer. Gut. 32:923–928.
47. McIntyre, A., P.R. Gibson, and G.P. Young. 1993. Butyrate production from dietary fibre and protection against large bowel cancer in rat model. Gut. 34:386–391.
48. D'Argenio, G., V. Cosenza, M. Delle Cave, P. Iovino, N. Della Valle, G. Lombardi, and G. Mazzacca. 1996. Butyrate enemas in experimental colitis and protection against bowel cancer in rat model. Gastroenterology. 110:1727–1734.
49. Medina, V., J.J. Afonso, H. Alvarez-Arguelles, C. Hernandez, and F. Gonzales. 1998. Sodium butyrate inhibits carcinoma development in a 1,2-dimethylhydrazine-induced rat colon cancer. J. Parenter Enteral. Nutr. 22:14–17.
50. Caderni, G., C. Luceri, L. Lancioni, L. Tessitore, and P. Dolora. 1998. Slow-release pellets of sodium butyrate increase apoptosis in the colon of rats treated with azoxymethane, without affecting aberrant crypt foci and colonic proliferation. Nutrition and Cancer. 30:175–181.
51. Scheppach, W., H. Sommer, T. Kirchner, G.M. Paganelli, P. Bartram, S. Christi, F. Richter, G. Dusel, and H. Kasper. 1992. Effect of butyrate enemas on the colonic mucosa in distal ulcerative colitis. Gastroenterology. 103:51–56.
52. Bradburn, D.M., J.C. Mathers, A. Gunn, J. Burn, P.D. Chapman, and I.D.A. Johnston. 1993. Colonic fermentation of complex carbohydrates in patients with familial adenomatous polyposis. Gut. 34:630–636.
53. Chapman, M.A.S., M.F. Grahn, M.A. Boyle, M. Hutton, J. Rogers, and N.S. Williams. 1994. Butyrate oxidation is impaired in colonic mucosa of suffers of quiescent ulcerative colitis. Gut. 35:73–76.
54. Scheppach, W., S.U. Christl, H.P. Bartram, F. Richter, and H. Kasper. 1997. Effects of short-chain fatty acids on the inflamed colonic mucosa. Scand. J. Gastroenterol. (suppl.). 222:53–57.
55. Scheppach, W., J.G. Muller, F. Boxberger, G. Dusel, F. Richter, H.P. Bartram, S.U. Christl, C.E. Dempfle, and H. Kasper. 1997. Histological changes in the colonic mucosa following irrigation with short-chain fatty acids. Eur. J. Gastroenterol. Hepatol. 9:163–168.
56. Mitsuyama, K., T. Saiki, O. Kanauchi, T. Iwanaga, N. Tomiyasu, T. Nishiyama, H. Tateishi, A. Shirachi, M. Ide, A. Suzyki, et al. 1998. Treatment of ulcerative colitis with germinated barley foodstuff feeding: a pilot study. Aliment. Pharmacol. Ther. 12:1225–1230.
57. Fernandez-Banares, F., J. Hinojosa, J.L. Sanchez-Lombrana, E. Navarro, J.F. Martinez-Salmeron, A. Garcia-Puges, F. Gonzalez-Huix, J. Riena, V. Gonzales-Lara, F. Dominguez-Abascal, et al. 1999. Randomized clinical trial of Plantago ovata seeds (dietary fiber) as compared with mesalamine in mantaining remission in ulcerative colitis. Am. J. Gastroenterol. 94:427–433.
58. Steinhart, A.H., T. Hiruki, A. Brzezinski, and J.P. Baker. 1996. Treatment of left-sided ulcerative colitis with butyrate enemas: a controlled trial. Aliment. Pharmacol. Ther. 10:729–736.
59. Velazquez, O.C., H.M. Lederer, and J.L. Rombeau. 1993. Butyrate and the colonocyte. Implication for neoplasia. Dig. Dis. Sci. 41:723–739.
60. Velazquez, O.C. and J.L. Rombeau. 1997. Butyrate. Potential role in colon cancer prevention and treatment. Adv. Exp. Med. Biol. 427:169–181.
61. Velazquez, O.C., H.M. Lederer, and J.L. Rombeau. 1997. Butyrate and the colonocyte. Production, absorption, metabolism, and therapeutic implication. Adv. Exp. Med. Biol. 427:123–134.
62. Barnard, J.A., J.A. Delzell, and N.M. Bulus. 1997. Short chain fatty acid regulation of intestinal gene expression. Adv. Exp. Med. Biol. 422:137–144.
63. Hague, A., A.J. Butt, and C. Paraskeva. 1996. The role of butyrate in human colonic epithelial cells: an energy source or inducer of differentiation and apoptosis. Proc. Nutr. Soc. 55:937–943.
64. Pouillart, P.R. 1998. Role of butyric acid and its derivatives in the treatment of colorectal cancer and hemoglobinopathies. Life Sci. 63:1739–1760.
65. Barnard, J.A. and G. Warwick. 1993. Butyrate rapidly induces growth inhibition and differentiation in HT-29 cells. Cell Growth Differ. 4:495–501.
66. Tanaka, Y., K.K. Bush, T.M. Klauck, and P.J. Higgins. 1989. Enhancement of butyrate-induced differentiation of HT-29 human colon carcinoma cells by 1,25-dihydroxyvitamin D3. Biochem. Pharmacol. 38:3859–3865.
67. Hodin, R.A., S. Meng, S. Archer, and R. Tang. 1996. Cellular growth state differentially regulates enterocyte gene expression in butyrate-treated HT-29 cells. Cell Growth Differ. 7:647–653.
68. Garcia-Bermejo, L., N.E. Vilaboa, C. Perez, A. Galan, E. De Blas, and P. Aller. 1997. Modulation of heat-shock protein 70 (HSP70) gene expression by sodium butyrate in U-937 promonocytic cells: relationship with differentiation and apoptosis. Exp. Cell Res. 236:268–274.

69. Yamamoto, H., J. Fujimoto, E. Okamoto, J. Furuyama, T. Tamaoki, and T. Hashimoto-Tamaoki. 1998. Suppression of growth of hepatocellular carcinoma by sodium butyrate in vitro and in vivo. Int. J. Cancer. 76:897–902.
70. Schroder, C., K. Eckert, and H.R. Maurer. 1998. Tributyrin induces growth inhibitory and differentiating effects on HT-29 colon cancer cells in vitro. Int. J. Oncol. 13:1335–1340.
71. Herz, F., A. Schermer, M. Halwer, and L.H. Bogart. 1981. Alkaline phosphatase in HT-29, a human colon cancer cell line: influence of sodium butyrate and hyperosmolality. Arch. Biochem. Biophys. 210:581–591.
72. Rabizadeh, E., M. Shaklai, A. Nudelman, L. Eisenbach, and A. Rephaeli. 1993. Rapid alteration of c-myc and c-jun expression in leukemic cells induced to differentiate by a butyric acid prodrug. FEBS Lett. 328:225–229.
73. Herutz, D.P., G.W. Zirnstein, J.F. Bradley, and P.G. Rothberg. 1993. Sodium butyrate causes an increase in the block to transcriptional elongation in the c-myc gene in SW837 rectal carcinoma cells. J. Biol. Chem. 268:20466–20472.
74. Sauleimani, A. and C. Asselin. 1993. Regulation of c-myc expression by sodium butyrate in the colon carcinoma cell line Caco-2. FEBS Lett. 326:45–50.
75. Sauleimani, A. and C. Asselin. 1993. Regulation of c-fos expression by sodium butyrate in the human colon carcinoma cell line Caco-2. Biochem. Biophys. Res. Commun. 193:330–336.
76. Czerniak, B., F. Herz, R.P. Wersto, and L.G. Koss. 1987. Modification of Ha-ras oncogene p21 expression and cell cycle progression in the human colonic cancer cell line HT-29. Cancer Res. 47:2826–2830.
77. Li, S., S. Ke, and R.J. Budde. 1996. The C-terminal Src kinase (Csk) is widely expressed, active in HT-29 cells that contain activated Src, and its expression is downregulated in butyrate-treated SW620 cells. Cell Biol. Int. 20:723–729.
78. Heerdt, B.G. and L.H. Augenlicht. 1991. Effect of fatty acids on expression of genes encoding subunits of cytochrome c oxidase and cytochrome c oxidase activity in HT29 human colonic adenocarcinoma cells. J. Biol. Chem. 266:19120–19125.
79. Moore-Hoon, M.L. and R.J. Turner. 1998. Increased expression of the secretory Na+-K+-2Cl-cotrasporter with differentiation of a human intestinal cell line. Biochem. Biophys. Res. Commun. 244:15–19.
80. Dangond, F. and S.R. Gullans. 1998. Differential expression of human histone deacetylase mRNAs in response to immune cell apoptosis induction by trichostatin A and butyrate. Biochem. Biophys. Res. Commun. 247:833–837.
81. Cuisset, L., L. Tichonicky, and M. Delpech. 1998. A protein phosphatase is involved in the inhibition of histone deacetylation by sodium butyrate. Biochem. Biophys. Res. Commun. 29:760–764.
82. Saunders, N., A. Dicker, C. Popa, S. Jones, and A. Dahler. 1999. Histone deacetylase inhibitors as potential antiskin cancer agent. Cancer Res. 59:399–404.
83. Bohm, L., F.A. Achneeweiss, R.N. Sharan, and L.E. Feinendegen. 1997. Influenze of histone acetylation on the modification of cytoplasmic and nuclear proteins by ADP-ribosylation in response to free radicals. Biochim. Biophys. Acta. 1334:149–154.
84. Toscani, A., D.R. Soprano, and K.J. Soprano. 1988. Molecular analysis of sodium butyrate-induced growth arrest. Oncogene Res. 3:223–238.
85. Russo, G.L., V. Della Pietra, C. Mercurio, F. Della Ragione, D.R. Marshak, A. Oliva, and V. Zappia. 1997. Down-regulation of protein kinase CKII activity by sodium butyrate. Biochem. Biophys. Res. Commun. 233:673–677.
86. Huet, C., C. Sahuquillo-Merino, E. Coudrier, and D. Louvard. 1987. Absorptive and mucus-secreting subclones isolated from a multipotent intestinal cell line (HT-29) provide new models for cell polarity and terminal differentiation. J. Cell Biol. 105:345–357.
87. Gum, J.R., W.K. Kam, J.C. Byrd, J.W. Hicks, M.H. Sleisenger, and Y.S. Kim. 1987. Effects of sodium butyrate on human colonic adenocarcinoma cells. J. Biol. Chem. 262:1092–1097.
88. Wice, B.M., G. Trugman, M. Pinto, M. Rousset, G. Chevalier, E. Dussaulx, B. Lacroix, and A. Zweibaum. 1985. The intracellular accumulation of UDP-N-acetylhexosamines is concomitant with the inability of human colon cancer cells to differentiate. J. Biol. Chem. 260:139–146.
89. Siavoshian, S., H.M. Blottiere, C. Cherbut, and J.-P. Galmiche. 1997. Butyrate stimulates cyclin D and p21 and inhibits cyclin-dependent kinase 2 expression in HT-29 colonic epithelial cells. Biochem. Biophys. Res. Commun. 232:169–172.
90. Lallemand, F., D. Courilleau, M. Sabbah, G. Redeuilh, and J. Mester. 1996. Direct inhibition of the expression of cyclin D1 gene by sodium butyrate. Biochem. Biophys. Res. Commun. 229:163–169.

91. Buquet-Fagot, C., F. Lallemand, L.H. Charollais, and J. Mester. 1996. Sodium butyrate inhibits the phosphorylation of the retinoblastoma gene product in mouse fibroblasts by a transcription-ependent mechanism. J. Cell Physiol. 166:631–636.
92. Yen, A. and R. Sturgill. 1998. Hypophosphorylation of the RB protein in S and G2 as well as G1 during growth arrest. Exp. Cell Res. 241:324–331.
93. Kitagawa, M., T. Okabe, H. Ogino, H. Matsumoto, I. Suzuki-Takahashi, T. Kokubo, H. Higashi, S. Saitoh, Y. Taya, H. Yasuda, et al. 1993. Butyrolactone I, a selective inhibitor of cdk2 and cdc2 kinase. Oncogene. 8:2425–2432.
94. Kitagawa, M., H. Higashi, I. Suzuki-Takahashi, T. Okabe, H. Ogino, Y. Taya, S. Nishimura, and A. Okuyama. 1994. A cyclin-dependent kinase inhibitor, butyrolactone I, inhibits phosphorylation of RB protein and cell cycle progression. Oncogene. 9:2549–2557.
95. Nasmyth, K. 1996. Viewpoint: putting the cell cycle in order. Science. 274:1643–1645.
96. King, R.W., R.J. Deshaies, J.-M. Peters, and M.W. Kirshner. 1996. How proteolysis drives the cell cycle. Science. 274:1652–1659.
97. Stillman, B. 1996. Cell cycle control of DNA replication. Science. 274:1659–1664.
98. Elledge, S.J. 1996. Cell cycle checkpoints: preventing an identity crisis. Science. 274:1664–1672.
99. Sherr, C.J. 1996. Cancer cell cycle. Science. 274:1672–1677.
100. Xiong, Y., G.J. Hannon, H. Zhang, D. Casso, R. Kobayashi, and D. Beach. 1993. p21 is a universal inhibitor of cyclin kinases. Nature. 366:701–704.
101. El-Deiry, W.S., T. Tokino, V.E. Velculescu, D.B. Levy, R. Parsons, J.M. Trent, D. Lin, W.E. Mercer, K.W. Kinzler, and B. Vogelstein. 1993. WAF1, a potential mediator of p53 tumor suppression. Cell. 75:817–825.
102. Xiao, H., T. Hasegawa, O. Miyaishi, K. Ohkusu, and K. Isobe. 1997. Sodium butyrate induces NIH3T3 cells to senescence-like state and enhances promoter activity of p21WAF/CIP1 in p53-independent manner. Biochem. Biophys. Res. Commun. 237:457–460.
103. Nakano, K., T. Mizuno, Y. Sowa, T. Orita, T. Yoshino, Y. Okuyama, T. Fujita, N. Ohtani-Fujita, Y. Matsukawa, T. Tokino, et al. 1997. Butyrate activates the WAF1/Cip1 gene promoter through Sp1 sites in a p53-negative human colon cancer cell line. J. Biol. Chem. 272:22199–22206.
104. Evers, B.M., T.C. Ko, J. Li, and E.A. Thompson. 1996. Cell cycle protein suppression and p21 induction in differentiating Caco-2 cells. Am. J. Physiol. 271:G722–G727.
105. Archer, S.Y., S. Meng, A. Shei, and R.A. Hodin. 1998. p21WAF1 is required for butyrate-mediated growth inhibition of human colon cancer cells. Proc. Natl. Acad. Sci. USA. 95:6791–6796.
106. Saini, K.S. and N.I. Walker. 1998. Biochemical and molecular mechanisms regulating apoptosis. Mol. Cell Biochem. 178:9–25.
107. Jacobson, M.D., M. Weil, and M.C. Raff. 1997. Programmed cell death in animal development. Cell. 88:347–354.
108. Nagata, S. 1997. Apoptosis by death factor. Cell. 88:355–365.
109. Evan, G. and T. Littlewood. 1998. A matter of life and cell death. Science. 281:1317–1322.
110. Green, D.R. and J.C. Reed. 1998. Mitochondria and apoptosis. Science. 281:1309–1312.
111. Adams, J.M. and S. Cory. 1998. The Bcl-2 protein family: arbiters of cell survival. Science. 281:13221326.
112. Thornberry, N.A. and Y. Lazebnik. 1998. Caspases: enemies within. Science. 281:1312–1316.
113. White, E. 1996. Life, death, and the pursuit of apoptosis. Genes & Dev. 10:1–15.
114. Hague, A., A.M. Manning, K.A. Hanlon, L.I. Huschtscha, D. Hart, and C. Paraskeva. 1993. Sodium butyrate induces apoptosis in human colonic tumour cell lines in a p53-independent pathway: implications for the possible role of dietary fibre in the prevention of large-bowel cancer. Int. J. Cancer. 55:498–505.
115. McBain, J.A., A. Eastman, C.S. Nobel, and G.C. Mueller. 1997. Apoptotic death in adenocarcinoma cell lines induced by butyrate and other histone deacetylase inhibitors. Biochem. Pharmacol. 53:1357–1368.
116. Marchetti, M.C., G. Migliorati, R. Moraca, C. Riccardi, I. Nicoletti, R. Fabiani, V. Mastrandrea, and G. Morozzi. 1997. Possible mechanisms involved in apoptosis of colon tumor cell lines induced by deoxycholic acid, short-chain fatty acids, and their mixtures. Nutr. Cancer. 28:74–80.
117. Marchetti, M.C., G. Migliorati, R. Moraca, C. Riccardi, I. Nicoletti, R. Fabiani, V. Mastrandrea, and G. Morozzi. 1997. Deoxycholic acid and SCFA-induced apoptosis in the human tumor cell-line HT-29 and possible mechanisms. Cancer Lett. 114:97–99.
118. Boisteau, O., B. Lieubeau, I. Barbieux, S. Cordel, K. Meflah, and M. Gregoire. 1996. Induction of apoptosis, in vitro and in vivo, on colonic tumor cells of the rat after sodium butyrate treatment. Bull. Cancer (Paris). 83:197–204.

119. Chang, S.T. and B.Y. Yung. 1996. Potentiation of sodium butyrate-induced apoptosis by vanadate in human promyelocytic leukemia cell line HL-60. Biochem. Biophys. Res. Commun. 221:594–601.
120. Heerdt, B.G., M.A. Houston, G.M. Anthony, and L.H. Augenlicht. 1998. Mitochondrial membrane potential (delta psi(mt)) in the coordination of p53-independent proliferation and apoptosis pathways in human colonic carcinoma cells. Cancer Res. 58:2869–2875.
121. Filippovich, I., N. Sorokina, K.K. Khanna, and M.F. Lavin. 1994. Butyrate induced apoptosis in lymphoid cells preceded by transient over-expression of HSP70 mRNA. Biochem. Biophys. Res. Commun. 198:257–265.
122. Mandal, M., X. Wu, and R. Kumar. 1997. Bcl-2 deregulation leads to inhibition of sodium butyrate-induced apoptosis in human colorectal carcinoma cells. Carcinogenesis. 18:229–232.
123. Palmer, D.G., C. Paraskeva, and A.C. Williams. 1997. Modulation of p53 expression in cultured colonic adenoma cell lines by naturally occurring lumenal factors butyrate and deoxycholate. Int. J. Cancer. 73:702–706.
124. Janson, W., G. Brandner, and J. Siegel. 1997. Butyrate modulates DNA-damage-induced p53 response by induction of p53-independent differentiation and apoptosis. Oncogene. 15:1395–1406.
125. Hass, R., R. Busche, L. Luciano, E. Reale, and W.V. Engelhardt. 1997. Lack of butyrate is associatyed with induction of Bax and subsequent apoptosis in the proximal colon of Guinea pig. Gastroenterology. 112:875–881.
126. Yamamoto, H., J.W. Soh, H. Shirin, W.Q. Xing, J.T. Lim, Y. Yao, E. Slosberg, N. Tomita, I. Schieren, and I.B. Weinsrein. 1999. Comparative effects of overexpression of p27Kip1 and p21Cip1/Waf1 on growth and differentiation in human colon carcinoma cells. Oncogene. 18:103–115.
127. Ellerhorst, J., T. Nguyen, D.N. Cooper, Y. Estrov, D. Lotan, and R. Lotan. 1999. Induction of differentiation and apoptosis in the prostate cancer cell line LNCaP by sodium butyrate and galectin-1. Int. J. Oncol. 14:225–232.
128. Thompson, M.A., M.A. Rosenthal, S.L. Ellis, A.J. Friend, M.I. Zorbas, R.H. Whitehead, and R.G. Ramsay. 1998. c-Myb down-regulation is associated with human colon cell differentiation, apoptosis, and decreased Bcl-2 expression. Cancer Res. 58:5168–5175.
129. Schwartz, B., C. Avivi-Green, and S. Polak-Charcon. 1998. Sodium butyrate induces retinoblastoma protein dephosphorylation, p16 expression and growth arrest of colon cancer cells. Mol. Cell Biochem. 188:21–30.
130. Hall, P.A., P.J. Coates, B. Ansari, and D. Hopwood. 1994. Regulation of cell number in the mammalian gastrointestinal tract: the importance of apoptosis. J. Cell Sci. 107:3569–3577.
131. Pinna, L.A. 1990. Casein kinase 2: an 'eminence grise' in cellular regulation? Biochim. Biophys. Acta. 1054:267–284.
132. Pinna, L.A. and F. Meggio. 1997. Protein kinase CK2 ("casein kinase-2") and its implication in cell division and proliferation, pp. 77–97. *In* L. Meijer, S. Guidet, and M. Philippe (ed.), Progress in cell cycle research, vol. 3. Plenum Press, New York, USA.
133. Tuazon, P.T. and J.A. Traugh. 1991. Casein kinase I and II—multipotential serine protein kinases: structure function and regulation. Adv. Second Messenger Phosphoprot. Res. 23:123–163.
134. Issinger, O.-G. 1993. Casein kinases: pleiotropic mediators of cellular regulation. Pharmac. Ther. 59:1–30.
135. Allende, J.E. and C.C. Allende. 1995. Protein kinase CK2: an enzyme with multiple substrates and a puzzling regulation. FASEB J. 9:313–323.
136. Litchfield, D.W., F.J. Lozeman, M.F. Cicirelli, M. Harrylock, L.H. Ericsson, C.J. Piening, and E.G. Krebs. 1991. Phoshorylation of the b subunit of casein kinase II in human A431 cells. J. Biol. Chem. 266:20380–20389.
137. Litchfield, D.W., D.G. Bosc, and E. Slominski. 1995. The protein kinase from mitotic cells that phosphorylates Ser-209 on the casein kinase II beta-subunit is p34cdc2. Biochim. Biophys. Acta. 1269:69–78.
138. Draetta, G. and D. Beach. 1988. Activation of cdc2 protein kinase during mitosis in human cells: cell cycle-dependent phosphorylation and subunit rearrangement. Cell. 54:17–26.
139. Russo, G.L., M.T. Vandenberg, I.J. Yu, Y.-S. Bae, B.R.J. Franza, and D.R. Marshak. 1992. Casein kinase II phosphorylates p34cdc2 kinase in G1 phase of the HeLa cell division cycle. J. Biol. Chem. 267:20317–20325.
140. Marshak, D.R. and G.L. Russo. 1994. Regulation of protein kinase CKII during the cell division cycle. Cell Mol. Biol. Res. 40:513–517.
141. Glover 3rd, C.V. 1998. On the physiological role of casein kinase II in Saccharomyces cerevisiae. Prog. Nucleic. Acid Res. Mol. Biol. 59:95–133.

142. Litchfield, D.W., G. Dobrowolska, and E.G. Krebs. 1994. Regulation of casein kinase II by growth factors—a reevaluation. Cell Mol. Biol. Res. 40:373–381.
143. Guerra, B., C. Gotz, P. Wagner, M. Montenarh, and O.G. Issinger. 1997. The carboxy terminus of p53 mimics the polylysine effect of protein kinase CK2-catalyzed MDM2 phosphorylation. Oncogene. 14:2683–2688.
144. Appel, K., P. Wagner, B. Boldyreff, O.G. Issinger, and M. Montenarh. 1995. Mapping of the interaction sites of the growth suppressor protein p53 with the regulatory beta-subunit of protein kinase CK2. Oncogene. 11:1971–1978.
145. Prowald, A., N. Schuster, and M. Montenarh. 1997. Regulation of the DNA binding of p53 by its interaction with protein kinase CK2. FEBS Lett. 408(1):99–104.
146. Gotz, C., P. Wagner, O.G. Issinger, and M. Montenarh. 1996. p21WAF1/CIP1 interacts with protein kinase CK2. Oncogene. 13:391–398.
147. Ashkenazi, A. and V.M. Dixit. 1998. Death receptors: signaling and modulation. Science. 281:1305–1308.
148. Carducci, M.A., J.B. Nelson, K.M. Chan-Tack, S.R. Ayyagari, W.H. Sweatt, P.A. Campbell, W.G. Nelson, and J.W. Simons. 1996. Phenylbutyrate induces apoptosis in human prostate cancer and is more potent than phenylacetate. Clin. Cancer Res. 2:379–387.
149. Siu, L.L., D.D. Von Hoff, A. Rephaeli, E. Izbicka, C. Cerna, L. Gomez, E.K. Rowinsky, and S.G. Eckhardt. 1998. Activity of pivaloyloxymethyl butyrate, a novel anticancer agent, on primary human tumor colony-forming units. Invest. New Drugs. 16:113–119.
150. Hague, A., B. Singh, and C. Paraskeva. 1997. Butyrate acts as a survival factor for colonic epithelial cells: further fuel for the in vivo versus in vitro debate. Gastroenterology. 112:1036–1039.
151. Calabresse, C., L. Venturini, G. Ronco, P. Villa, L. Degos, D. Belpomme, and C. Chomienne. 1994. Selective induction of apoptosis in myeloid leukemic cell lines by monoacetone glucose-3 butyrate. Biochem. Biophys. Res. Commun. 201:266–283.

13

SHORT-CHAIN FATTY ACID IN THE HUMAN COLON

Relation to Inflammatory Bowel Diseases and Colon Cancer

Giuseppe D'Argenio and Gabriele Mazzacca

Gastrointestinal Unit, School of Medicine
Federico II University, Naples Italy

ABSTRACT

Short chain fatty acids (SCFAs) are the end products of anaerobic bacteria break down of carbohydrates in the large bowel. This process, namely fermentation, is an important function of the large bowel; SCFAs, mainly acetate, propionate and butyrate account for approximately 80% of the colonic anion concentration and are produced in nearly constant molar ratio 60:25:15. Among their various properties, SCFAs are readily absorbed by intestinal mucosa, are relatively high in caloric content, are metabolized by colonocytes and epatocytes, stimulate sodium and water absorption in the colon and are trophic to the intestinal mucosa. While the fermentative production of SCFAs has been acknowledged as a principal mechanism of intestinal digestion in ruminants, the interest in the effects of SCFAs production on the human organism has been raising in the last ten years. SCFAs are of major importance in understanding the physiological function of dietary fibers and their possible role in intestinal neoplasia. SCFAs production and absorption are closely related to the nourishment of colonic mucosa, its production from dietary carbohydrates is a mechanism whereby considerable amounts of calories can be produced in short-bowel patients with remaining colonic function and kept on an appropriate dietary regimen. SCFAs enemas or oral probiotics are a new and promising treatment for ulcerative colitis. The effects have been attributed to the oxidation of SCFAs in the colonocytes and to the ability of butyrate to induce enzymes (i.e. transglutaminase) promoting mucosal restitution. Evidence is mounting regarding the effects of butyrate on various cell functions the significance of which needs further considerations. Up until now, attention has been related especially to cancer prophylaxis and treatment. This article briefly reviews the role of

Advances in Nutrition and Cancer 2, edited by Zappia *et al.*
Kluwer Academic / Plenum Publishers, New York, 1999.

SCFAs, particularly butyrate, in intestinal mucosal growth and potential clinical applications in inflammatory and neoplastic processes of the large bowel.

INTRODUCTION

Short chain fatty acids (SCFA) are produced in the mammalian intestinal tract by anaerobic bacterial fermentation of carbohydrate. In man, the production of SCFA in healthy individuals contribute to the energy needs of the body, although much less than in ruminants. In these animals, fermentation occurs primarily in the forestomach as well as in the large intestine,[1] whereas in nonruminants the main fermentation sites are the cecum and the colon. The difference in energy provided to the body by SCFA in ruminant and nonruminant (70–80% vs 5–10%) explains why the role of SCFA as nutrients has previously been neglected in humans; however recent reports have demonstrated that the colonic digestion of carbohydrates is an important energy source in patients with short bowel and severe malabsorption.[2–4] Besides being a source of calories, SCFA provide energy-yielding substrates for colonocytes,[5,6] influence intestinal mucosal growth,[7] increase colonic blood flow,[8] and stimulate sodium and water absorption in the colon.[9] In the last decade, it has become evident that SCFA play a more important role in the colonic mucosa welfare than previously suspected, and several studies have now focused on the possible role of SCFA in colonic diseases such as diversion colitis,[10] ulcerative colitis,[11–12] pouchitis,[13–14] and colonic neoplasia.[13,15,16] It is now known that fiber rich diets have a number of effects on intestinal physiology and on the large bowel luminal environment, by which they mediate a protective effect. Fibers increase the intestinal transit rate[17,18]modify the intestinal microflora altering carcinogen metabolism,[19] absorb carcinogens and mutagens,[20] modify fecal bile salt excretion,[21] lower the colonic pH, and increase colonic and fecal SCFA concentration.[22] One characteristic property of fibers is fermentability. Fermentation, the process whereby anaerobic bacteria break down carbohydrates to obtain energy for growth and maintenance of cellular function, is an important component of normal large bowel activity. Hydrogen, carbon dioxide, methane, lactate and SCFA, predominantly acetate, propionate and butyrate, are the principal end products of microbial carbohydrate fermentation in man and in all animal species. One product of fermentation, butyrate, is particularly interesting as it has been found to induce transglutaminase activity stimulating mucosal repair in rat induced colitis[23] and to slow proliferation and promote expression of phenotypic markers of differentiation in several large bowel cancer cell lines.[24,25] In vivo studies have demonstrated that a high fiber diet, which is associated with a decreased risk of colorectal cancer, leads to a high colonic production of SCFA.[26] Among dietrary fibers, only those producing high butyrate concentrations in the distal colon are protective against colon cancer.[27] What has been apparent is that highly fermentable fibers are less protective than less fermentable fibers. These studies suggest that it may be possible to select a dietary fiber source that stimulates the production of individual SCFA, i.e. butyrate. An increased production of butyrate in the colon may be beneficial, as it provides energy to colonocytes, regulates differentiation of cells, and inhibits inflammation and tumor growth.

Short Chain Fatty Acids and Colonic Inflammation

In the early 1980's, Roediger showed that SCFAs are a luminal nutrient source for the colonocyte.[28] More recently Breuer et al, demonstrated the efficacy of rectal irrigation with SCFAs in patients with distal ulcerative colitis;[11] An SCFA mixture was also successfully used in the treatment of diversion colitis.[10] Among SCFAs, sodium butyrate is the preferred oxidative fuel for the colonocytes.[29] Once inside the cells, SCFAs are metabolized via a β-oxidation pathway requiring the presence of coenzyme A (CoA) as an essential cofactor and produce energy entering the Krebs cycle. Low CoA levels have been found in human UC,[30] reducing SCFAs activation and oxidation to CO_2 and ketone bodies. On these bases, Sheppach et al used sodium butyrate enemas in the treatment of UC.[12] The treatment consists of sodium butyrate 100 mM b.i.d., for 2 weeks enemas in patients with active distal UC in a single-blind cross-over design. Clinical symptoms regressed under butyrate but not under placebo; endoscopic and histologic scores fell significantly only after butyrate irrigation. In a multicentric follow-up study, a SCFAs mixture, butyrate monotherapy or saline placebo were tested in 47 patients receiving enemas b.i.d. for 8 weeks. A disease activity index ameliorated significantly after SCFAs and butyrate, but the statistical significance was not reached. However there was a trend towards more complete responses following SCFAs or butyrate enemas than placebo.[31] Steinart et al,[32] in an open trial, treated patients with refractory ulcerative proctosigmoiditis using 80 mM butyrate, single dose for 6 weeks. Six of 10 patients responded to therapy and 4 of them had a complete response. In a combined treatment of butyrate with 5-Aminosolicylic Acid, histologic improvement was observed in 7 of 9 patients with refractory distal UC. Enemas containing a SCFAs mixture at various concentration were used; Breuer et al[11] demonstrated a beneficial effect of SCFAs enemas administration in 9 of 10 patients with distal colitis. In a large follow-up study, the authors randomized 103 patients with left-sided UC to rectal SCFAs or saline enemas: in a subgroup patients with a short current episode (<6 months) a dramatic response to SCFAs was seen that was significantly different from placebo. In an extension trial for those who did not improve on placebo in the blind trial, 65% improved on SCFAs.[33] Thus, local application of a mixture of SCFAs or butyrate alone seems to ameliorate inflammation and does indeed appear to improve colonic function in patients with ulcerative colitis. Regarding the mechanism of action, the prevaling hypothesis involves an improved energy production in colonocytes from SCFAs because of or despite the suggested inhibited mitocondrial β-oxidation of butyrate, which should be achieved by restoring or increasing luminal concentration of SCFAs, possibly to levels higher than normal. However, the luminal deficiency of SCFAs in patients with ulcerative colitis has not been universally accepted. Although some authors have demonstrated that concentrations were in fact reduced in faeces of patients with active ulcerative colitis, other have found that levels were normal, or even increased. Therefore, the mechanism behind the clinical efficacy of SCFAs and butyrate enemas is at present not clarified, and may not even be related to the oxidative and energy creating process of SCFAs metabolism.

In recent years we observed that serum and tissue transglutaminase (TG) activity correlates with the severity of inflammation in trinitrobenzensulfonic acid (TNB)-induced colitis in the rat.[34] It is also known that butyrate stimulate TG activity in several types of cultured cells.[35,36] These findings prompted us to determine whether irrigation of the colon with sodium butyrate, mesalamine or both, could affect the histology and tissue TG activity in a model of chronic colitis induced in the rat. Rats were treated twice a day for 2 weeks with intraluminal administration of 40 mmol/l sodium butyrate solution (pH 7.0), 25 mg/kg mesalamine or both. Four paramenters, each scored on a 0–5 scale of severity, were considered: extent of

Table 1. Criteria for histological assessment of colonic damage

	Score
Extention of ulceration	
No ulcer	0
Small ulcer (<3 cm)	1–2
Large ulcer (>3 cm)	3–5
Submucosal infiltration	
None	0
Mild	1
Moderate	2–3
Severe	4–5
Crypt abscesses	
None	0
Rare	1–2
Diffuse	3–5
Wall thickness (μm)	
<470 (normal)	0
<600	1
<700	2
<800	3
<900	4
>900	5

ulceration, submucosal infiltration, crypt abscesses, and wall thickness (Table 1). The sum of assigned scores determined a rating of slight (score 1–5), moderate (score 6–10), or severe (score 11–20) colonic inflammation. Following the period of treatment, we found TG activity higher than in mesalamine and saline control groups in the colon of rats given sodium butyrate and sodium butyrate plus mesalamine (Table 2). Furthermore, the rats in these two groups showed a slight histological activity, which instead was moderate in the mesalamine and severe in the saline group (Fig. 1). On the basis of these results we hypothesized a role for colon TG in ulcer healing; it is likely that the enzyme, increased by sodium butyrate treatment, migrates to the extracellular matrix and provides a protein network by promoing fibronectin cross-linking, which is important for re-epithelization and increased tensile strength of the damaged tissues. This study shows that SCFAs, namely sodium butyrate, affect ulcerative colitis by different mechanisms one of whom involves the complex interaction among various molecules in tissue remodeling and indicates an interesting field for future research.

Table 2. Effects of mesalamine, sodium butyrate, and mesalamine plus sodium butyrate treatment on transglutaminase and on histological activity in colon of rats with TNB chronic colitis

	Normal	Mesalamine	Sodium butyrate	Mesalamine plus sodium butyrate	Saline
Transglutaminase (mU/g)	1800 + 192*	658 + 152	1390 + 228*	1226 + 172*	783 + 157
Histological activity	0*	6.2 + 1.4*	2.5 + 1.3*	2.3 + 1.1*	13.7 + 1.7

*p < 0.01 vs saline.

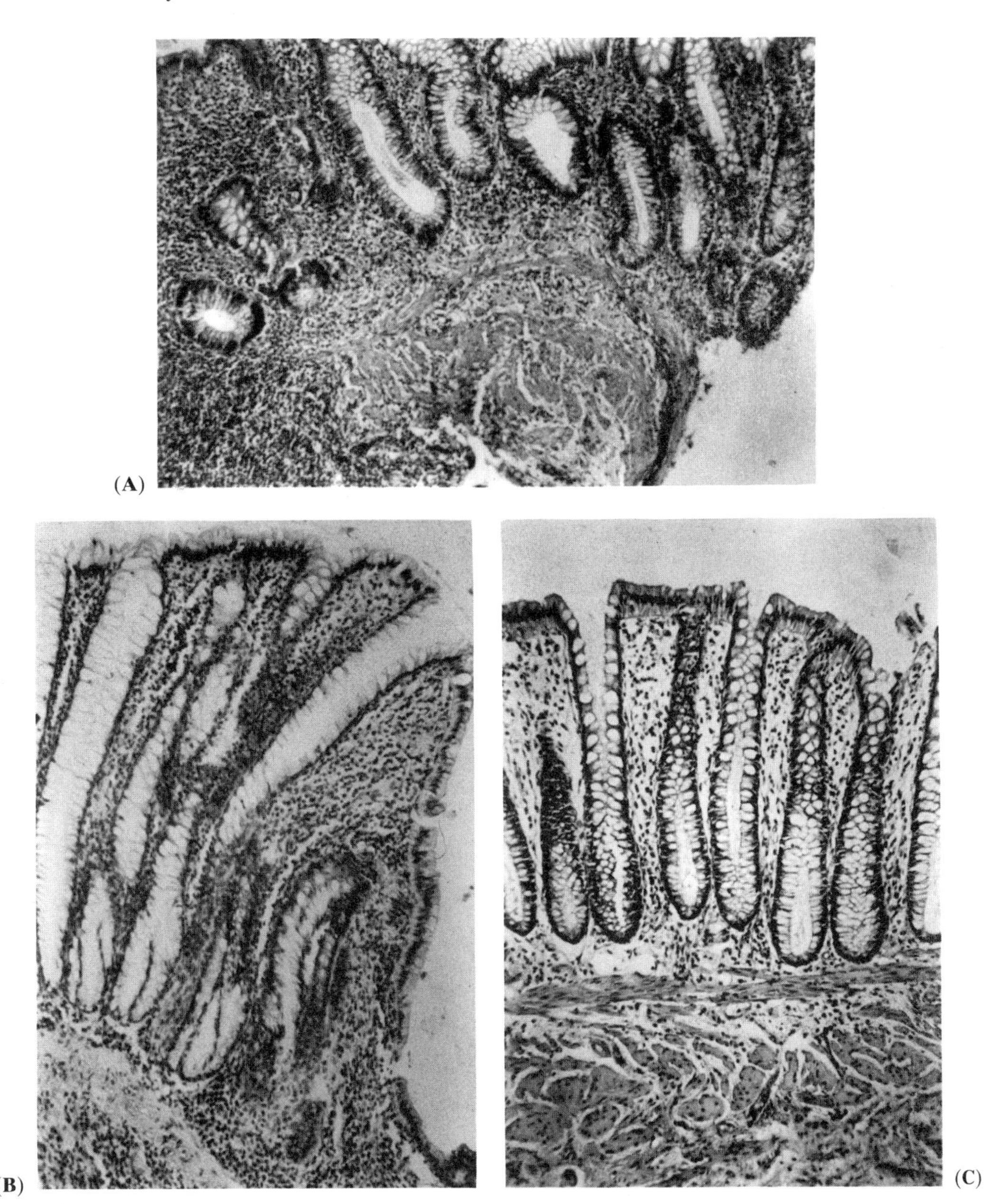

Figure 1. Histological picture of severe colitis after TNB administration in the rat (A). The above colitis improved to moderate after treatment with mesalamine (B) and was slight after butyrate enemas (C). (original magnification X39).

SHORT CHAIN FATTY ACIDS AND COLON CANCER

Diet is suspected to have a considerable influence on large bowel cancer risk. Attempts to classify components of the diet that inhibit or enhance colon carcinogenesis have been made; among these components, dietary fibers have been considered to inhibit carcinogenesis, although the mechanisms by which fibers may protect against colon cancer development are not fully clarified. Originally, fibers were seen simply as providing bulk to dilute potential carcinogens and speed their transit through the colon; recent data support the idea that many of the beneficial effects of dietary fibers may be due to their fermentation to SCFAs. Among these, sodium butyrate is known as an effective inducer of cell differentiation[37,38] and a more differentiated cell state is consistent with a marked effect of butyrate on human colorectal cell lines.[39] The protective effect of dietary fibers has thus been associated with their production of butyrate into the colonic contents, which possibly decreases the risk of neoplasia in colonocytes. A suggested mechanism is the regulation of acetylation of histones which is important in gene regulation. Evidence is mounting regarding different effects of butyrate on colonic and non colonic cells, but how each of these effects is connected to function in the cell is at present not clear. McIntyre et al. fed three different fibers to rats with dimethylhydrazine induced large-bowel cancer and demonstrated that only fibers producing butyrate in the distal colon affect tumor mass.[27] Results indicated also that fibers associated with high butyrate concentrations in the colon were protective against cancer, in contrast to fiber not increasing butyrate levels. Fecal levels of butyrate in patients with colonic neoplasia have been investigated by different studies. Vernia et al.[40] compared 20 patients with colorectal cancer, 8 patients colon polyps, and 32 healthy controls. No significant differences were found, although patients with rectal cancer showed slightly lower levels of propionate and butyrate than those with more proximal cancer. SCFAs distributions in enema samples from a sigmoidoscopy population has been screened and a significantly lower ratio of butyrate to total SCFAs was found. The fecal concentration of butyrate has been found lower in patients with previous colonic adenomas compared with controls, but this difference was not observed when the patients were placed on a balanced diet. The investigation of fecal concentration and in vitro production of SCFAs in patients with familial adenomatous polyposis with and without polyps at the time of examination have shown normal concentration and ratios of SCFAs, including butyrate. Clausen et al.[15] showed that patients with polyps produced less butyrate than healthy controls and patients without polyps.

To date there are no simple answers to the question concerning which type of fiber to be considered optimal in minimizing the risk of colorectal cancer, but different physiological responses to different sources of dietary fiber indicate that data are required not only in the actual amount of fiber present, but also concerning the properties of fibre sources, such as fermentability and production of SCFAs, particularly butyrate. Our recent experience relates to the protective effect of butyrate against large bowel cancer in experimental colitis suggesting its usefulness in long-term therapy to decrease colon cancer risk in ulcerative colitis.[41] There is a strong link between colonic proliferative rate and colon cancer risk. In humans and animals, expansion of the proliferative compartment and increased proliferative rate seem to occur in the histologically normal-appearing colonic mucosa that is at increased risk for the subsequent development of colon cancer.[42,43] In this study, we investigated whether the TNB model of chronic colitis affects the development

Table 3. Characteristics of intestinal tumors (number and surface area) after azoxymethane exposure in rats with (Group A) or without (Group B) trinitrobenzensulphonic acid induced colitis

	Small Bowel	Proximal Colon	Distal Colon (last 8 cm)	
Rats	N. of tumors	N. of tumors	N. of tumors	Surface area (mm^2)
Group A (TNB + AOM)	1.16 ± 0.98	0.34 ± 0.51	2.16 ± 0.40[a]	635 ± 417[a]
Group B (AOM)	1.20 ± 0.44	0.50 ± 1.00	1.00 ± 0.70	164 ± 167

[a] $p < 0.05$.

Table 4. Characteristics of intestinal tumors (number and surface area) after azoxymethane exposure in rats with trinitrobenzensulphonic acid induced colitis. Effects of butyrate or mesalamine treatments

	Small Bowel	Proximal Colon	Distal Colon (last 8 cm)	
Treatment	N. of tumors	N. of tumors	N. of tumors	Surface area (mm^2)
Butyrate	0.66 ± 0.51	0.71 ± 0.48	0.83 ± 1.17[a]	47 ± 70[a]
Mesalamine	1.50 ± 1.07	0.57 ± 0.53	2.62 ± 0.82	452 ± 426
Saline	1.20 ± 0.89	0.45 ± 0.31	2.22 ± 0.51	598 ± 388

[a] $p < 0.05$ vs Saline.

of intestinal tumors produced by azoxymethane in rats and whether butyrate treatment of colitis reduces mucosal sensitivity to colon cancer development. The effects of colitis on tumor development are reported in Table 3 while the effects of butyrate or mesalamine treatments of colitis are summarized in Table 4. Accordingly, proliferative indices are increased by colitis induction and restored by butyrate treatment (Table 5). The study shows that experimental colitis induced by TNB enhances colonic neoplasia in rats exposed to azoximethane, and that this model may be useful in the assessment of potentially chemopreventive agents in colonic precencerous conditions.

Table 5. Proliferative indices (mean ± SD) of colonic epithelial cells in normal rats, and 27 weeks after the first azoxymethane injection in rats with trinitrobenzensulphonic acid induced colitis. Effects of butyrate or mesalamine treatments

	Normal rats	Group A (saline)	Group B (mesalamine)	Group C (butyrate)
Number of rats	10	10	10	10
Columns per rat (mean)	36.9	37.2	31.8	30.1
Total cells per column	35.8 ± 1.6	35.0 ± 1.1	40.2 ± 1.9[a]	37.5 ± 1.8
Labeled cells per column	3.1 ± 0.6	6.0 ± 1.5[a]	6.5 ± 1.7[a]	5.6 ± 1.1[a]
Whole Labeling Index, Labeled/total cells %	8.2 ± 1.2	16.8 ± 1.8[a]	16.3 ± 2.8[a]	14.9 ± 1.9[a]
Labeling Index by crypt compartment				
Compartment 1	16.4 ± 2.3	24.8 ± 2.7[a]	26.8 ± 3.9[a]	30.2 ± 4.3[a]
Compartment 2	14.2 ± 1.9	27.2 ± 3.3[a]	33.4 ± 3.5[a]	28.5 ± 2.6[a]
Compartment 3	9.1 ± 1.4	22.6 ± 3.1[a]	15.7 ± 1.8[a]	11.9 ± 2.0[b]
Compartment 4 ± 5	0.6 ± 0.08	4.4 ± 0.7[a]	2.7 ± 0.5[a, b]	1.6 ± 0.7[a, b]

[a] $p < 0.01$, significantly different from normal rats; [b] $p < 0.01$, significantly different from Group A according to one way ANOVA ($p = 0.0001$) followed by Newman Keuls' test.

The study was divided into two parts, the former showing that, in the distal colon of animals with colitis, azoxymethane produced a greater number of tumors. These tumors were also larger, and developed where the TNB mucosal damage had been induced. In the second part of the study, we demonstrated that treatment of colitis with butyrate reduced the incidence of tumors in the distal colon; moreover, their surface area was approximately 10 times lower than that of tumors developing in saline controls.

Expansion of cell proliferation in the epithelial lining has been found in animal models and in familial polyposis,[44] sporadic adenomas,[45] and ulcerative colitis.[46] This altered proliferative pattern, with a shift of proliferating cells in the upper crypt compartment, has been indicated as a marker of increased cancer risk.[12] In our studies, autoradiographic evaluations showed that azoxymethane increases proliferation in the upper compartments of the crypts; moreover, the association of azoxymethane with colitis leads to a further increase of the upper crypts labelling index. Treatment of colitis with butyrate reduced epithelial cell hyperproliferation, improves the induced colitis and induces transgluatminase activity.[36] These effects indicate that the reduced susceptibility of colonic mucosa to carcinogens in colitis caused by butyrate may depend on multifactorial effects: first, the trophic effect of butyrate on altered mucosa; second, the ability of butyrate in modifying cell kinetics, restoring the physiological proliferative pattern; and third, the induction of transglutaminase activity that is involved in apoptosis and that catalyzes reactions between extracellular matrix proteins to each other, promoting tissue restitution and remodeling. Our data raises the possibility that butyrate, by influencing colonic homeostasis, may represent a therapeutic tool useful in long-term therapy to decrease disease relapses and to prevent colon cancer onset in ulcerative colitis.

SCFAs DELIVERY

As the therapeutic role of SCFAs in colorectal diseases becomes more defined, modes of delivery of the drug to the colon need to be improved. Oral SCFAs are absorbed in the small intestine, minimizing the effects in the colon. One method of delivery would be in the form of fibers, but different types of fiber led to different amounts and quality of SCFAs. Soluble fibers are almost 100% fermented in the colon, whereas lignin and cellulose are only partially fermented in the colon. Variations in the type of fiber may result in differences in the ratio of SCFAs produced. Another way of delivery would involve lactulose or similar compounds which are not absorbed in the small intestine and are fermented in the colon. The use of gastro-resistant capsules could be a suitable route of oral SCFAs administration, with bacterial degradation providing targeted release in the colon similar to sulfasalazine. Finally, the preparations of SCFA enemas can be used for distal colonic diseases. SCFAs have multiple effects on the colon, from electrolytes and water absorption to cell growth and differentiation; in spite of several reports on the therapeutic benefits of SCFA in a variety of colonic inflammatory diseases, a final, convincing demonstration of the efficacy of SCFAs in inflammatory bowel diseases as well as in colon cancer development and/or prevention has not been obtained. The efforts made in recent years to understand the basic physiology of SCFAs should perhaps be made in the near future to better clarify the clinical role of SCFAs.

REFERENCES

1. Czerkawski J.W. Energetics of rumen fermentation. In: An introduction to rumen studies. Pergamon Press, Oxford, 1986, pp. 85–101.
2. Nordgaard I., Hansen B.S., and Mortensen P.B. Colon as a digestive organ in patients with short bowel. Lancet 1994;343:373–376.
3. Nordgaard I. and Mortensen P.B. The digestive process in the human colon. Nutrition 1995;11:37–45.
4. Cummings J.H. and Branch W.J. Fermentation and the production of short chain fatty acids in the human large intestine. In Vahouny G.B. and Kritchevsky D., eds. Dietary fiber: basic and clinical aspect. New York: Plenum, 1986:131–152.
5. Clausen M.R. and Mortensen P.B. Kinetic studies on the metabolism of short-chain fatty acids and glucose by isolated rat colonocytes. Gastroenterology 1994;106:423–432.
6. Ruppin H., Bar-Mier S., Soergel K.H., Wood C.M., and Schmitt M.G. Absorption of short-chain fatty acids by the colon. Gastroenterology 1980;78:1500–1507.
7. Sheppach W. Effects of short chain fatty acids on gut morphology and function. Gut 1994;35(suppl. 1):S35–38.
8. Mortensen F.V., Nielsen H., Mulvany M.J., and Hessov I. Short chain fatty acids dilate isolated human colonic resistence arteries. Gut 1990;31:1391–1394.
9. Binder J.H. and Metha P. Short chain fatty acids stimulate active sodium and chloride absorption in vitro in the rat distal colon. Gastroenterology 1989;96:989–996.
10. Harig J.M., Soergel K.H., Komorowski R.A., and Wood C.M. Treatment of diversion colitis with short chain fatty acids irrigation. N Engl J Med 1989;320:23–28.
11. Breuer R.I., Buto S.K., Christ M.L. et al. Rectal irrigation with short-chain fatty acids for distal ulcerative colitis: preliminary report. Dig Dis Sci 1991;36:185–187.
12. Sheppech W., Sommer H., Kirchner T. et al. Effect of butyrate enemas on the colonic mucosa in distal ulcerative colitis. Gastroenterology 1992;103:51–56.
13. Clausen M.R., Tvede M., and Mortensen P.B. Short chain fatty acids in pouch contents from patients with and without pouchitis after ileal pouch-anal anastomosis. Gastroenterology 1992;103:1144–1153.
14. Nordgaard—Andersen I., Clausen M.R., and Mortensen P.B. Short-chain fatty acids, lactate and ammonia in ileorectal and ileal pouch contents. A model of cecal fermentation. J Parent Ent Nutr 1993;34:324–331.
15. Clausen M.R., Bonnen H., and Mortensen P.B. Colonic fernebtation of dietary fiber to short chain fatty acids in patients with adenomatous polyps and colonic cancer. Gut 1991;32:923–928.
16. Hove H., Clausen M.R., and Mortensen P.B. Lactate and pH in faeces from patients with colonic adenomas or cancer. Gut 1993;34:625–629.
17. Burkitt D.P. Epidemiology of cancer of the colon and rectum. Cancer 1971;28:3–13.
18. Cummings J.H. Constipation, dietary fibre, and the control of the large bowel function. Postgrad Med J 1984;60:811–819.
19. Jacobs L.R. Relationship between dietary fibre and cancer: metabolic, physiologic, and cellular mechanisms. Proc Soc Exp Biol Med. 1986;183:299–310.
20. Smith-Barbaro P., Hanson D., and Reddy B.S. Carcinogen binding to various type of dietary fiber. J Natl Cancer Inst. 1991;67:495–497.
21. Reddy B.S., Watanabe K., and Sheinfil A. Effects of dietary weat bran, α pectin and carrageenan on plasma cholesterol and fecal bile acid and neutral sterol in rats. J Nutr 1980;110:1247–1254.
22. Fleming S.E., Fitch M.D., and Chansler M.W. High-fiber diets: influence on characteristics of cecal digesta including short chain fatty acid concentration and pH. Am J Clin Nutr 1989;50:93–99.
23 D'Argenio G., Cosenza V., Sorrentini I. et al. Butyrate, mesalamine, and Factor XIII in experimental colitis in the rat: effects on transglutaminase activity. Gastroenterology 1994;106:399–404.
24. Whitehead R.H., Young G.P., and Bhathal P.S. Effects of short chain fatty acids on a new human colon carcinoma cell line (LIM 1215). Gut 1986;27:1457–1463.
25. Kim Y.S., Tsao D., and Siddiqui B. Effects of sodium butyrate and dimethylsulfoxide on biochemical properties of human colon cancer cells. Cancer 1980;45:1189–1192.
26. McIntire Young G.P., Taranto T., Gibson P.R., and Ward P.B. Different fibers have different regional effects on luminal contents of rat colon. Gastroenterology 1991;101:1274–1281.
27. McIntire A., Gibson P.R., and Young G.P. Butyrate production from dietary fibers and protection against large bowel cancer in a rat model. Gut 1993;34:386–391.

28. Roediger W.E. Role of anaerobic bacteria in the metabolic welfare of the colonic mucose in man. Gut 1980;21:793–798.
29. Roediger W.E. Utilization of nutrients by isolated epithelial cells of the rat colon. Gastroenterology 1982;83:424–429.
30. Ellestad-Sayad J.J., Nelson R.A., Adson M.A., Palmer W.M., and Soule E.H. Pantothenic acid, coenzyme A and human chronic ulcerative colitis and granulomatous colitis. Am J Clin Nutr 1976;29:1333–1338.
31. Sheppach W. For the German-Austrian SCFA Study Group. Are Short-chain fatty acids effective in the local treatment of ulcerative colitis? Gastroenterology 1996;110:A1010.
32. Steinart A.H., Brzezinski A., and Baker J.P. Treatment of refractory ulcerative proctosigmoiditis with butyrate enemas. Am J Gastroenterol 1994;89:179–183.
33. Breuer R.I., Soergel K.H., Lashner B.A., Christ M.L., Hanauer S.B., Vanagunas A. et al. Short chain fatty acid rectal irrigation for left-sided ulcerative colitis: a randomized, placebo controlled trial. gastroenterology 1996;110:A873.
34. D'Argenio G., Sorrentini I., Cosenza V. et al. Serum and tissue transglutaminase correlates with the severity of inflammation in induced colitis in the rat. Scand J Gastroenterol 1982;82:673–679.
35. Shmidt R., Cathelineau C., Cavey M.T. et al. Sodium butyrate selectively antagonizes the inhibitory effect of retinoids on cornified envelope formationin cultured human hepatocytes. J Cell Physiol 1989;140:281–287.
36. Birckbiekler P.J., Orr J.R., Patterson M.K. Jr et al. Enhanced transglutaminase activity in transformed human lung fibroblast cells after exposure to sodium butyrate. Biochim Biophis Acta 1983;723:27–34.
37. Kruh J. Effects of sodium butyrate, a new pharmacological agent, on cells in culture. Mol Cell Biochem 1982;42:65–82.
38. Gum J.R., Kam W.K., Byrd J.C. et al. Effects of sodium butyrate on human colonic adenocarcinoma cells: induction of placental-like alkaline phosphatase. J Biol Chem 1987;262:1092–1097.
39. Hague A., Manning A.M., Harlon K.A.et al. Sodium butyrate induces apoptosis in human colonic tumor cell lines in a p53-independent pathway. Implications for the possible role of dietary fibre in the prevention of large bowel cancer. Int J Cancer 1993;55;498–505.
40. Vernia P. and Cittadini M. Short chain fatty acids and colorectal cancer. Eur J Clin Nutr 1995;49:18–20.
41. D'Argenio G., Cosenza V., Delle Cave M. et al. Butyrate enemas in experimental colitis and protection against large bowel cancer in a rat model.
42. Deschner E.E. and Moshen A.P. Significance of labeling index and labeling distribution on kinetic parameters in colorectal mucosa of cancer patients and DMH-treated animals. Cancer 1982; 50:1136–1141.
43. Eilers G.A. Abnormal pattern of cell proliferation in the entire colonic mucosa of patients with colon adenoma or cancer. Gastroenterology 1987;92:704–708.
44. Lipkin M., Blattner W.A., Gardner E.J. et al. Classification and risk assessment of individuals with familiar polyposis, Gardners syndrome and familial non-polyposis colon cancer from ^{3}H thymidine labeling patterns in colonic epithelial cells. Cancer Res 1984;44:4201–4207.
45. Maskenzs A.P. and Deschner E.E. Tritiated thymidine incorporation into epithelial cells of normal appearing colorectal mucosa of cancer patients. J Natl Cancer Inst 1977;58:1221–1224.
46. Bleiberg H., Maingurt P. Galand P. et al. Cell renewal in the human rectum: in vitro autoradiographic study on active ulcerative colitis. Gastroenterology 1970;58:851–855.

14

BRASSICA VEGETABLES AND CANCER PREVENTION

Epidemiology and Mechanisms

Geert van Poppel, Dorette T. H. Verhoeven, Hans Verhagen, and R. Alexandra Goldbohm

TNO Nutrition and Food Research Institute
Zeist, The Netherlands

ABSTRACT

This paper first gives an overview of the epidemiological data concerning the cancer-preventive effect of brassica vegetables, including cabbages, kale, broccoli, Brussels sprouts, and cauliflower. A protective effect of brassicas against cancer may be plausible due to their relatively high content of glucosinolates. Certain hydrolysis products of glucosinolates have shown anticarcinogenic properties. The results of six cohort studies and 74 case-control studies on the association between brassica consumption and cancer risk are summarized. The cohort studies showed inverse associations between the consumption of brassica's and risk of lung cancer, stomach cancer, all cancers taken together. Of the case-control studies 64% showed an inverse association between consumption of one or more brassica vegetables and risk of cancer at various sites. Although the measured effects might have been distorted by various types of bias, it is concluded that a high consumption of brassica vegetables is associated with a decreased risk of cancer. This association appears to be most consistent for lung, stomach, colon and rectal cancer, and least consistent for prostatic, endometrial and ovarian cancer. It is not yet possible to resolve whether associations are to be attributed to brassica vegetables per se or to vegetables in general. Further epidemiological research should separate the anticarcinogenic effect of brassica vegetables from the effect of vegetables in general.

The mechanisms by which brassica vegetables might decrease the risk of cancer are reviewed in the second part of this paper. Brassicas, including all types of cabbages, broccoli, cauliflower, and Brussels sprouts, may be protective against cancer due to their

Advances in Nutrition and Cancer 2, edited by Zappia *et al.*
Kluwer Academic / Plenum Publishers, New York, 1999.

glucosinolate content. Glucosinolates are usually broken down through hydrolysis catalysed by myrosinase, an enzyme that is released from damaged plant cells. Some of the hydrolysis products, viz. indoles, and isothiocyanates, are able to influence phase 1 and phase 2 biotransformation enzyme activities, thereby possibly influencing several processes related to chemical carcinogenesis, e.g. the metabolism, DNA-binding, and mutagenic activity of promutagens. Most evidence concerning anticarcinogenic effects of glucosinolate hydrolysis products and brassica vegetables has come from studies in animals. In addition, studies carried out in humans using high but still realistic human consumption levels of indoles and brassica vegetables have shown putative positive effects on health. The combination of epidemiological and experimental data provide suggestive evidence for a cancer preventive effect of a high intake of brassica vegetables.

INTRODUCTION

The consumption of vegetables and fruits has always been seen as health-promoting. In bygone days, certain vegetables and fruits were used as medicine, and nowadays vegetables and fruits are supposed to play a protective role in the etiology of various diseases, such as cancer and coronary heart diseases. The protective effect against cancer of the consumption of a wide variety of vegetables and fruits has been examined in many epidemiological studies and has been reviewed by Block et al.[1] and Steinmetz and Potter[2] who concluded that there is a consistent, inverse association between the consumption of a wide variety of vegetables and fruits and the risk of cancer at most sites.

One group of vegetables that was already used for medicinal purposes in ancient times and is nowadays seen as possibly cancer-protective are vegetables of the family *Cruciferae*.[3] The protective effect of cruciferous vegetables against cancer has been suggested to be partly due to their relatively high content of glucosinolates which distinguishes them from other vegetables. Vegetables of the Brassica genus, including cabbages, kale, broccoli, cauliflower, Brussels sprouts, kohlrabi, rape, black, and brown mustard, and root crops such as turnip and rutabaga (swede), contribute most to our intake of glucosinolates (Table 1).[3]

Certain hydrolysis products of glucosinolates, viz. indoles, and isothiocyanates, have shown anticarcinogenic properties.[4-6] The enzyme myrosinase, found in plant cells in a compartment separated from glucosinolates, catalyses the hydrolysis of glucosinolates.[3] When the plant cells are damaged, e.g by cutting or chewing, the myrosinase comes in contact with the glucosinolates and hydrolysis occurs. All glucosinolates share a common basic skeleton, but differ in side chain (R):

$$S - \beta - D\ glucose$$

$$R - C$$

$$N - OSO_3$$

with R = alkyl, alkenyl, alkylthioalkyl, aryl, β–hydroxyalkyl or indolylmethyl. The glucosinolate hydrolysis products consist of equimolar amounts of an aglucon, glucose, and

Table 1. Estimated consumption of some cruciferous vegetables in the Netherlands (source: 11)

	Estimated NL consumption (g/day)	Glucosinolates (mg/kg)
Cabbages	11	400–1200
Brussels sprouts	4	600–2500
Cauliflower	10	100–1200
Broccoli	2	300–700

sulphate. The aglucones are unstable and undergo further reactions to form for instance thiocyanates, nitriles, isothiocyanates or indoles (Table 2). Besides the conditions of the hydrolysis and the presence of any cofactors, the nature of the hydrolysis products depends primarily on the side chain of the glucosinolate. Many experimental studies have shown that indoles and isothiocyanates given to animals after a carcinogen insult reduced tumour incidence and multiplicity e.g.[7–10] A possible inhibitory activity of isothiocyanates and indoles against tumorigenesis appears to stem mainly from their ability to influence phase 1 and phase 2 biotransformation enzyme activities, thereby influencing several processes related to chemical carcinogenesis, such as the metabolism and DNA-binding of carcinogens.[4–6]

This paper reviews both the epidemiological and experimental evidence on brassica vegetables, glucosinolates and their breakdown products in relation to cancer risk. First, the epidemiological evidence will be discussed. Subsequently, the mechanisms by which glucosinolates from brassica's may influence cancer risk and the experimental evidence are reviewed.

Table 2. Some glucosinolates and their hydrolysis products found in brassica vegetables[a]

		Myrosinase hydrolysis products			
Side chain (R) =	Trivial name	N[b]	I[b]	T[b]	Others
2-Butyl	Glucocochlearin	+	+	–	
2-Propenyl	Sinigrin	+	+	–	Epithionitrile[c]
3-Butenyl	Gluconapin	+	+	–	Epithionitrile[c]
4-Pentenyl	Glucobrassicanapin	+	+	–	Epithionitrile[c]
2-Hydroxy-3-butenyl	Progoitrin	+	–	–	L-5-vinyl-oxazolidine-2-thione (goitrin), Epithionitrile[c]
2-Hydroxy-4-pentenyl	Gluconapoleiferin	+	–	–	L-5-allyl-oxazolidine-2-thione, Epithionitrile[c]
3-Methylthiopropyl	Glucoiberverin	+	+	–	
3-Methylsulfinylpropyl	Glucoiberin	+	+	–	
3-Methylsulfonylpropyl	Glucocheirolin	+	+	–	
4-Methylthiobutyl	Glucoerucin	+	+	–	
5-Methylthiopentyl	Glucoberteroin	+	+	–	
Benzyl	Glucotropaeolin	+	+	–	
2-Phenylethyl	Gluconasturtiin	+	+	–	
3-Indolylmethyl	Glucobrassicin	+	–	+	Indoles
1-Methoxy-3-indolylmethyl	Neo-glucobrassicin	+	–	+	Indoles

[a]Reproduced with permission from De Vos and Blijleven (34).
[b]*N* Nitrile, R—C N; *I* isothiocyanate, R—N=C=S; *T* thiocyanate ion, S—C N.
[c]General structure: R—C N with sulphur inserted in the terminal double bond of R.

EPIDEMIOLOGY

Data on epidemiological studies were taken from a recent exhaustive review by Verhoeven et al.[11] This covers all epidemiological studies reviewed by Steinmetz and Potter[2] and by Block et al.[1] and studies, additionally, cited in MEDLINE on CD-ROM up to 1995. The current paper limits itself to an overview of epidemiological data on cancers of the lung, stomach, colon, rectum, prostate, oral cavity, endometrium, and ovarium. Two types of epidemiological studies are reviewed. In *prospective cohort studies*, the diets of large groups of people are recorded, mostly using questionnaires, and these people are subsequently followed up for disease occurrence. Cohort studies normally consist of ten thousands of people with a follow-up of well over ten years. *Retrospective case-control studies* use interviews or questionnaires to estimate dietary patterns in the near or distant past. Data from patients (usually well over 100) are then being compared with data from disease free controls. In nutritional epidemiology, comparisons of disease incidence are made between groups of similar size with different intake levels (high-middle-low = tertiles; four levels = quartiles, five levels = quintiles). A common way to quantify associations of intake levels with disease rate is the Relative Risk (RR); RR of 0.25 implies that the risk of disease at the particular intake level is 25% of the risk in the comparison group (usually the group with the lowest intake level). Multivariate techniques are used to adjust the Relative Risks for other factors known to influence the disease risk.

In Fig. 1, *prospective cohort studies* of brassica consumption and cancer risk are summarized. Five studies report protective effects of the consumption of one or more brassicas, one study shows no association and one study reports a somewhat increased risk. Of all these associations, either positive or inverse, none was statistically significant. The results of the case control studies are given in Table 3. Of the 74 studies given here 46 (62%) found lower cancer risk to be associated with consumption of at least one type of brassica. Protective effects are most consistent for lung, colon, stomach, and rectal cancer, and least convincing for prostate, endometrium, and ovary cancer.

The following should be taken into account when interpreting the results of the studies. In case-control studies the measured effect can be distorted by selection bias, resulting from procedures used to select subjects (both cases and controls). For example, health conscious controls who eat vegetables on a regular basis may be more inclined to participate in studies on cancer than less health consciousness controls who eat less vegetables. Furthermore, in case-control studies dietary information is collected retrospectively from cases. The disease process may have influenced consumption, or knowledge of the disease status may have resulted in recall bias in patients. In prospec-

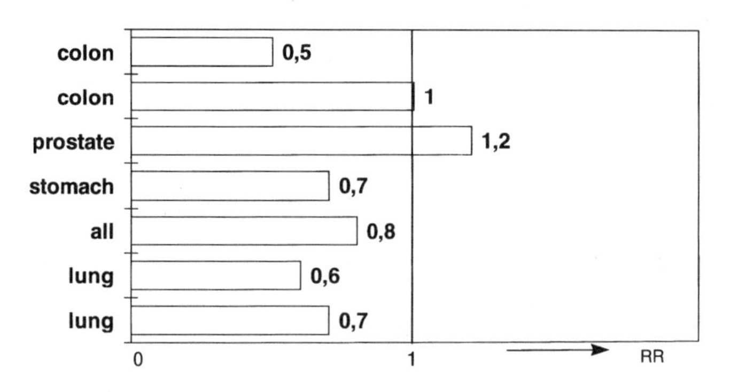

Figure 1. Associations between high brassica consumption and risk of cancer at several sites in prospective cohort studies. (source: 11)

Table 3. Overview of case control studies on brassica consumption and cancer risk (data given are number of studies; decreased associations with at least one type of brassica are reported). (source: 11)

Cancer site	Decreased risk, statistically significant	Decreased risk,[1] nonsignificant	No effect[2]	Increased risk,[3] nonsignificant	Increased risk, statistically significant
Lung	6	3	2	0	0
Stomach	4	3	4	1	1
Colon	8	2	5	3	0
Oral Cavity, Pharynx	5	5	3	1	0
Rectum	5	2	2	1	0
Prostate	0	1	1	1	0
Endometrium	0	1	1	1	0
Ovary	0	1	0	1	0

[1] $RR \leq 0.8$, but not statistically significant.
[2] $0.9 \leq RR \leq 1.1$, not statistically significant.
[3] $RR \leq 1.2$, but not statistically significant.

tive studies, such bias is avoided since dietary intake is measured a number of years prior to diagnosis. The results of both case-control and cohort studies may be influenced by a rather crude assessment of vegetable consumption. This can result in underestimation of the strength of that association between brassica consumption and cancer risk. It should also be realised that the studies mostly did not sort out whether the observed association was attributable to brassica vegetables per se, to vegetables as a whole, or to other factors associated with vegetable consumption. Moreover, the observed protective effects of brassicas may be caused by other compounds than glucosinolates[12]

In general, there seems to be an inverse association between the consumption of brassica vegetables and the risk of cancer. Although case-control studies may carry some risk of bias, the consistent inverse results suggest that (part of) the association is real. This association appears to be most consistent for lung, stomach, colon, and rectal cancer, and least consistent for prostatic, endometrial, and ovarian cancer. The few cohort studies conducted confirm the inverse association, in particular for lung and stomach cancer. However, it is not yet possible to decide whether the protective effect is attributable to brassica vegetables or to vegetables in general.

MECHANISMS

Carcinogenesis is a multistage process; cells susceptible to genetic changes (initiation) and epigenetic changes (promotion) may gain a growth advantage and undergo clonal expansion. Genetic changes are considered to result from interactions between DNA and a carcinogen/mutagen, which can be metabolised into an electrophilic intermediate and bind to DNA. If repair of the damage does not occur, replication of DNA can lead to permanent DNA lesion, and, in presence of a tumour promontor, to preneoplastic cells, neoplastic cells, and finally metastases.[13]

At each stage of the carcinogenic process a possibility of intervention exists. One anticarcinogenic action is the modulation of metabolism of carcinogenic/mutagenic compounds, thereby preventing the formation of electrophilic intermediates. This modulation of metabolism comprises inhibition of activation of promutagens/

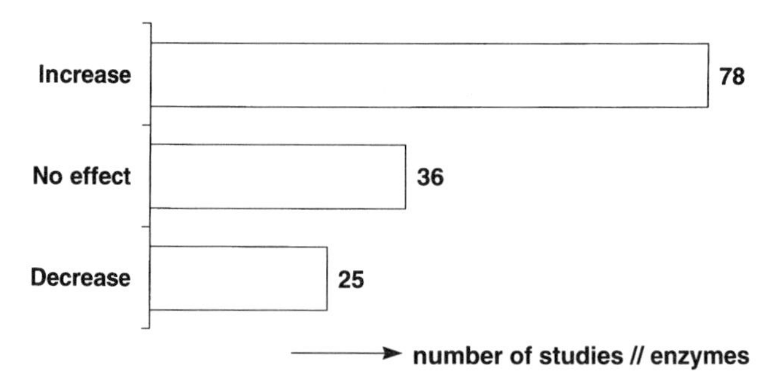

Figure 2. Overview of studies on modulation of Phase I enzymes by isothiocyanates and indoles (*in vivo* and *in vitro*). (source: 17)

procarcinogens, induction of detoxifying mechanisms, and stimulation of activation coordinated with detoxification and blocking of reactive metabolites.[14] Involved in this modulation are phase 1 and phase 2 biotransformation enzymes.[15] Phase 1 involves oxidation, reduction and hydrolysis reactions, thereby making xenobiotics more hydrophilic (which can result in inhibition of activation but also in activation of the compound) as well as susceptible to detoxification. The most important phase 1 enzymes are the cytochrome P450 enzymes. Phase 2 metabolism, a detoxifying mechanism, comprises conjugation reactions making phase 1 metabolites more polar and more readily excretable. Examples of phase 2 enzymes are glutathione S-transferases and UDP-glucuronyl transferases. Alteration of biotransformation enzyme activities is supposed to be involved, at least partly, in the alteration of the toxicity, mutagenicity, and tumorigenicity of specific chemicals.[16] Isothiocyanates and indoles, both glucosinolate hydrolysis products, are considered to be able to modulate biotransformation enzyme activities and could thereby modulate cancer risk. Recently, Verhoeven et al.[17] have reviewed 181 experimental studies that have used isothyocyanates, indoles or brassica vegetables to evaluate their effects on various phase 1 and phase 2 enzymes, mutagenicity, and carcinogenicity in several experimental systems, including animals and humans. Though test systems, experimental conditions, and compounds differ it is clear from this review that isolated indoles and isothyocyanates lead to an increase in Phase 1 enzymes in most but certainly not all experiments (Fig. 2). In vivo animal studies using whole brassica vegetables confirm this picture (Fig. 3). For phase 2 enzymes, decreases are not reported and the reported inductions thus seem be more consistent than for phase 1 enzymes (Figs. 4 and 5).

The overall picture thus seems to be favourable for prevention of DNA damage and cancer, since the induction of phase 1 enzymes can result in inhibition of activation of compounds, but also in bioactivation of compounds. When phase 2 enzymes are

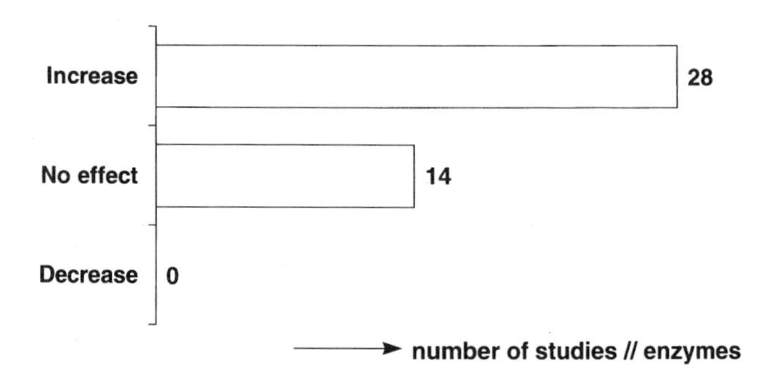

Figure 3. Overview of studies on modulation of Phase I enzymes by brassica vegetables (*in vivo*, mouse and rat). (source: 17)

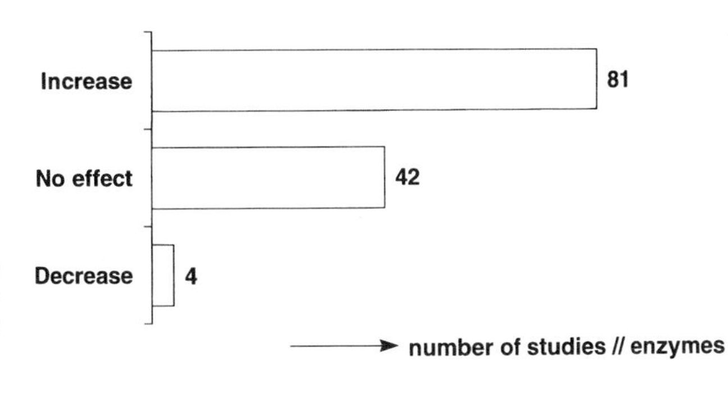

Figure 4. Overview of studies on modulation of Phase II enzymes by isothiocyanates and indoles (*in vivo* and *in vitro*). (source: 17)

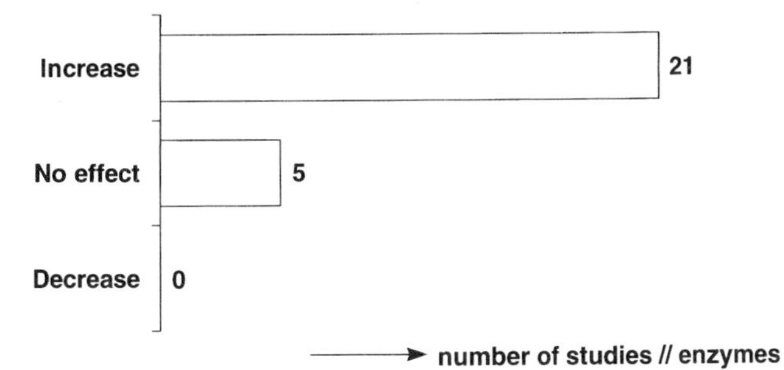

Figure 5. Overview of studies on modulation of Phase II enzymes by brassica vegetables (*in vivo*, mouse and rat). (source: 17)

induced besides phase 1 enzymes, the activated compounds can be detoxified. Jongen et al.[18] have found that modulating effects of indoles on DNA damage are not directly related to induction of phase 1 enzymes, but to a shift in the balance between phase 1 and phase 2 enzymes in favour of phase 2 enzymes. Indeed, a favourable picture arises when the experiments on mutagenicity are viewed (Fig. 6). The shifts in phase 1 and phase 2 enzymes given in Figs. 2–5 thus seem to result in more protection against DNA damage in the test systems used. A favourable picture with regard to cancer prevention also emerges form the data on experimental carcinogenesis where aninals are exposed to known carcinogens (Figs. 7 and 8). For isolated isothiocyanates and indoles, there is a decrease in cancer occurrence in the vast majority, but not in all experiments (Fig. 7). For the effects of whole brassica's on carcinogenesis in animal models, the picture indicates consistent protection (Fig. 7), though the number of experiments is more limited.

Most evidence concerning anticarcinogenic effects of glucosinolate hydrolysis products and brassica vegetables results from *in vivo* studies in animals. In animal

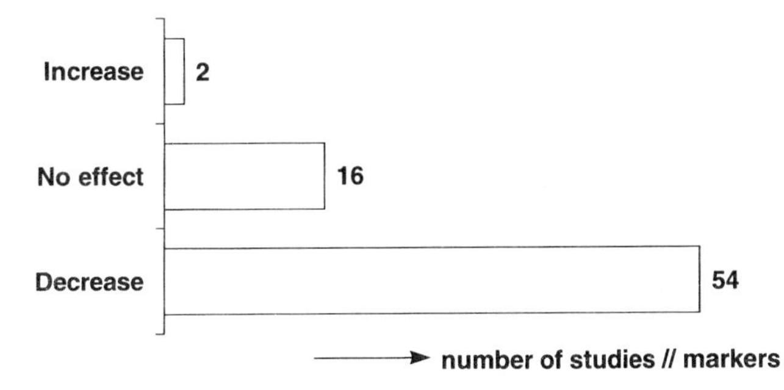

Figure 6. Overview of studies on modulation of mutagenicity of various agents by isothiocyanates and indoles (*in vivo* and *in vitro*). (source: 17)

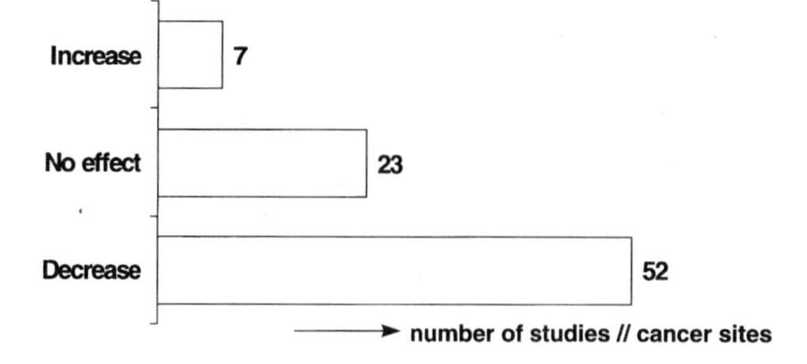

Figure 7. Overview of animal studies on *in vivo* modulation of carcinogenicity of various agents by sothiocyanates and indoles (source: 17)

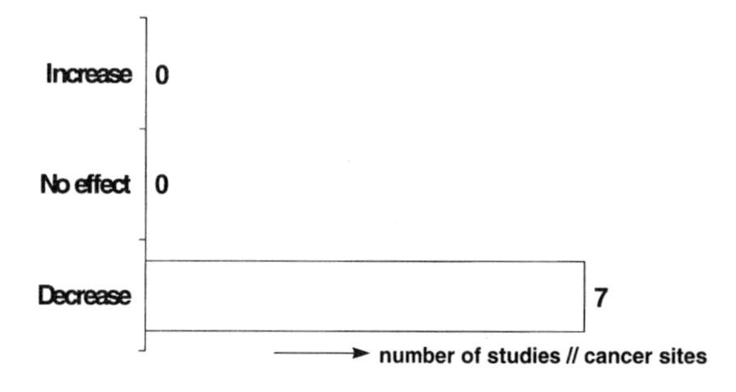

Figure 8. Overview of modulation of carcinogenicity of various agents by brassica vegetables (liver, mammary, respiratory). (source: 17)

studies differences between species are found which makes it difficult to extrapolate the results to humans. The dose of both the anticarcinogenic and the carcinogenic compound used in animal studies is usually much higher than those to which man is exposed via a normal diet. However, isothiocyanates, indoles and brassicas may have beneficial effects at high but realistic consumption levels, in human studies[19–28] (Table 4). Besides beneficial effects, glucosinolates, and their hydrolysis products have also shown some adverse effects in high doses. A few isothiocyanates and indoles showed mutagenic and carcinogenic properties in some studies (Figs. 6 and 7), and very high intake of glucosinolates or their hydrolysis products resulted in goitrogenic effects, growth decrease, and liver enlargement in animals.[29–31] However, no adverse effects were observed when humans were given a diet with a high but realistic content of Brussels sprouts[32,33] and the doses used in the animal studies exceed the normal human daily consumption by far.

Overall, the experimental data in Figs. 2 to 8 support the theory that isolates and isothiocyanates in brassica vegetables induce detoxifying enzymes resulting in increased resistance against DNA damage and, as a consequence, increased resistance against carcinogenesis. The limited data from experiments in humans point in the same direction.

Table 4. Overview of human studies on the effect of indoles and brassicas on biomarkers

	Increase	No effect	Decrease
Phase 1 enzymes	6	—	—
Phase 2 enzymes	4	2	—
DNA damage (80 × odG)	—	—	1

CONCLUSION

There is now much evidence to suggest that brassica vegetables may possess anticarcinogenic properties. Components of brassicas possibly responsible for these properties are glucosinolates that are converted by the plant enzyme myrosinase to isothiocyanates or indoles as the active principles. The combination of epidemiological and experimental data provide evidence for a cancer preventive effect of brassica vegetables.

REFERENCES

1. G. Block, B. Patterson, and A. Subar. Fruit, vegetables, and cancer prevention: a review of the epidemiological evidence. Nutr. Cancer, 18 (1992) 1–29.
2. K.A. Steinmetz and J.D. Potter. Vegetables, fruit, and cancer. I. Epidemiology. Cancer Causes Contr., 2 (1991) 325–357.
3. G.R. Fenwick, R.K. Heany, and W.J. Mullin. Glucosinolates and their breakdown products in food and food plants. CRC Crit. Rev. Food Sci. Nutr., 18 (1983) 123–201.
4. Y. Zhang and P. Talalay. Anticarcinogenic activities of organic isothiocyanates: chemistry and mechanisms. Cancer Res., 54 (1994) 1976s–1981s.
5. C.W. Boone, G.J. Kelloff, and W.E. Malone. Identification of candidate cancer chemopreventive agents and their evaluation in animal models and human clinical trials: a review. Cancer Res., 50 (1990) 2–9.
6. R. McDanell and A.E.M. McLean. Chemical and biological properties of indole glucosinolates (glucobrassicins): a review. Food Chem. Toxicol., 26 (1988) 59–70.
7. M.A. Morse, K.I. Eklind, S.S. Hecht, and F.-L. Chung. Inhibition of tobacco-specific nitrosamine 4-(*N*-nitrosomethylamino)-1-(3-pyridyl)-1-butanone (NNK) tumorigenesis with aromatic isothiocyanates. IARC Sci. Publ., 105 (1991) 529–534.
8. S. Sugie, A. Okumura, T. Tanaka, and H. Mori. Inhibitory effects of benzyl isothiocyanate and benzyl thiocyanate on diethylnitrosamine-induced hepatocarcinogenesis in rats. Jpn. J. Cancer Res., 84 (1993) 865–870.
9. R.H. Dashwood, D.N. Arbogast, A.T. Fong, C. Pereira, J.D. Hendricks, and G.S. Bailey. Quantitative inter-relationships between aflatoxin B1 carcinogen dose, indole-3-carbinol anti-carcinogen dose, target organ DNA adduction and final tumor response. Carcinogenesis, 10 (1989) 175–181.
10. T. Tanaka, Y. Mori, Y. Morishita, A. Hara, T. Ohno, T. Kojima, and H. Mori. Inhibitory effect of sinigrin and indole-3-carbinol on diethylnitrosamine-induced hepatocarcinogenesis in male ACI/N rats. Carcinogenesis, 11 (1990) 1403–1406.
11. D.T.H. Verhoeven, R.A. Goldbohm, G. van Poppel, H. Verhagen, and P.A. van den Brandt. Review: Epidemiological studies on brassica vegetables and cancer risk. Cancer Epidemiology, Biomarkers and Prevention 5 (1996) 733–748.
12. K.A. Steinmetz and J.D. Potter. Vegetables, fruit, and cancer. II. Mechanisms. Cancer Causes Contr., 2 (1991) 427–442.
13. C.C. Harris. Chemical and physical carcinogenesis: advances and perspectives for the 1990s, Cancer Res., 51 (1991) 5023s–5044s.
14. S. de Flora, A. Izzotti, and C. Bennicelli. Mechanisms of antimutagenesis and anticarcinogenesis: role in primary prevention. In: G. Bronzetti, H. Hayatsu, S. de Flora, M.D. Waters, and D.M. Shankel (Eds.), Antimutagenesis and anticarcinogenesis mechanisms III, Plenum Press, New York, 1993, pp. 1–16.
15. W.B. Jakoby. Enzymatic basis of detoxification, Vol I and II, Academic press, London, 1980.
16. Nordic Council of Ministers, Naturally occurring antitumourigens. II Organic isothiocyanates, TemaNord, Copenhagen, 1994.
17. D.T.H. Verhoeven, H. Verhagen, R.A. Goldbohm, P.A. van den Brandt, and G. van Poppel. A review of mechanisms underlying anticarcinogenicity by brassica vegetables. Chem-Biol. Interactions 103 (1997) 79–129.
18. W.M.F. Jongen, R.J. Topp, P.J. van Bladeren, J. Lapre, K.J.H. Wienk, and R. Leenen. Modulating effects of indoles on benzo[*a*]pyrene-induced sister chromatid exchanges and the balance between drug-metabolizing enzymes, Toxic. in Vitro, 3 (1989) 207–213.

19. J.J. Michnovicz and H.L. Bradlow. Induction of estradiol metabolism by dietary indole-3-carbinol in humans, J. Natl Cancer Inst., 82 (1990) 947–949.
20. J.J. Michnovicz and H.L. Bradlow. Altered estrogen metabolism and excretion in humans following consumption of indole-3-carbinol, Nutr. Cancer, 16 (1991) 59–66.
21. H.L. Bradlow, J.J. Michnovicz, M. Halper, D.G. Miller, G.Y.C. Wong, and M.P. Osborne. Long-term responses of women to indole-3-carbinol or a high fiber diet, Cancer Epidemiol., Biomarkers & Prev., 3 (1994) 591–595.
22. H. Verhagen, H.E. Poulsen, S. Loft, G. van Poppel, M.I. Willems, and P.J. van Bladeren. Reduction of oxidative DNA-damage in humans by Brussels sprouts, Carcinogenesis, 16 (1995) 969–970.
23. J.J. Bogaards, H. Verhagen, M.I. Willems, G. van Poppel, and P.J. van Bladeren. Consumption of Brussels sprouts results in elevated α-class glutathione S-transferase levels in human blood plasma, Carcinogenesis, 15 (1994) 1073–1075.
24. E.J. Pantuck, C.B. Pantuck, W.A. Garland, B.H. Min, L.W. Wattenberg, K.E. Anderson, A. Kappas, and A.H. Conney. Stimulatory effect of Brussels sprouts and cabbage on human drug metabolism, Clin. Pharmacol. Ther., 25 (1979) 88–95.
25. W.A. Nijhoff, T.P.J. Mulder, H. Verhagen, G. van Poppel, and W.H.M. Peters. Effects of consumption of Brussels sprouts on plasma and urinary glutathione S-transferase class α- and -π in humans, Carcinogenesis, 16 (1995) 955–958.
26. W.A. Nijhoff, F.M. Nagengast, M.J.A.L. Grubben, J.B.M.J. Jansen, H. Verhagen, G. van Poppel, and W.H.M. Peters. Effects of consumption of Brussels sprouts on intestinal and lymphocytic glutathione and glutathione S-transferases in humans, Carcinogenesis, 16 (1995) 2125–2128.
27. S. Yannai, H.L. Bradlow, J. Westin, and E.D. Richter. Consumption of cruciferous vegetables—probable protection against breast cancer, In: H. Kozlowska, J. Fornal, and Z. Zdu czyk (Eds.), Proceedings of the International Conference Euro Food Tox IV "Bioactive substances in food of plant origin", Vol.2 "Dietary Cancer Prevention", Centre for Agrotechnology and Veterinary Sciences, Olszytn, Poland, 1994, pp. 480–483.
28. M.A. Kall, O. Vang, and J. Clausen. Effects of dietary broccoli on human *in vivo* drug metabolizing enzymes: evaluation of caffeine, oestrone, and chlorzoxazone metabolism, Carcinogenesis, 17 (1996) 793–799.
29. R. Mawson, R.K. Heany, Z. Zdunczyk, and H. Kozlowska. Rapeseed meal-glucosinolates and their antinutritional effects. Part 3. Animal growth and performance, Nahrung, 38 (1994) 167–177.
30. R. Mawson, R.K. Heany, Z. Zdunczyk, and H. Kozlowska. Rapeseed meal-glucosinolates and their antinutritional effects. Part 4. Goitrogenicity and internal organs abnormalities in animals, Nahrung, 38 (1994) 178–191.
31. L. Nugon-Baudon, S. Rabot, O. Szylit, and P. Raibaud. Glucosinolates toxicity in growing rats: interactions with the hepatic detoxification system, Xenobiotica, 20 (1990) 223–230.
32. G. van Poppel, W.I. Willems, and H. Verhagen. Thyroid function after consumption of Brussels sprouts: a controlled experiment in humans, TNO Nutrition and Food Research Institute, Zeist, Netherlands, 1993.
33. M. McMillan, E.A. Spinks, and G.R. Fenwick. Preliminary observations on the effect of dietary Brussels sprouts on thyroid function, Hum. Toxicol., 5 (1986) 15–19.
34. R.H. Vos and W.G.H. Blijleven. The effect of processing conditions on glucosinolates in cruceferous vegetables, Zeitschrift für Lebensmitteluntersuchung und -Forschung 187,6 (1988) 525–529.

15

STILBENES AND BIBENZYLS WITH POTENTIAL ANTICANCER OR CHEMOPREVENTIVE ACTIVITY

Fulvia Orsini and Luisella Verotta

Dipartimento di Chimica Organica e Industriale
Università degli Studi di Milano
Via Venezian 21, 20133 Milan
Italy

1. OCCURRENCE

Stilbenes and bibenzyls are widely distributed in the plant kingdom both in lower plants (liverworts mostly) and in higher plants (gymnosperms and angiosperms).

Biogenetically the stilbenoids derive from cinnamic acid (via the shikimic pathway) and three acetate units from malonyl coenzyme A.[1] The first part of the pathway is common to stilbenoids and flavonoids. They diverge at the font of a styryl-3, 5,7-triketoeptanoic acid **1**: an aldol condensation gives a stilbene 2-carboxylic acid usually unstable and intermediate to several structures known as stilbenoids (bibenzyls, bis-bibenzyls, stilbenes, phenantrenes, 9,10-dihydrophenantrenes, phenyldihydroisocoumarins); a acylation produces a calchone which, subsequently modified, gives flavonoids (flavones, flavonols, flavanones, iso-flavonoids, anthocyanins, tannins, etc . . .) (Scheme 1).

The stilbenoid and flavonoid pathways occur together in liverworts as well as in higher plants. In the higher plants, angiosperms produce a greater variety of stilbenoids than any other plant. These compounds may be simple bibenzyls and stilbenes, stilbene oligomers and also mixed stilbene-flavonoid derivatives.

Their biological role in plants is not always clear: however antibiotic and growth-regulatory activities have been reported *in vitro*. In most cases they are phytoalexins as, for example, resveratrol (3,4′,5-trihydroxystilbene, **18**, Fig. 3) and its oligomers present in grapevines.

Advances in Nutrition and Cancer 2, edited by Zappia *et al.*
Kluwer Academic / Plenum Publishers, New York, 1999.

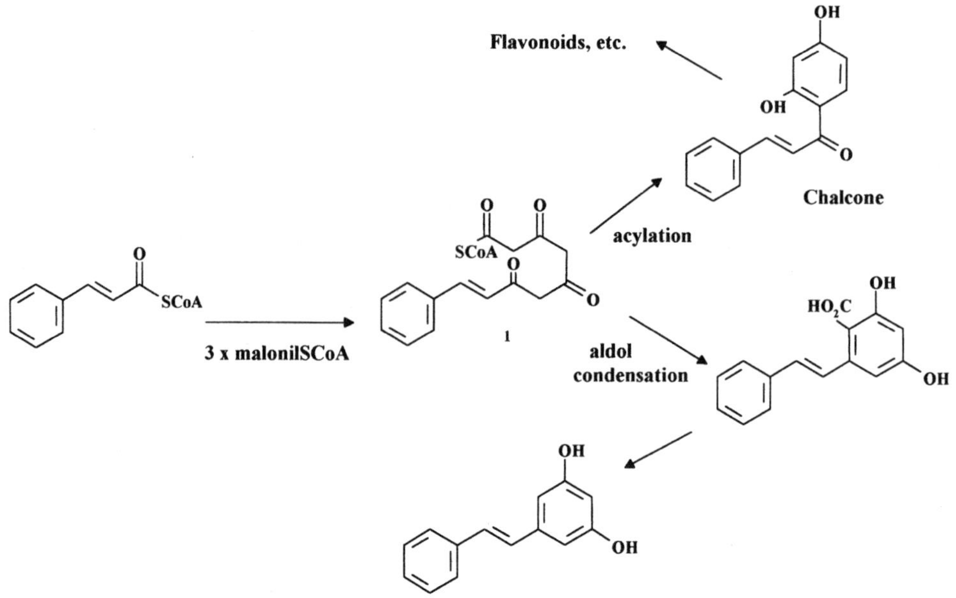

Scheme 1. Biosynthetic pathway.

1.1. Bibenzyls

Several hydroxylated bibenzyls (the parent hydrocarbon has not been found in plants) occur in higher plants.[1] Among potentially edible species, 2,3′,4-trihydroxybibenzyl **2**, dihydro-oxyresveratrol **3**, and lunularin **4** were found in *Morus* species (Fig. 1).[2]

Prenylated bibenzyls **5**, **6**, **7** have been found in *Glycyrrhiza* species.[3]

Bibenzyls with plant growth regulatory activity, batatasins III **8**, IV **9**, and V **10** have been found in *Dioscorea* species.[4] Lunularic acid **11** (widespread in liverworts) and other bibenzyl-2-carboxylic acids have been found in A*llium*[5] and in *Glycyrrhiza* species.[6]

Partially O-methylated 3,4,5,3′,4′-pentahydroxy- and 3,4,5,2′,3′,4′-exahydroxy derivatives such as **12** (combretastatin) and **13** (combretastatin B-1) form a remarkable group of bibenzyls. To our knowledge such bibenzyls have not yet been isolated from edible plants, but only from *Combretum* species.[7] Nevertheless they are worth mentioning for their biological properties.[8]

Bibenzyl glycosides are also found in higher plants. Examples are 2,2′,4,4′-tetrahydroxybibenzyl-2-O-xylopyranoside together with icaraside A-4 **15** and icaraside A-6, **16** (Fig. 2).[9] The fruit of *Combretum kraussi* yielded combretastatin B-1 **13** and its 2′-O-β-D-glucopyranoside **17**.[10]

Differently from bibenzyl, the parent compound, *trans*-stilbene has been found in some species such as A*llium* species. Stilbenes usually have the *trans-(E)* configuration and are 3,5-dihydroxy substituted in agreement to their biogenetic pathway. However, *cis*-(Z)-stilbenes have also been found. Strong heat or light may induce isomerization between the two forms. 3,5-Dihydroxystilbene and its methyl derivative are found in *Pinus* genus and *Alnus* species. 3,4′,5-Trihydroxystilbene (resveratrol) **18** has a wider distribution and is found in various members of several species, for example: *Morus*,

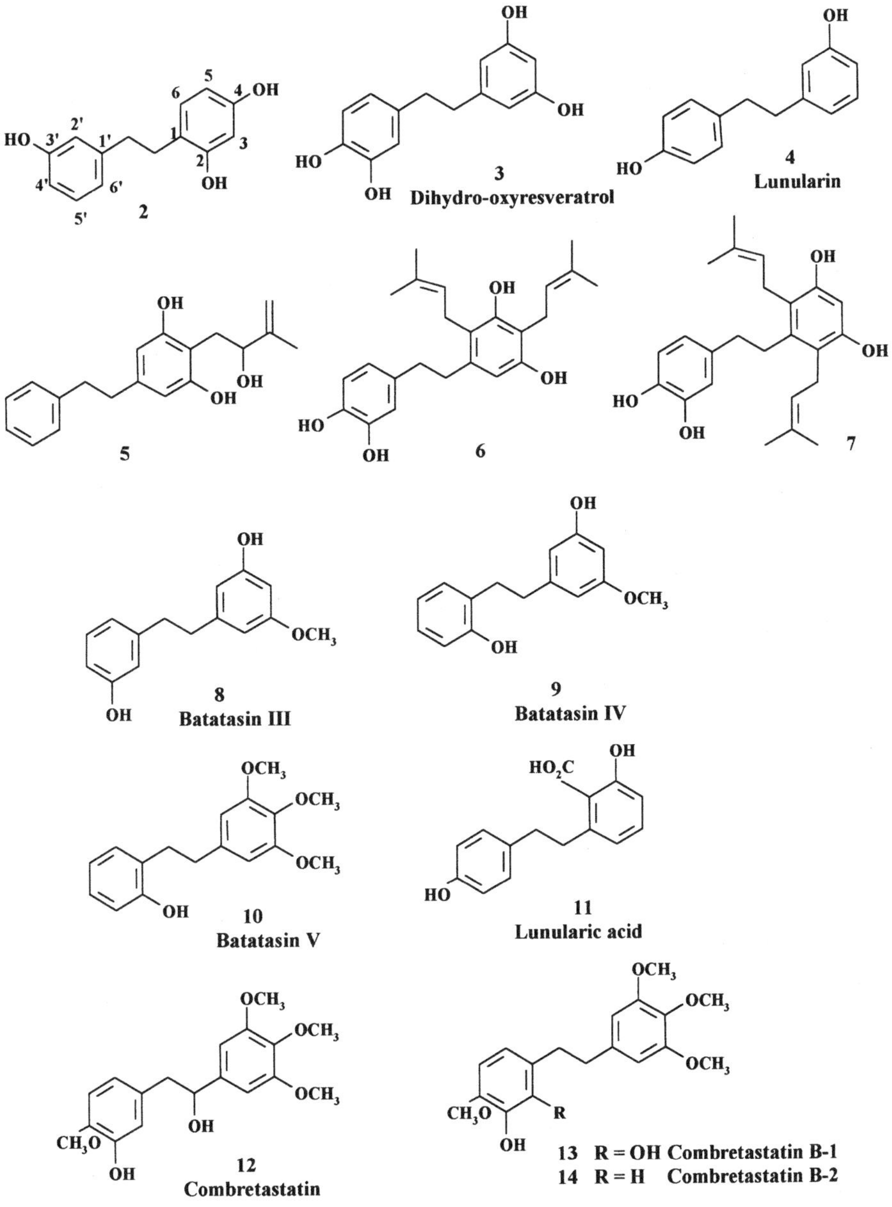

Figure 1. Examples of bibenzyls.

Vitis, *Rheum* species (Fig. 3). Resveratrol is also precursor to many oligostilbenes. Methyl ethers of resveratrol are also found as in *Vitis*[11] and *Rheum* species.[12]

The 3-O-β-D-glucopyranoside of resveratrol, piceid, **19**, common in *Eucalyptus* species, has also been found in *Erythrophleum* lasianthum.[13] The 4′-O-glucoside **20** (resveratroloside) has been found in rhubarb (*Rheum rhaponticum*). In rhubarb another, unusual, glycoside of resveratrol has been found: the 4′-O-(6″-O-galloyl) glucopyranoside 21.[14]

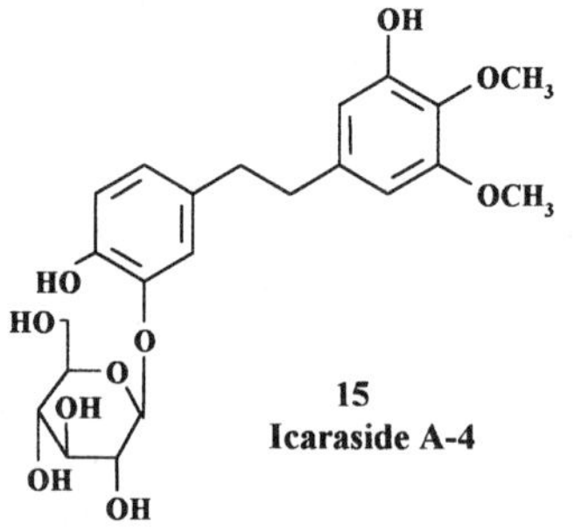

15
Icaraside A-4

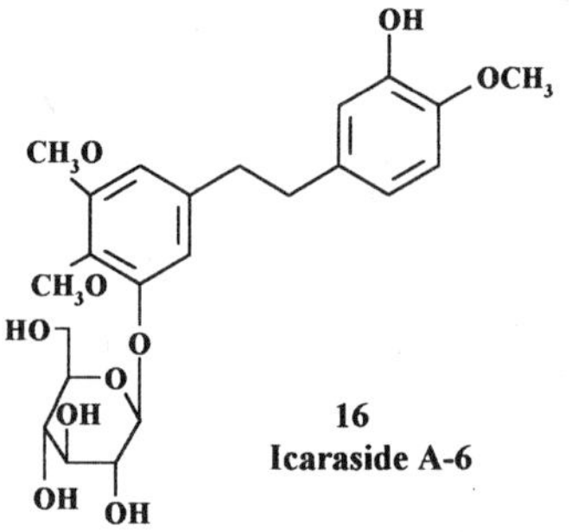

16
Icaraside A-6

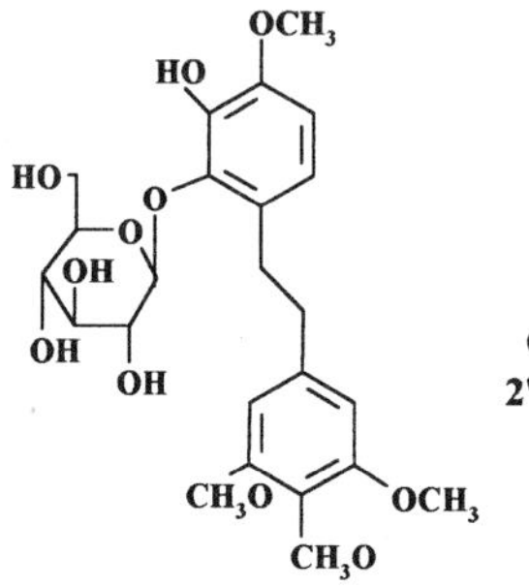

17
Combretastatin B-1
2'-O-β-D-glucopyranoside

Figure 2. Examples of bibenzyl glycosides.

1.2. Stilbenes3,3′, 4′, 5-Tetrahydroxystilbene, piceatannol **22**, present in several species, is also present in Morus,[2] *Rheum*[14], and *Saccharum*[15] (sugarcane) species and a prenylated derivative of piceatannol occurs in groundnuts (*Arachis hypogaea*) (Fig. 4). The 3-O-β-D-glucopyranoside of piceatannol, astringin **23** is present in several plants, among them rhubarb (*Rheum rhaponticum*) together with the 3′-O-β-D-glucopyranoside **24**, the 3′-O-β-D-xylopyranoside **25**, and the 3-O-(6″-O-galloyl)glucopyranoside 26.[14b] Rhapontigenin (4′-methoxy-3,3′,5-trihydroxystilbene) **27**, the 3′-O-β-D-glucopyranoside **28**, the 3-O-β-D-glucopyranoside, rhapontin, **29**, and its

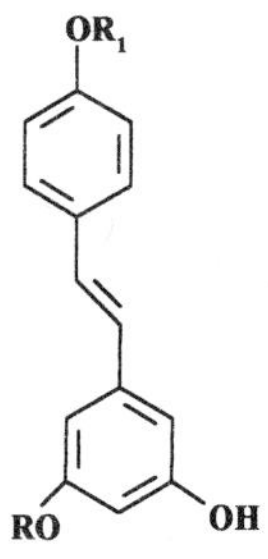

18 R, R_1 = H **Resveratrol**
19 R = Glc; R_1 = H **Piceid**
20 R = H; R_1 = Glc **Resveratroloside**
21 R = H; R_1 = Glc(6"-O-galloyl)

Glc =

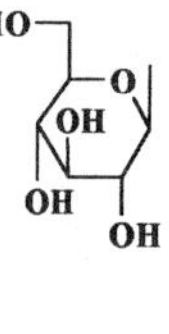

Glc(6"-O-galloyl) =

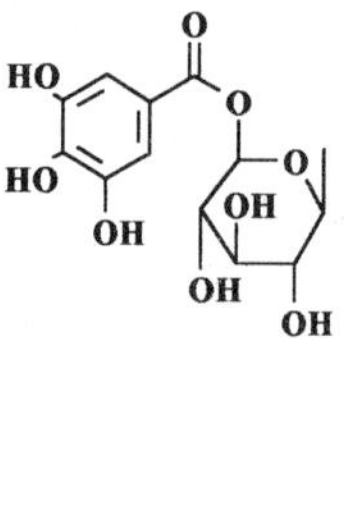

Figure 3. Examples of tri-hydroxystilbenes and their glycosides.

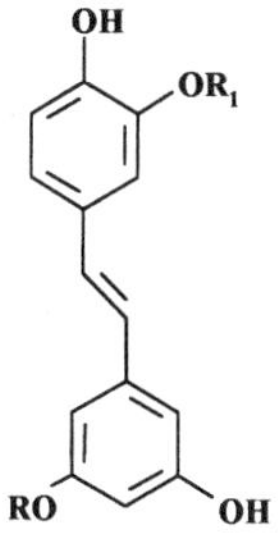

22 R, R_1 = H (E)-Piceatannol
23 R = Glc, R_1 = H (E)-Astringin
24 R = H, R_1 = Glc
25 R = H, R_1 = Xyl
26 R = Glc(6"-O-galloyl), R_1 = H

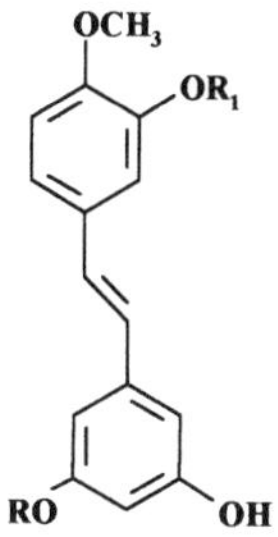

27 R, R_1 = H Rhapontigenin
28 R = H, R_1 = Glc
29 R = Glc, R_1 = H Rhapontin
30 R = Glc(2"-O-galloyl), R_1 = H
31 R = Glc(6"-O-galloyl), R_1 = H
32 R = Glc(2"-O-coumaroyl), R_1 = H

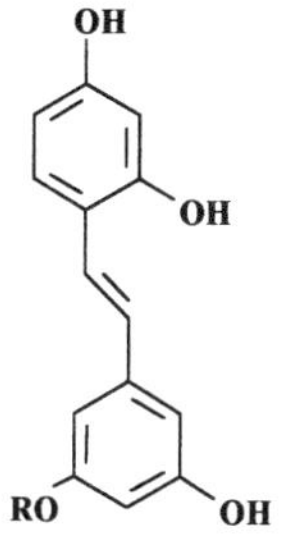

33 R = H Oxyresveratrol
34 R = Glc

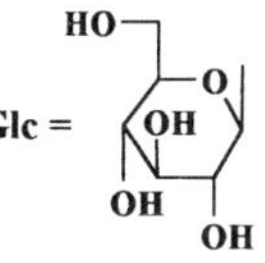

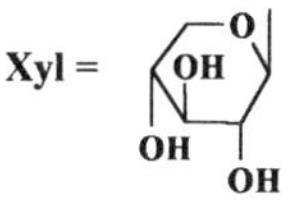

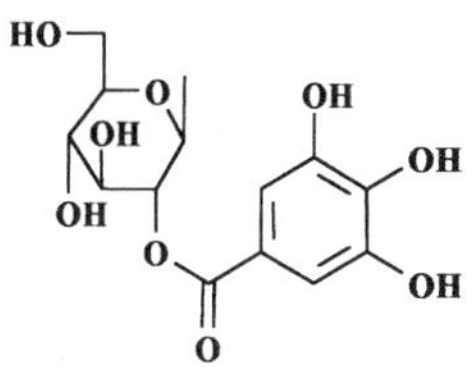

Glc(2"-O-galloyl) 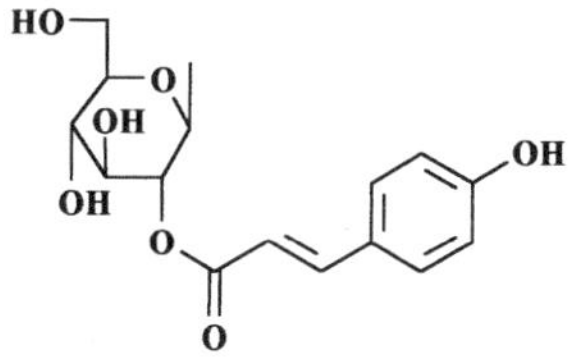 Glc(2"-O-coumaroyl) 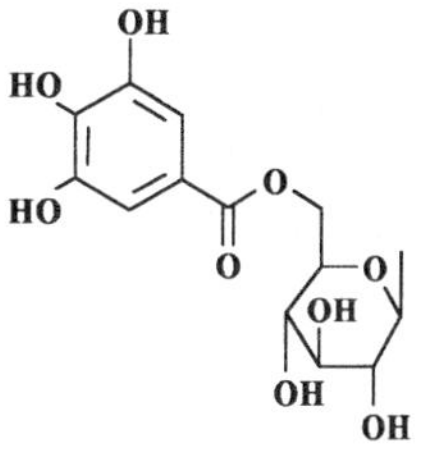 Glc(6"-O-galloyl)

Figure 4. Examples of tetra-hydroxystilbenes and their glycosides.

corresponding 2″-O-galloyl- **30**, 6″-O-galloyl- **31**, and 2″-O-p-coumaroyl derivatives **32** were also found in *Rheum* species. Oxyresveratrol **33** (2′,3,4′,5-tetrahydrostilbene) and its 3-O-β-D-glucopyranoside **34** were found in members of the Moraceae.[16]

(E)-piceid **19** (Fig. 3), (E)-resveratroloside **20**, (E)-astringin **23** (Fig. 4), and the corresponding (Z)-isomers were isolated from cell cultures of *Vitis vinifera* L. (Vitaceace).[17]

Resveratrol together with *trans* and *cis* 4-methoxy 3,3′,5,5′-tetrahydroxystilbene **35** and **36** (an example of pentahydroxystilbenes different from combretastatins reported in Fig. 6) occur in the date palm (*Phoenix dactylifera*)[18] (Fig. 5).

Several prenylated stilbenes with a *trans* configuration are produced as phytoalexins in groundnuts: among them are arachidin I-III **37–39** (Fig. 5).[19] Some

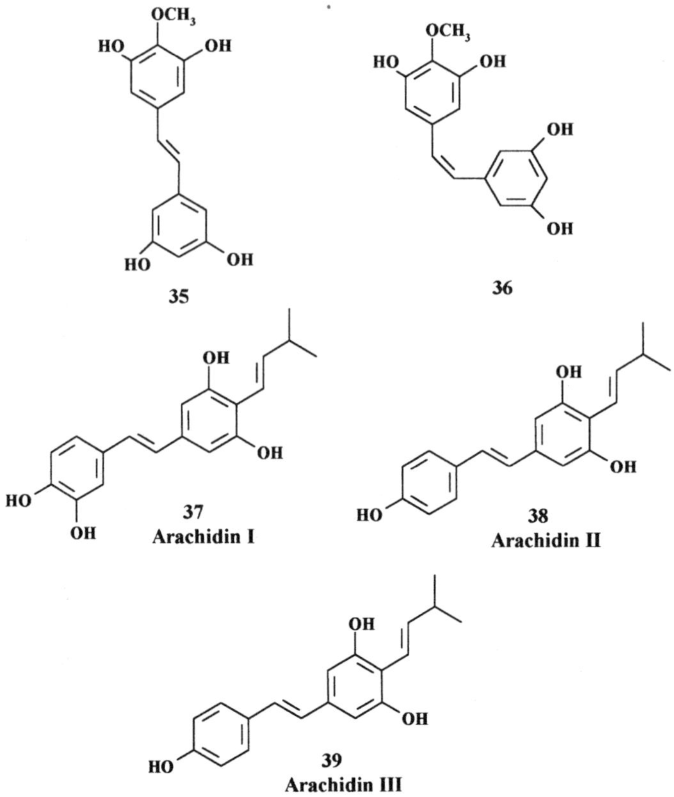

35

36

37
Arachidin I

38
Arachidin II

39
Arachidin III

Figure 5. Examples of penta-hydroxystilbenes and their glycosides.

prenylated derivatives of oxyresveratrol, active as tyrosine inhibitors, have been recently found in Moraceae.[20]

As already mentioned for bibenzyls, some partially O-methylated 3,4,5,3′,4′-pentahydroxy- and 3,4,5,2′,3′,4′-exahydroxy-*cis*-stilbenes, at present isolated only from the non-edible *combretum* species, showed quite interesting biological properties (Fig. 6).[7,8] They include combretastatin A-1 (*cis* 2′, 3′-dihydroxy-3,4,4′-5-tetramethoxystilbene) **40**, combretastatin A-3 (*cis* 3, 3′-dihydroxy-4,4′,5-trimethoxystilbene) **41**, combretastatin A-4 (*cis* 3′-hydroxy-3,4,4′,5-tetramethoxystilbene) **42**, combretastatin A-5 (3-hydroxy-3′,4,4′,5-tetramethoxystilbene **43**, and combretastatin A-6 (*trans* 3,3′-dihydroxy-4,4′,5-trimethoxystilbene) **45**. The 2′-O-β-D-glucopyranoside of combretastatin A-1 **46** has been found in the fruit of *Combretum krauts*.[10]

2. ISOLATION FROM NATURAL SOURCES

The chemical nature of stilbenoids and their glycosides refers to phenolic structures where phenol groups could be free or “masked”. Masked phenolic hydroxyls mean that these groups exist in the molecule derivatized as ether, ester or acetal functions. Among acetal-type derivatizations, glycosides represent a diffused way in plants to detoxify or protect a biologically active aglycone.

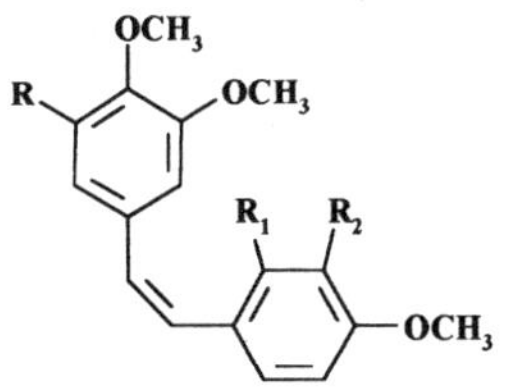

40 R = OCH_3; R_1 = OH; R_2 = OH **Combretastatin A-1**
41 R = OH; R_1 = H; R_2 = OH **Combretastatin A-3**
42 R = OCH_3; R1 = H R_2 = OH **Combretastatin A-4**
43 R = OH; R_1 = H; R_2 = OCH_3 **Combretastatin A-5**

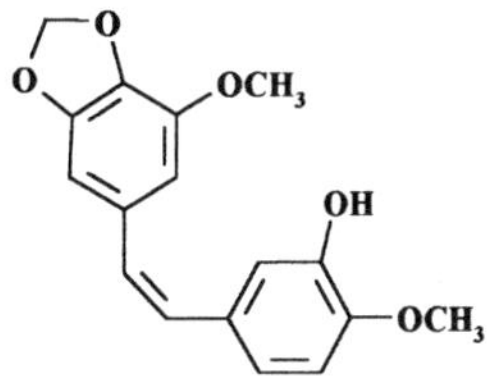

44 Combretastatin A-2

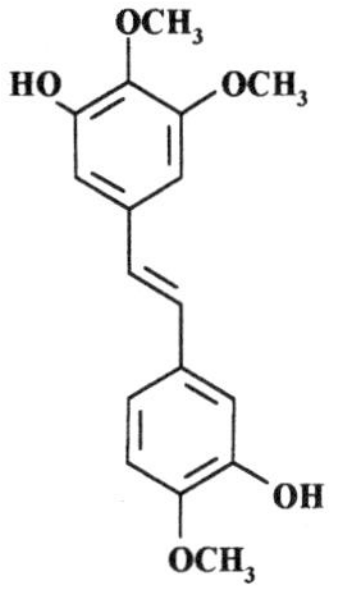

45 Combretastatin A-6

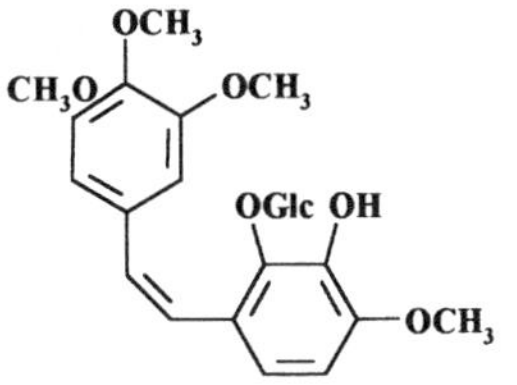

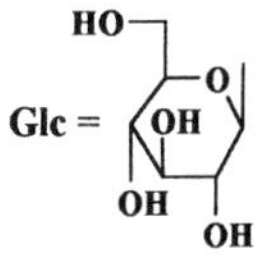

46 Combretastatin A-1, 2'-β-O-D-glucopyranoside

Figure 6. Penta-hydroxystilbenes (Combretastatins) and their glycosides.

Due to the contemporary presence of a diaryl-ethyl (or ethylene) moiety and a number of hydroxyls, stilbenoids are considered low-medium polar compounds; other way is the contemporary presence of one or more sugar moieties which give the molecule polar characteristics.

The isolation and purification of stilbenoids and their glycosides, thus, should follow different approaches. Two, three, four hydroxyls bearing stilbenoids can be obtained by dichloromethane extraction of the vegetable material.[21] Glycosides can only be obtained from a total alcoholic extraction of the plants. A n-butanol counter extraction is necessary to avoid most of mono, disaccharides which make difficult further separations.[13,21] Stilbenoids can be easily separated and purified through silica gel chromatographies by using alcohol modified chloformic mobile phases.

The n-butanol extract usually undergoes different purification steps. A preliminary gel filtration chromatography allows us to separate glycosides of different molecular sizes, and to obtain crops of fractions enriched in different classes of glycosides. Triterpene glycosides are separated from phenol and stilbenoids glycosides, which are themselves separated from flavonoid glycosides.[22]

A subsequent step is the use of reversed phase column chromatographies at medium or high pressures (HPLC)[12,21] and/or counter-current chromatography.[23] Both the techniques take into account the polar characteristics of stilbenoid glycosides, avoiding irreversible interactions of phenolic-alcoholic hydroxyl groups with silanol groups of underivatized silica gel.

Due to be mentioned are the counter-current methods that give quantitative recovery of samples by the use of liquid-liquid chromatography without a sorbent. Three, four components solvents which give two immiscible phases are the base of this chromatography. Solutes are partitioned between the two phases with respect to their distribution coefficients, thus allowing separation.

Typical solvent systems for the separation of phenolic glycosides are $CHCl_3$/MeOH/H_2O (medium polar characteristics) or n-BuOH/n-PrOH/H_2O (strong polar characteristics).[23]

3. SYNTHETIC METHODS

3.1. Aglycones

Several synthetic methods (some general other quite specific) have been developed for the synthesis of stilbenes and bibenzyls. Stilbenes may be prepared by condensation of a substituted sodium phenylacetate with a substituted benzaldehyde as exemplified for lunularin **3** (Scheme 2).

A valid and quite general alternative to the condensation reported above is a Wittig reaction between a properly substituted benzaldehyde and a benzyl triphenylphosphonium halide as exemplified in the Scheme 3 for several stilbenoids related to resveratrol and combretastatins.[12,13]

Several variations of this approach concerning the protecting group chosen for hydroxyl groups[24] and the type of the organolithium used[25,24c] have been proposed.

A mixture of *cis* and *trans* isomers is obtained when a benzyl triphenylphosphonium halide is used. On the contrary only the *trans*-isomer is obtained when the substituted benzaldehyde is reacted with a diethyl benzylphosphonate.[26]

Other methods include: condensation of a 1,3-dithiane intermediate with an appropriate bromo derivative;[27] base-catalysed condensation of a substituted ben-

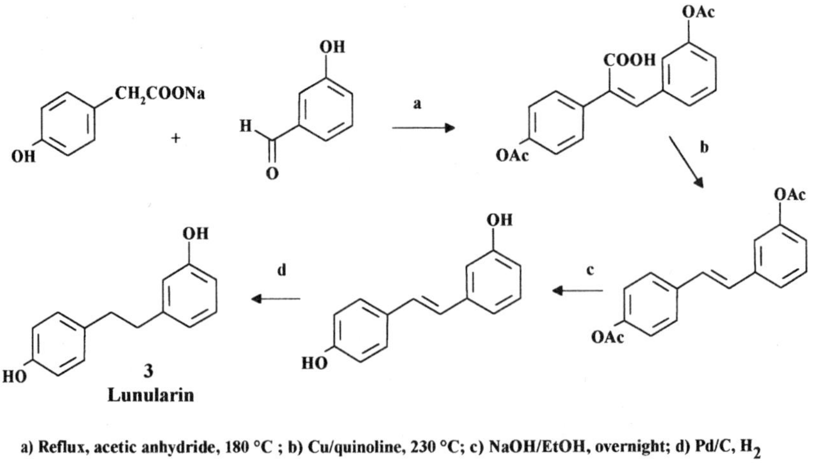

a) Reflux, acetic anhydride, 180 °C ; b) Cu/quinoline, 230 °C; c) NaOH/EtOH, overnight; d) Pd/C, H_2

Scheme 2. Synthetic methods of stilbenes and bibenzyls.

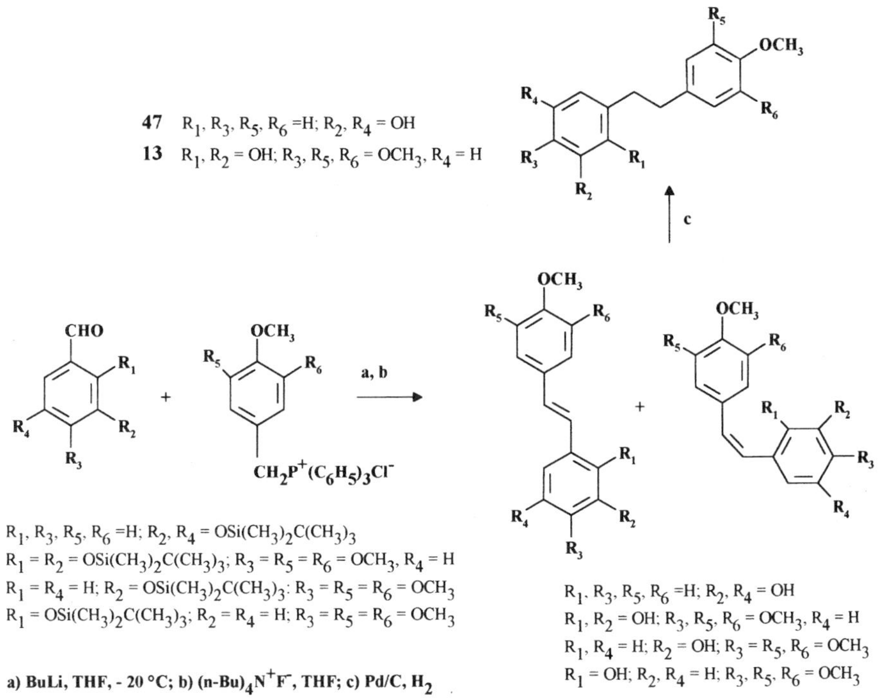

47 $R_1, R_3, R_5, R_6 = H; R_2, R_4 = OH$
13 $R_1, R_2 = OH; R_3, R_5, R_6 = OCH_3, R_4 = H$

$R_1, R_3, R_5, R_6 = H; R_2, R_4 = OSi(CH_3)_2C(CH_3)_3$
$R_1 = R_2 = OSi(CH_3)_2C(CH_3)_3; R_3 = R_5 = R_6 = OCH_3, R_4 = H$
$R_1 = R_4 = H; R_2 = OSi(CH_3)_2C(CH_3)_3; R_3 = R_5 = R_6 = OCH_3$
$R_1 = OSi(CH_3)_2C(CH_3)_3; R_2 = R_4 = H; R_3 = R_5 = R_6 = OCH_3$

$R_1, R_3, R_5, R_6 = H; R_2, R_4 = OH$
$R_1, R_2 = OH; R_3, R_5, R_6 = OCH_3, R_4 = H$
$R_1, R_4 = H; R_2 = OH; R_3 = R_5, R_6 = OCH_3$
$R_1 = OH; R_2, R_4 = H; R_3, R_5, R_6 = OCH_3$

a) BuLi, THF, - 20 °C; b) $(n\text{-}Bu)_4N^+F^-$, THF; c) Pd/C, H_2

Scheme 3. Synthetic methods of stilbenes and bibenzyls.

zaldehyde with a phenylacetic acid or with a phenylacetonitrile[28] (a modification of the procedure reported in the scheme 2, used to prepare a series of 3′-hydroxy-3,4,4′,5-tetramethoxystilbene analogues); lithium-mediated condensation of a benzyltrimethyl silyl ether with a benzaldehyde (a protocol used to synthesise resveratrol and piceatannol) (Scheme 4).[29]

Bibenzyls can be obtained from the corresponding *cis-* and *trans* stilbene by catalytic hydrogenation on Pd/C as exemplified for the synthesis of 4′-methyl-dihydroresveratrol **47** and 2′,3′-dihydroxy-3,4,4′,5-tetramethoxybibenzyl **13** (combretastatin B-1)[10,13] in the scheme 3.

However, direct methods have also been reported. For example 3,5-dihydroxybibenzyl **48** was synthesised from 3,5-dimethoxybenzoic acid and phenetyl bromide[30] and by condensation of 3,5-dimethoxybenzyl trimetylsilyl ether with benzaldehyde, followed by dehydroxylation (Scheme 4).[29]

3.2. Glycosides

The 3-O-β-D-glucopyranoside of resveratrol **19** piceid, was synthesised by reaction of 4′-O-methyl resveratrol with tetra-O-acetyl-α-D-glucopyranosyl bromide under phase transfer catalysis (Scheme 5).[13]

The protocol used for the synthesis of piceid was applied to the synthesis of the glucosides of combretastatin A-1 **49**, combretastatin A-4 **50** (the most potent inhibitor of tubulin assembly), combretastatin *iso*-A-4 **51**, and combretastatin B-1 **52** (Scheme 5).[10] A different approach, via a C-1 imidate, was used for the synthesis of the β-D-meliobioside and the β-D-glucuronide of 3′-hydroxy-3,4,4′,5-tetramethoxystilbene.[31]

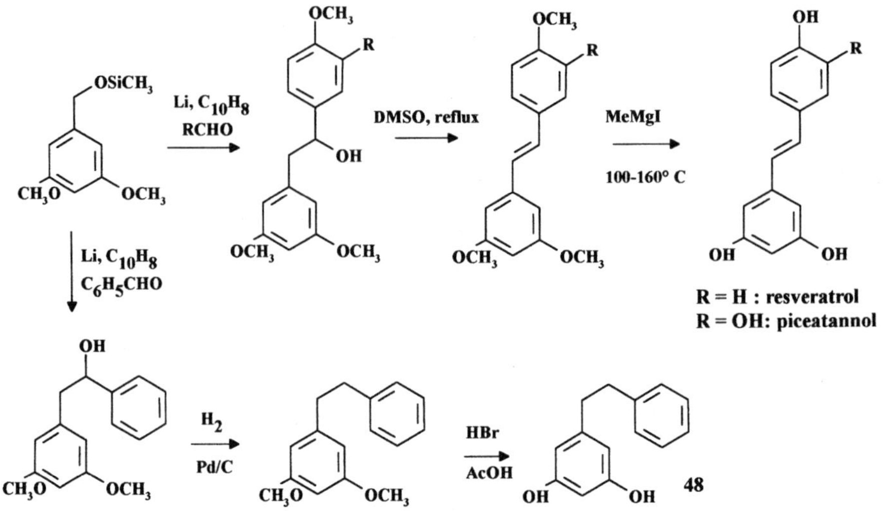

Scheme 4. Synthetic methods of stilbene and bibenzyls.

More recently the 3-O-β-D-glucopyranoside of resveratrol, piceid, **19** has been obtained by biotransformation of resveratrol with whole-cell suspension *of Bacillus cereus* UI 1477.[32]

A major problem often encountered in the pharmacological testing of the stilbenoids is their limited solubility that makes it difficult to prepare suitable formulations for clinical trials. Water-soluble derivatives, as glycosides could have been useful as prodrugs. Unfortunately the glycosides were in general less active than the corresponding aglycones. To obtain other practical and water- soluble prodrugs of polyhydroxystilbenoids, several derivatives (reported in the section 3) were synthesised by simple conversion of the stilbenoids to the corresponding sodium and potassium salts or by incorporating various basic or acidic units.[33]

Water-soluble derivatives with a nitrogen-containing group (reported in the section 3) were also designed.[28]

4. BIOLOGICAL ACTIVITY

Stilbenoids have been reported to show several and quite diverse biological effects[1] such as antifungal, antibacterial, cytotoxic, anti-inflammatory, cardiotonic, oestrogenic[34] activities. Most of these biological activities are considered to be due to their phenolic structures.

Stilbenoids, for example, influence the arachidonate metabolism exerting their inhibitory effect either on cyclooxygenase (COX) or on 5-lipoxygenase 5-(LOX), the key enzymes of the arachidonic acid cascade. These effects may have quite important consequences for human health such as the prevention of heart diseases and cancer.

Such effects have been reported both for bibenzyls, prenyl bibenzyl bis(bibenzyl)s as marchantin A and paleatin B (Fig. 7) (from liverworts, nevertheless worth mentioning for their peculiar structures)[35] which showed 5-lipoxygenase inhibitory activity and for stilbenes such as resveratrol **18** (Fig. 8) which alters eicosanoid synthesis,[36]

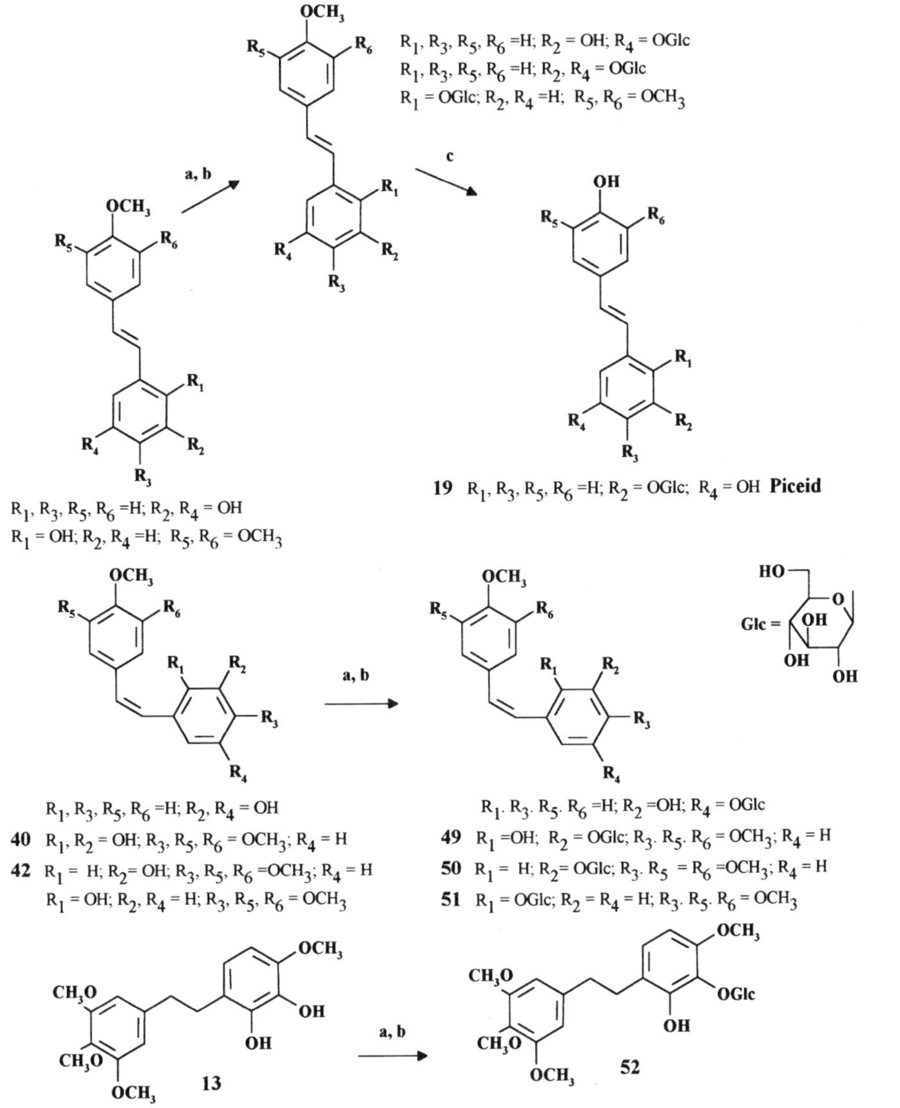

R_1, R_3, R_5, R_6 =H; R_2 = OH; R_4 = OGlc
R_1, R_3, R_5, R_6 =H; R_2, R_4 = OGlc
R_1 = OGlc; R_2, R_4 =H; R_5, R_6 = OCH_3

R_1, R_3, R_5, R_6 =H; R_2, R_4 = OH
R_1 = OH; R_2, R_4 =H; R_5, R_6 = OCH_3

19 R_1, R_3, R_5, R_6 =H; R_2 = OGlc; R_4 = OH **Piceid**

R_1, R_3, R_5, R_6 =H; R_2, R_4 = OH
40 R_1, R_2 = OH; R_3, R_5, R_6 = OCH_3; R_4 = H
42 R_1 = H; R_2= OH; R_3, R_5, R_6 =OCH_3; R_4 = H
R_1 = OH; R_2, R_4 = H; R_3, R_5, R_6 = OCH_3

R_1, R_3, R_5, R_6 =H; R_2 =OH; R_4 = OGlc
49 R_1 =OH; R_2 = OGlc; R_3, R_5, R_6 = OCH_3; R_4 = H
50 R_1 = H; R_2= OGlc; R_3, R_5 = R_6 =OCH_3; R_4 = H
51 R_1 = OGlc; R_2 = R_4 = H; R_3, R_5, R_6 = OCH_3

13 → **52**

a) $Et_3(PhCH_2)N^+Br^-$, **α-bromo-tetra-O-acetyl-glucose, NaOH,** $CHCl_3$**, 60° C; b) MeONa, MeOH, 20 °C. c) EtSNa, DMF.**

Scheme 5. Synthetic methods of stilbenes' and bibenzyls' glucosides.

influences the lipid metabolism,[37] inhibits the human low-density lipoprotein oxidation,[38] influences human polymorphonuclear leukocyte function,[39] and inhibits platelet aggregation.[13,40,41] It is conceivable that resveratrol, present in high quantities in *Vitis vinifera* as a response to fungal infections, plays a role in the prevention of heart disease because of its effects on lipid and arachidonate metabolism.

Resveratrol inhibits platelet aggregation at a molar concentration in the same order of magnitude of quercetin, but presumably with a different mechanism: the latter directly inhibiting phospholipase C activity, resveratrol inhibiting TXA_2 production.[40b] Resveratrol 3-O-β-D-glucopyranoside **19**, tested in vitro on human platelet-rich plasma also showed inhibitory effect on platelet aggregation induced by collagen, adrenaline

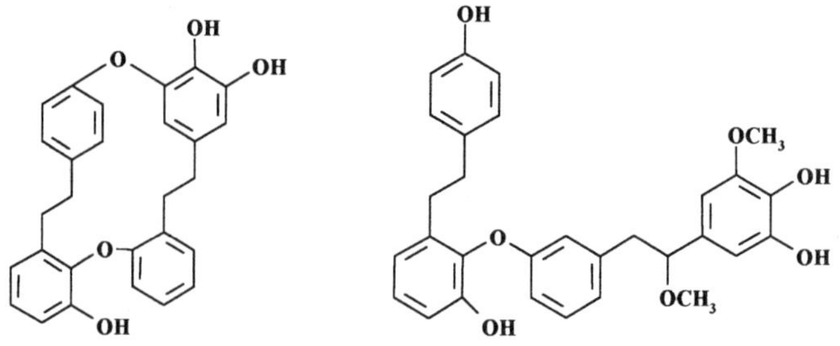

Marchantin A

Paleatin B

Figure 7. Examples of bis(bibenzyl)s with biological activity.

and, to a minor extent, by arachidonic acid and ADP. Several stilbenoids of the resveratrol and combretastatin series were tested as antiplatelet aggregation agents (aggregation induced by collagen and ADP). The study suggested a relationship between antiplatelet aggregation activity and a) the number, the location and the derivatization of the hydroxyl groups; b) the stereochemistry of the double bond. The maximum activity appeared to be related to the (E) configuration of the double bond and to the presence of two free hydroxyl groups. Glycosides were less active than the corresponding aglycones.[13]

A further interesting activity, potentially related to the prevention of cardiac diseases, was recently observed in combretastatin B-1 **13**: it selectively and reversibly increased the duration of the action potential in sensory neurones by inhibiting the repolarizing K^+ currents.[42] Combretastatin A-1 **40** and the corresponding (E)-isomer were less active. The 2′-O-β-D-glucopyranoside **17** was inactive, whereas a 50% loss of activity resulted from acetylation of the phenolic groups in combretastatin B-1 **13**; bibenzyl and stilbene were totally inactive. Further investigation is therefore currently in progress on other stilbenoids, in particular of the resveratrol series to investigate the structure/activity relationship and design new drugs. Preliminary results have evidenced for 4′-O-methyl-dihydro-resveratrol **47** a greater activity with respect to combretatstain B-1 **13** and to resveratrol itself.[43] This type of K^+ channel inhibition has been described, in other tissues, in relation to some polyunsaturated fatty acids, cardiac antiarrhythmic agents (such as tedisamil) and antitumour drugs (as tamoxifen).

Both the lipoxygenase and the cycloxygenase reactions and the non enzymatic lipid oxidations form free radicals and all these reactions are prone to inhibition by antioxidant and/or radical scavengers. Stilbenoids can act as structural analogues of arachidonic acid and substrates of 5-(LOX) and COX and/or act as radical scavengers. Both effects may be relevant to cancer chemoprevention. Resveratrol was found to inhibit free-radical formation when promyelocitic leukemia (HL-60) cells were treated with 12-O-tetradecanoylphorbol-13-acetate[41] Piceid **19** protected rats fed with peroxidized oil from liver injury.[44] (E)-piceid **19**; (E)-resveratroloside **20**, (E)-astringin **23**, and the corresponding (Z)-isomers, extracted from *Vitis vinifera* cell cultures, act as scavengers of free radicals and inhibit the Cu^{++} induced lipid peroxidation in low-density lipoprotein.[17]

Antioxidant and/or radical scavenger's properties and COX inhibition are also relevant to cancer chemoprevention. COX inhibition inhibits the conversion of arachidonic acid to substances such as prostaglandins, which can suppress immune defences and stimulate cell growth.

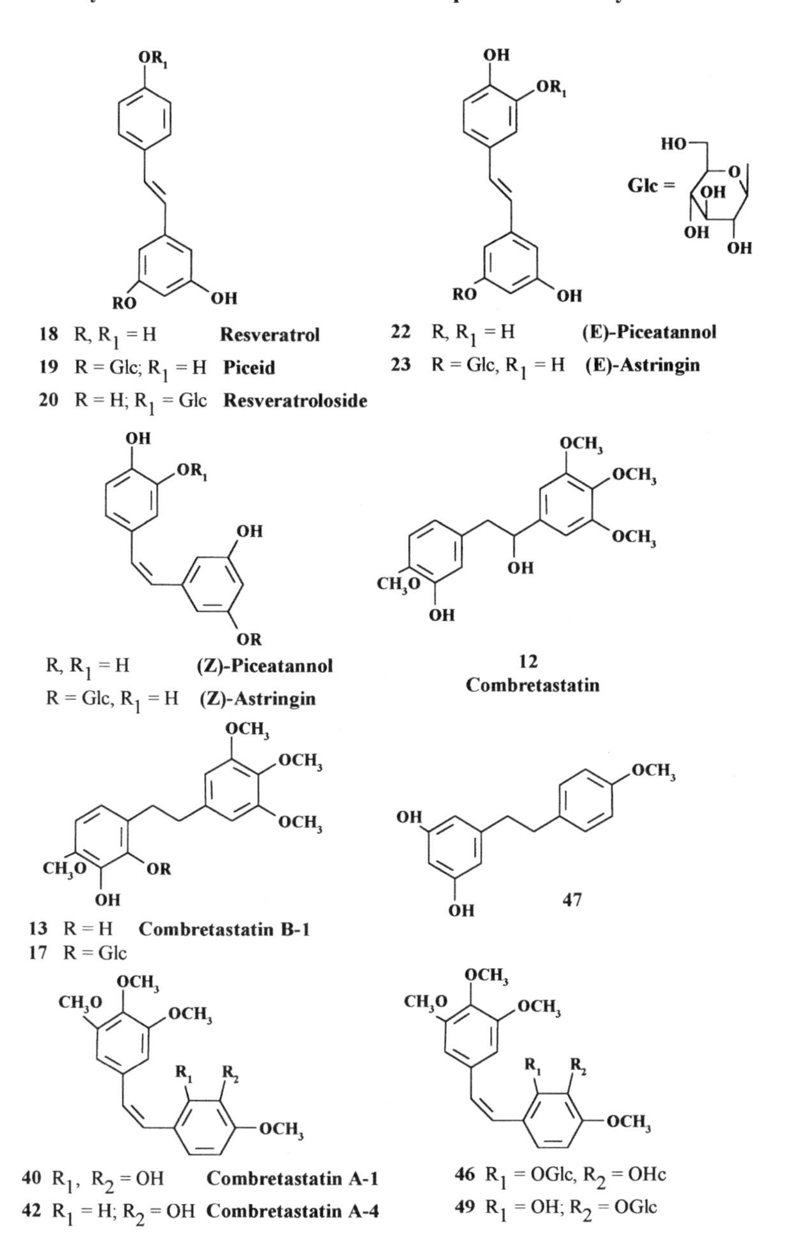

Figure 8. Examples of stilbenoids with biological activity.

Recently resveratrol **18**, which has been shown to inhibit the hydroperoxidase activity of COX-1 free radical formation, was reported to inhibit cellular events associated with tumour initiation, promotion and progression[41] and, due to its very weak toxicity, it was suggested that this compound should be investigated as a cancer chemopreventive agent in humans. More recently resveratrol was shown to inhibit ribonucleotide reductase[45] and DNA polymerase[46] and to arrest the cell division cycle at S/G2 phase transition.[47]

Resveratrol **18** and the corresponding 3- and 4′-O-glucosides were also reported to inhibit bovine protein-tyrosine kinase.[48] The glucosides were less active than the corresponding aglycone. Piceatannol **22** and related compounds also inhibits the FcERI-induced protein-tyrosine phosphorylation.[1,49]

Several resveratrol-like stilbenes and bibenzyls were reported to inhibit mouse leukaemia L1210, the most active being astringin **23** and 3,3′,4,5′-tetrahydrobibenzyl.[50]

Concerning antileukaemic activity, a screening of plant extracts in the murine P388 lymphocytic leukaemia system by the NCI resulted in the identification of several cytotoxic stilbenoids, at present extracted only from the non edible *Combretum* species, nevertheless worth mentioning for their biological activities. Combretastatin **12**, the first of the series, isolated from *Combretum caffrum* caused pronounced astrocyte reversal in the NCI astrocytoma bioassay and inhibition for the murine P-388 lymphocytic leukaemia cell line.[7a] Further investigations of *Combretum* metabolites afforded other pentahydroxy- and exahydroxystilbenoids such as combretastatin A-1 **40**, the corresponding 2′-β-O-D-glucopyranoside, combretastatin A-4 **42**, combretastatin B-1 **13**, and the corresponding 2′-β-O-D-glucopyranoside **17**.

Combretastatin B-1 **13** and its 2′-O-β-D-glucopyranoside **17** were tested *in vitro* on the growth inhibition of mouse leukaemia L1210. Although the active concentrations were of the order of 10 μM, the effect was not reversed after drug washout, thus indicating that both compounds were not only cytostatic but also cytotoxic. Both drugs induced a drastic change in the distribution of cells in the various phase of the cell cycles (most cells at the end of the treatment were in late S-G2M phase whereas most control cells were in G1) and induced the appearance of a new tetraploid cell population, an effect already reported for antimitotics and other DNA interacting agents.[21]

Both combretastatin A-1 **40** and combretastatin B-1 **13** were potent inhibitors of microtubule assembly in vitro and of the binding of colchicin to tubulin. But the most potent cancer cell growth inhibitory stilbene is combretastatin A-4 **42**, which is a potent cytotoxic agent against several cancer cell lines including MDR ones and strongly inhibits the polymerization of tubulin by binding to the colchicine site.

The glucosides of stilbenoids, due to their improved water-solubility with respect to the corresponding aglycones, could have acted as useful prodrugs for clinical trials. Unfortunately the 2′-O-β-D-glucopyranoside **46** (the natural glucoside), the 3′-O-β-D-glucopyranoside **49** (the synthetic glucoside) of combretastatin A-1; the β-D-glucopyranoside 50,[10] the β-D-glucuronide **53** and the β-D-meliobioside **54**[31] (Fig. 9) of combretastatin A-4, tested for *in vitro* cytotoxicities were not as active as the corresponding aglycones. Of relevance, however, is the behaviour of combretastatin A-1 3′-O-β-D-glucopyranoside **49** (non natural) which is five times more active than the naturally occurring 2′-O-β-D-glucopyranoside **46** against a cancer cell-line[10] In the bibenzyl series, the 3′-β-O-D-glucopyranoside of combretastatin B-1 **17** showed selectivity towards LoVoDX colon cancer, the IC50 values being twice more active than the parent phenol.[51]

To overcome the problem of water-solubility, common to several polyhydroxylated stilbenoids, several water-soluble prodrugs **55** (Fig. 9) have been formulated choosing 3′-hydroxy-3,4,4′,5-tetramethoxystilbene **40** as a model compound.[33]

The more suitable compounds in term of water solubility and stability were the disodium and dipotassium phosphate derivatives: the former, tested *in vitro* against several cancer cell-lines proved to be from about 2 to 10-fold more active than the potassium salt.[33]

The requirement of water-soluble and *in vivo* active prodrugs, also supported by structure-activity studies,[52] stimulated the preparation of either natural or synthetic

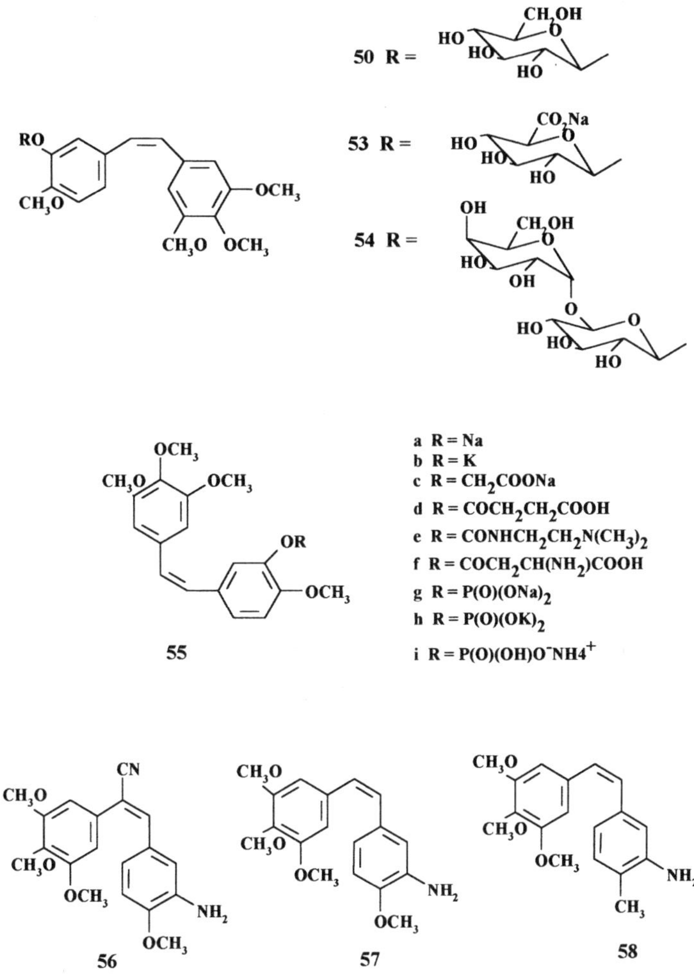

50 R =

53 R =

54 R =

55

a R = Na
b R = K
c R = CH_2COONa
d R = $COCH_2CH_2COOH$
e R = $CONHCH_2CH_2N(CH_3)_2$
f R = $COCH_2CH(NH_2)COOH$
g R = $P(O)(ONa)_2$
h R = $P(O)(OK)_2$
i R = $P(O)(OH)O^{-}NH4^{+}$

56 57 58

Figure 9. Examples of water-soluble derivatives of stilbenes.

non-natural stilbenoids Water soluble derivatives with a nitrogen-containing group were also designed. Among them, those with an amino moiety in place of the phenolic group showed potent antitubulin activity and cytotoxicity against murine Colon 26 adenocarcinoma *in vitro*. Furthermore **56**, **57**, and **58** showed significant antitumour activity in the animal model, while combretatstain A-4 **42** was ineffective.[28]

5. CONCLUSIONS

Stilbenes and bibenzyls (stilbenoids in general) are a group of quite interesting compounds because of their biochemistry, taxonomic distribution, and biological activities.

Our present knowledge of their distribution in plants can be regarded as fragmentary due to the number of plants investigated and the number of new reports that continue to appear in the literature.

Much work is therefore still to be done with particular reference to stilbenoids that are present as minor constituents of edible plants and could have a significant impact on human health.

ACKNOWLEDGMENTS

CNR and Ministero dell' Universita' e della Ricerca Scientifica e Tecnologica are gratefully acknowledged for financial support.

REFERENCES

1. Gorham John (1995). The biochemistry of stilbenoids Chapman & Hall London.
2. Deshpande V.H., Srinivasan R., and Rao A.V.R. (1975). Wood phenolics of *Morus* species. IV. Phenolics of the heartwood of five *Morus* species. Ind. J. Chem. 13, 453–457.
3. a) Gollapudi S.R., Telikepalli H., and Keshavarz-Shokri A. (1989). Glepidotin C: a minor antimicrobial bibenzyl from *Glycyrrhiza lepidota.* Phytochemistry 28, 3556–2357; b) Fukai T., Wang Q.W., and Nomura T. (1991). Six prenylated phenols from *Glycyrrhiza uralensis.* Phytochemistry 30, 1245–1250.
4. Hashimoto T., Hasegawa K., and Yamaguchi H. (1974). Structure and synthesis of batatasins, dormancy-inducing substances of yam bulbils. Phytochemistry 13, 2849–2852.
5. Goda Y. and Sankawa U. (1985). Symposium Papers, 105th Annual Meeting of the Pharmacological Society of Japan, Kanazawa, Japan. 468.
6. Ghisalberti E., Jefferies P.R., and McAdam P. (1981). Isoprenylated resorcinol derivatives from *Glycyrrhiza acanthocarpa.* Phytochemistry 20, 1059–1061.
7. a) Rogers C.B. and Verotta L. (1996). Chemistry and biological properties of the African Combretaceae. In: Hostettmann K., Chinyanganya F., Maillard M., and Wolfender J.L. (eds) Chemistry, Biological and Pharmacological Properties of African Medicinal Plant pag. 121–141. University of Zimbabwe Publications; b) Verotta L. and Rogers C.B. (1997). The Hiccup Nut. *Combretum* species as source of bioactive compounds. In: Verotta L. (eds) Virtual Activity, Real Pharmacology. Different Approaches to the Search for Bioactive Compounds from Natural Sources. (1997). pag. 209–225. Research Signpost, Trivandrum, India,
8. a) Pettit G.R., Cragg G.M., Herald D.L., Schmidt J.M., and Lohavanijaya P. (1982). Isolation and structure of combetastatin. Can. J. Chem. 60, 1374–1376; b) Pettit G.R., Cragg G.M., and Singh S.B. (1987). Antineoplastic agents, 122. Constituents of *Combretum caffrum.* J. Nat. Prod. 50, 386–391; c) Pettit G.R. and Singh S.B. (1987). Isolation, structure, and synthesis of combretastin A-2, A-3, and B-2. Can. J. Chem. 65, 2390–2396; d) Malan E. and Swinny E. (1993). Substituted bibenzyls, phenanthrenes, and 9,10-dihydrophenanthrenes from the heartwood of *Combretum apiculatum.* Phytochemistry, 33, 1139–1142; e) Pettit G.R., Singh S.B., Hamel E., Lin C.M., Alberts D.S., and Garcia-Kendall D. (1989). Isolation and structure od the strong cell growth and tubulin inhibitor combretastatin A-4. Experientia 45, 209–211.
9. Miyase T, Ueno A., Takizawa N., Kobayashi H., and Karasawa H. (1988). Studies on the glycosides of *Epimedium grandiflorum* Morr. Var. *thunbergianum* (Miq.) *Nakai III.* Chem. Pharm. Bull. 36, 2475–2484.
10. Orsini F., Pelizzoni F., Bellini B., and Miglierini G. (1997). Synthesis of biologically active polyphenolic glycosides (combretastatin and resveratrol series). Carbohydrate Research 301, 95–109.
11. Langcake P. (1981). Disease resistance of *Vitis* spp. and the production of the stress metabolites resveratrol, ε-viniferin and pterostilbene. Physiol. Plant Pathol. 18, 213–226.
12. a) Klimek B. (1973). Evaluation of the usefulness of some *Rheum L.* species in medical treatment. Part I. Occurence of anthraquinone compounds and stilbene derivatives. Ann. Acad. Med. Lodz, 14, 133–148; b) Banks H.J., and Cameron D.W. (1971). A new natural stilbene glucoside from *Rheum rhaponticum* (Polygonaceae). Aust. J. Chem. 24, 2427–2430.
13. Orsini F., Pelizzoni F., Verotta L., Aburjai T., and Rogers C.B. (1997). Isolation, synthesis, and antiplatelet aggregation activity of resveratrol 3-O-β-D-glucopyranoside and related compounds. J. Nat. Prod. 60, 1082–1087.

14. a) Nonaka G., Minami M., and Nishioka I. (1977). Studies on rhubarb (*Rhei rhizoma*). III. Stilbenes glycosides. Chem. Pharm. Bull. 25, 2300–2305; b) Kashiwada Y., Nonaka G.I., and Nishioka I. (1984). Studies on rhubarb (*Rhei rhizoma*). VI. Isolation and characterisation of stilbenes Chem. Pharm. Bull. 32, 3501–3517.
15. Brinker A.M. and Seigler D.S. (1991). Isolation and identification of piceatannol as a phytoalexin from sugarcane. Phytochemistry 30, 3229–3232; Brinker A.M. and Seigler D.S. (1993). Time course of piceatannol accumulation in resistant and susceptible sugarcane stalks after inoculation with *Colletotrichum falcatum* Physiol. Mol. Plant Pathol. 42, 169–176.
16. Qiu F., Komatsu K., Kawasaki K., Saito K., Yao X., and Kano Y. (1996). A novel stilbene glucoside, oxyresveratrol 3′-O-β-glucopyranoside, from the root bark of *Morus alba*. Planta Med. 62, 559–561.
17. Teguo P.W., Fauconneau B., Deffieux G., Huguet F., Vercauteren J., and Merillon J.M. (1998). Isolation, identification, and antioxidant activity of three stilbene glucosides newly extracted from *Vitis vinifera* cell cultures. J. Nat. Prod. 61, 655–657.
18. Fernandez M.A., Pedro J.R., and Seoane E. (1983). Two polyhydroxystilbenes from stems of *Phoenix dactylifera*. Phytochemistry 22, 2819–2821.
19. a) Aguamah E., Langcake P., Leworthy D.P., Page J.A., Pryce R.J., and Strange R.N. (1985). Two novel stilbene phytoalexins from *Arachis hypogaea*. Phytochemistry 20, 1381–1383; b) Wotton H.R. and Strange R.N. (1985) Circumstantial evidence for phytoalexin involvement in the resistance of peanuts to *Aspergillus flavus*. J. Gen. Microbiol. 131, 487–494; Cooksey C.J., Garratt P.J., Richards S.E., and Strange R.N. (1988). A dienyl stilbene phytoalexin from *Arachis hypogaea*. Phytochemistry 27, 1015–1016
20. Shimuzu K., Kondo R., Sakai K., Lee S.O., and Sato H. (1998). The inhibitory components from *Artocarpus incisus* on melanin biosynthesis. Planta Medica 64, 408–412.
21 a) Pelizzoni F., Verotta L., Rogers C.B., Colombo R., Pedrotti B., Balconi G., Erba E., and D'Incalci M. (1993). Cell growth inhibitor constituents from *Combretum kraussii* Nat. Prod. Lett. 1 (4), 273; b) Pelizzoni F., Colombo R., D'Incalci M., and Verotta L. (1992). Combretastatin derivatives with antitumour activity, and process for the preparation thereof. P. WO 9405682 A1 940317. Priority IT 92-MI2033 920831. CA 121:73872.
22. Fossati S. Thesis dissertation in Biological Sciences. University of Milano Italy (A.a. 1996–1997)
23. Marston, A. and Hostettmann K. (1994). Counter-current chromatography as a preparative tool-applications and perspectives. J. Chromatogr. A, 658, 315–341.
24. a) Crombie L., Crombie W.M.L., and Jamieson S.V. (1980). Extractives of Thailand *Cannabis*: synthesis of canniprene and isolation of new geranylated and prenylated chrysoeriols. Tetrahedron Lett. 3607–3610; b) Reimann E. (1969). Natural stilbenes. Synthesis of polyhydroxystilbene ethers by the Wittig reaction. Chem. Ber. 102, 2881–2888; c) Reimann E. (1970). Naturliche polyhydroxystilbene. Die synthese von oxyresveratrol, piceatannol, and rhapontigenin" Tetrahedron Lett. 47, 4051–4053; d) Reimann E. (1971). Natural stilbenes. II. Synthesis of polyhydroxystilbenes. Justus Liebigs Ann. Chem. 750, 109–127.
25. Cardona M.L., Fernandez M.I., Garcia M.I., and Pedro J.R. (1986). Synthesis of natural polyhydroxystilbenes. Tetrahedron 42, 2725–2730.
26. a) Bachelor F.W., Loman A.A., and Snowdon L.R., (1970). Synthesis of pinosylvin and related heartwood stilbenes. Can. J. Chem. 48, 1554–1557; b) Wheeler O.H. and Battle de Pabon H.N. (1965). Synthesis of stilbenes. A comparative study. J. Org. Chem. 30, 1473–1477.
27. Medarde M., Pelaez-Lamamie de Clirac R., Lopez J.L., and San Feliciano A. (1994). A versatile approach to the synthesis of Combretastatins. J. Nat. Prod. 57, 1136–1144.
28. Ohsumi K., Nakagawa R., Fukuda Y., Hatanaka T., Morinaga Y., Nihei Y., Ohishi K., Suga Y., Akiyama Y., and Tsuji T. (1998). Novel combretastatin analogues effective against murine solid tumors: design and structure-activity relationships. J. Med. Chem. 41, 3022–3032.
29. Alonso E., Ramón D.J., and Yus M. (1997). Simple synthesis of 5-substituted resorcinols: a revisited family of interesting molecules. J. Org. Chem. 62, 417–421.
30. Crombie L.W., Crombie W.M., and Firth D.F. (1988). Synthesis of bibenzyl cannabinoids, hybrids of two biogenetic series found in *Cannabis sativa*. J. Chem. Soc. Perkin Trans. 1, 1263–1270.
31. Brown R.T., Fox B.W., Hadfield J. A, McGown A.T., Mayalarp S.P., Pettit G.R., and Woods J.A. (1995). Synthesis of water-soluble derivatives of Combretastatin A-4. J. Chem. Soc. Perkin Trans. I, 577–581.
32. Cichewicz, R.H. and Kouzi Samir A. (1998). Biotransformation of resveratrol to Piceid by *Bacillus cereus*. J. Nat. Prod. 61, 1313–1314.
33. Pettit G.R., Temple C., Narayanan Ven L., Varma R., Simpson M.J., Boyd M.R., Rener G.A., and Bansal N. (1995). Anti-Cancer Drug Des. 10, 299–309.

34. a) Williams R. and Rutledge R. (1998). Recent phytoestrogen research" Chemistry and Industry, 14–16; b) Gehm B.D., McAndrews J.M., Chien P.Y., and Jameson J.L. (1997). Resveratrol, a polyphenolic compound found in grapes and wine, is an agonist of the estrogens receptor. Proc. Natl. Acad. Sci. USA 94, 14138–14143.
35. Schwartner C., Bors W., Michel C., Franck U., Muller-Jakic B., Nenninger A., Asakawa Y., and Wagner H. (1995). Effects of Marchantins and related compounds on 5-lipoxygenase and cycloxygenase and their antioxidant properties: a structure activity relationship study. Phytomedicine 2, 113–117.
36. Kimura Y., Okuda H., and Arichi S. (1985). Effects of stilbenes on arachidonate metabolism in leukocytes. Biochim. Biophys. Acta 834, 275–278.
37. Arichi H., Kimura Y., and Okuda H. (1982). Effects of stilbene components of the roots of *Polygonum cuspidatum* Sieb. Et Zucc. on lipid metabolism. Chem. Pharm. Bull. 30, 1766–1770.
38. Frankel E.N., Waterhouse A.L., and Kinsella J.E. (1993). Inhibition of human LDL oxidation by resveratrol. Lancet 341, 1103–1104.
39. Rotondo S., Rajtar G., Manarini S., Celardo A., Rotilio D., De Gaetano G., Evangelista V., and Cerletti C. (1998). Effect of *trans*-resveratrol, a natural polyphenolic compound, on human polymorphonuclear leukocyte function. British J. Pharmacol. 123, 1691–1699.
40. a) Pace-Asciak C.R., Rounova O., Hahn S.E., Diamandis E.P., and Goldberg D.M. (1996). Wines and grape juice as modulators of platelet aggregation in healthy human subjects. Clin. Chim. Acta. 246, 163–182; b) Pace-Asciak C.R., Hahn S.E., Diamandis E.P., Soleas G., and Goldberg D.M. (1995). The red wine phenolics trans-resveratrol and quercetin block human platelet aggregation and eicosanoid synthesis: Implication for protection against coronary hearth disease. Clin. Chim. Acta 235, 207–219.
41. Jang M., Cai L., Udeani G.O., Slowing C.V., Thomas C.F., Beecher C.W.W., Fong H.H.S., Farnsworth N.R., Kinghorn A.D., Metha G.R., Moon R.C., and Pezzuto J.M. (1997). Cancer chemopreventive activity of resveratrol, a natural product derived from grapes. Science 275, 218–220.
42. Guatteo E., Bianchi L., Faravelli L., Verotta L., Pelizzoni F., Rogers C.B., and Wanke E. (1996). A novel K^+ channel blocker isolated from hiccup nut toxin. Neuroreport 2575–2579.
43. Orsini F., Verotta L., and Wanke E. Unpublished results.
44. Kimura Y., Ohminami H., Okuda H., Baba K., Kozawa M., and Arichi S. (1983). Effects of stilbene components of roots of *Polygonum* spp. on liver injury in peroxidized oil fed rats. Planta Med. 49, 51–54.
45. Fontecave M., Lepoivre M., Elleingand E., Gerez C., and Guitter O. (1998). Resveratrol, a remarkable inhibitor of ribonucleotide reductase. FEBS Lett. 421, 277–279
46. Sun N.J., Woe S.H., Cassady J.M., and Snapka R.M. (1998). DNA Polymerase and topoisomerase II inhibitors from *Psoralea coryfolia*. J. Nat. Prod. 61, 362–366.
47. Della Ragione F., Cucciolla V., Borriello A., Della Pietra V., Raciuppi L., Soldati G., Manna C., Galletti P., and Zappia V. (1998). Resveratrol arrests the cell division cycle at S/G2 phase transition. Biochem. Biophys. Res. Commun. 250, 53–58.
48. Jayatilake G.S., Jayasuriya H., Lee E.S., Koonchanok L.N., Geahlen R.L., Ashendel C.L., McLauglin J.L., and Chang C.J. (1993). Kinase inhibitors from *Polygonum cuspidatum*. J. Nat. Prod. 56, 1805–1810.
49. Deanin G.G., Oliver J.M., and Burg D.L. (1993). Piceatannol is a selective inhibitor of Fc epsilon-R1-induced protein tyrosine phosphorylation and a strong inhibitor of receptor-mediated signal transduction in RBL-2H3 rat-tumor mast-cells. J. Immunol. 150, 221.
50. a) Mannila E. and Talvitie A. (1992). Stilbenes from *Picea abies* bark. Phytochemistry 31, 3288–3289; b) Mannila E. and Talvitie A. (1993). Combretastatin analogs via hydration of stilbene derivatives. Justus Liebigs Ann. Chem. 1037. Mannila E., Talvitie A., and Kolehmainen E. (1993). Anti-leukaemic compounds derived from stilbenes in *Picea abies* bark. Phytochemistry 33, 813–816.
51. Pelizzoni F., Bellini B., and Miglierini G. It. Pat. MI94A/000921 (1994).
52. Cushman M., Nagarathnam D., Gopal D., Chakraborti A.K., Lin C.M., and Hamel E. (1991). Synthesis and evaluation of stilbene and dihydrostilbene derivatives as potential anticancer agents that inhibit tubulin polymerisation. J. Med. Chem. 34, 2579–2588

16

POST-TRANSLATIONAL MODIFICATIONS OF EUKARYOTIC INITIATION FACTOR-5A (eIF-5A) AS A NEW TARGET FOR ANTI-CANCER THERAPY

Michele Caraglia,[1] Pierosandro Tagliaferri,*[2] Alfredo Budillon,[3] and Alberto Abbruzzese[1]

[1] Department of Biochemistry and Biophysics "F. Cedrangolo"
Second University of Naples
[2] Department of Endocrinology and Molecular and Clinical Oncology
University "Federico II" of Naples
[3] Division of Experimental Oncology C
National Tumour Institute
Foundation "G. Pascale," Naples
Italy

1. ABSTRACT

Eukaryotic translation initiation factor 5A (eIF-5A) is the only cell protein that contains the unusual basic amino acid hypusine [N^{ε}-(4-amino-2-hydroxybutyl)lysine]. Hypusine is formed by the transfer of the butylamine portion from spermidine to the ε-amino group of a specific lysine residue of eIF-5A precursor and the subsequent hydroxylation at carbon 2 of the incoming 4-aminobutyl moiety. Agents that reduce cell hypusine levels inhibit the growth of mammalian cells. These observations suggest that hypusine is crucial for proliferation and transformation of eukaryotic cells. Here we have studied whether the inhibition of hypusine synthesis can potentiate the anti-cancer activity of the anti-tumour agents interferon-α (IFNα) and cytosine arabinoside (ara-C). We have found that IFNα increased epidermal growth factor receptor (EGF-R) expression, but reduced S phase and proliferative marker expression in human

---

*Present address: Department of Experimental and Clinical Medicine, University "Magna Graecia" of Catanzaro

Advances in Nutrition and Cancer 2, edited by Zappia *et al.*
Kluwer Academic / Plenum Publishers, New York, 1999.

epidermoid KB cells and that this effect was antagonised by epidermal growth factor (EGF). Growth inhibition induced by IFNα was paralleled by decreased hypusine synthesis and, when EGF counteracted anti-proliferative effects, a reconstitution of hypusine levels was recorded. We also studied the effects of IFNα on the cytotoxicity of the recombinant toxin TP40 which inhibits elongation factor 2, another step of protein synthesis, through EGF-R binding and internalisation; IFNα induced an about 27-fold increase of TP40 cytotoxicity in KB cells.

Ara-C, another antineoplastic agent commonly used in haematologic malignancies, induced both apoptosis and iron depletion in human acute myeloid leukaemic cells. The combination of ara-C and of the iron chelator desferioxamine, a strong inhibitor of hypusine synthesis, had a synergistic activity on apoptosis in these cells. The data strongly suggest that the post-translational modifications of eIF-5A could be a suitable target for the potentiation of the activity of anti-cancer agents.

2. INTRODUCTION

Epidermal growth factor (EGF) activates polyamine uptake and synthesis in cancer cells.[1,2] Polyamines play an important role in the regulation of cell growth and are involved in the biosynthesis of hypusine [N^{ε}-(4-amino-2-hydroxybutyl)lysine], an unusual basic amino acid found only in the eukaryotic translation initiation factor 5A (eIF-5A, previously designated eIF-4D).[3–5] Hypusine is formed by the transfer of the butylamine portion from spermidine to the ε-amino group of a specific lysine residue of eIF-5A[6–10] and its subsequent hydroxylation at carbon 2 of the incoming 4-aminobutyl moiety.[11–12] Recently we have also found that eIF-5A metabolic pathway is further correlated to the transglutaminase expression and activity.[13,14] eIF-5A probably acts in the final stage of the initiation phase of protein synthesis by promoting the formation of the first peptide bond.[15] Hypusine plays a key role in the regulation of eIF-5A function because eIF-5A precursors,[16] which do not contain hypusine, have little, if any, activity.[17] The correlation between hypusine content and cell proliferation suggests that hypusine might play a role in cell growth and differentiation.[18–21] Moreover, agents that reduce cell hypusine levels[21–23] inhibit growth of mammalian cells thus demonstrating that hypusine is crucial for the proliferation of eukaryotic cells.[24]

We previously demonstrated that interferon α_2 recombinant (IFNα) inhibits growth, upregulates the expression of the receptor for epidermal growth factor (EGF-R) and enhances the proliferative response of human epidermoid cancer KB cells to EGF.[25,26] The latter effect is paralleled by an increased tyrosine phosphorylation of cellular proteins and of EGF-R itself.[26] Moreover, we have reported that cytostatic concentrations of cytosine arabinoside (ara-C) increase the surface expression of EGF-R and TRF-R in epithelial cancer and melanoma cells and we have hypothesised that the TRF-R upregulation could mediate a protective effect against the growth inhibitory activity of ara-C in these cells.[27–29] These results suggested that increased peptide growth factor receptor (PGF-R) expression could be a general homeostatic reaction of tumour cells to several growth inhibitory stimuli.[29] In the study described herein we evaluated if EGF could reverse IFNα-induced KB cell cycle perturbation. We also studied whether the recovery of IFNα-treated KB cell proliferation induced by EGF is paralleled by restoration of hypusine synthesis. Inhibition of protein synthesis is a new strategy in tumour cell killing and, in this context, the contemporaneous inhibition of initiation and elongation factors of protein synthesis is an attractive perspective. Therefore, we evaluated the effects of IFNα on KB cell cytotoxicity induced by the recombinant toxin TP40, which inhibits Elongation Factor-2 (EF2) after binding to EGF-R.[27] Furthermore, we have evaluated the effects of ara-C on the TRF-R expression and iron metabolism in human leukaemic cells. The effects of the combination between ara-C and the hypusine synthesis inhibitor desferioxamine on the apoptosis of human leukaemic cells were also investigated.

2. MATERIALS AND METHODS

2.1. Cell Culture and Cell Proliferation Assays

The human oropharyngeal epidermoid carcinoma KB cell line, obtained from ATCC (Rockville, MD), was grown in DMEM supplemented with heat inactivated 10% FBS, 20 mM HEPES, 100 U/ml penicillin, 100 μg/ml streptomycin, 1% L-glutamine, and 1% sodium pyruvate. Leukaemia cells were derived from 16 adult patients affected by AML. Primary cultures were obtained from peripheral blast cells after Ficoll-Hypaque separation and subsequent washing in PBS. Leukemia cells were grown in RPMI 1640 supplemented with 10% FBS, 2 mM L-glutamine, 10 mM HEPES, 100 U/ml penicillin, and 100 μg/ml streptomycin. For cell growth experiments leukemia cells were seeded at 4×10^5 cells per well in 6-well culture plates (Becton Dickinson Co.) Different concentrations of ara-C and/or desferioxamine were added 24 h later. Cell growth assessment was performed by haemocytometric cell count and Trypan Blue viability assay. The cells were grown in a humidified atmosphere of 95% air/5% CO_2 at 37°C.

2.2. TRF-R Live-Cell Radioimmunoassay

Leukaemic cells were seeded in 96-well microtiter plates at 5×10^4 cells per well inoculum size and were exposed for 48 h to 100 nM and 1000 nM ara-C. After over-night incubation in serum-free medium (RPMI 1640 with non essential amino acids and vitamins added), the growth medium was removed. Then 100 μl 5% BSA (w/v) in RPMI 1640 were added to each well. After 60 min of incubation at 37°C, cells were washed with RPMI 1640 containing 5% BSA. Samples containing 50 μl of appropriately diluted MAb OKT9 were then added to each well. After incubation for 3 h at 4°C, the cells were washed twice with 5% PBS/BSA (w/v). Then, ^{125}I-labelled sheep anti-mouse IgG was added to each well. Following a 60 min incubation (at 37°C), the cells were washed three times with PBS/BSA 5% (w/v) and adsorbed with a cotton swab; radioactivity was counted in a Beckmann γ-counter, as previously described.[27]

2.3. Determination of Intracellular Non Haem Iron Content

AML cells, grown in 100-mm dishes, were exposed for different times to ara-C. For cell extract preparation the cells were washed twice with PBS and counted with the haemocytometer. Then the cells were incubated for 30 min at 4°C with 1 ml of lysis buffer (1% Triton, 10% Glycerol, 25 mM Hepes, pH 7.4 in PBS) and passed through a 21 G needle. The lysates were incubated for 20 min at 22°C, in 0.3 N HCl, TCA-precipitated and finally centrifuged at 1200 *g* for 15 min. The supernatants were mixed with chromogen (0.51 mM disulfonate bathophenantroline, 1.3 M sodium acetate, pH 4.6, 30 mM sodium pyrosulfite, 1.83 mM p-(N-methyl)aminophenol) and the absorbancy was spectrophotometrically determined at the wavelength of 546 nm. The non haem iron concentration was calculated by interpolation, as previously described.[31]

2.4. FACS Analysis of Ki67 and PCNA

KB cells were treated with IFNα and/or EGF as described above and then harvested in EDTA solution (0.1% EDTA in PBS without Ca^{++}/Mg^{++}, pH 7.4). 2×10^6 cells were centrifuged and resuspended in washing medium (PBS, 1% human serum and 0.1% NaN_3, pH 7.4). Appropriately diluted anti-Ki67 or anti-PCNA MAbs (Dako, Bastrup, Denmark) were added to the pellet, while an irrelevant IgG_1 MAb was used as control. After 30 min at 4°C, KB cells were washed twice in washing medium and cell pellets were incubated with fluorescein-isothyocianate (FITC)-labelled goat anti-mouse Ig (Technogenetics, Milan, Italy) diluted 1:20 at 4°C for 30 min in the dark. After two washings with PBS KB cells were analysed by a FACSCAN flow cytometer (Beckton Dickinson, USA).

2.5. FACS Analysis of Cell Cycle Distribution

KB or human leukaemic cells were treated with IFNα and/or EGF or with ara-C and or desferioxamine, respectively, as described above and trypsinised, washed twice with PBS without Ca^{++}/Mg^{++} and fixed in 70% ethanol. 10^6 KB or human leukaemic cells were incubated at room temperature for 30 min in 1 ml of a propidium iodide staining solution (50 μg/ml in PBS without Ca^{++}/Mg^{++}, pH 7.4). DNA analysis was performed in duplicate with a FACSCAN flow cytometer (Becton Dickinson) linked to a Hewlett-Packard computer. Cell cycle data analysis was performed with the CELL-FIT program (Becton Dickinson). Pulse area versus pulse width gating was performed to avoid doublets from the G_2/M region. The per cent proliferation index was determined as the ratio between the sum of cell percentage in G_1, G_2, S, and M phase and the sum of cell percentage in G_0, G_1, G_2, S, and M phase of cell cycle. The ipodyploid peak was considered as expression of apoptosis.

2.6. Protein Synthesis Inhibition Assay

KB cells were grown with and without 1000 IU/ml IFNα for 48 h in 24-multiwell plates. Different concentrations of TP40 were added to the cells that were incubated at 4°C for 60 min. KB cells were washed three times with PBS/BSA 0.1% (w/v) and incubated at 37°C for 5 h in complete medium. The medium was withdrawn and complete medium containing [^{3}H]L-Leucine (45–70 Ci/mmol, Amersham, Arlington Heights, IL) was added to KB cells which were then incubated at 37°C for 60 min. Then, the cells were washed three times with PBS without Ca^{++} or Mg^{++} and dissolved in 1 ml 0.1 N NaOH. After 5 min of incubation at 37°C, the proteins were precipitated in 12% ice-cold trichloro-acetic acid (TCA) for 60 min. The precipitated TCA was washed twice in 6% TCA and digested in 1.5 ml 0.1 N NaOH at 56°C for 30 min. Aliquots were assayed with the Lowry method for protein quantitation and counted in scintillation liquid in a β-counter (Beckmann) to detect incorporation of [^{3}H]L-Leucine, as previously described.[30]

2.7. Isolation, Purification, and Identification of Hypusine

KB cells were seeded in 100-mm dishes and 24 h before processing, 8 μl [terminal methylenes ^{3}H]spermidine 3 HCl (15 Ci/mmol) were added to each dish. Cell lysates were prepared using cells from 10 dishes by suspending the cells in 4 ml of PBS, sonicating, and finally centrifuging for 30 min at 25,000 × *g*. The lysates were treated with solid ammonium sulphate and the precipitate hydrolysed in 6 N HCl at 110°C for 18 h. The hydrolysates were applied to 0.5 × 4 cm columns of AG 50 × 2 (H^+ form, 200–400 mesh) and eluted with 30 ml of 1 N HCl, 20 ml 3 N HCl, and 30 ml of 6 N HCl. The hypusine contained in the 3 N HCl fraction was determined by using a reversed-phase high performance liquid chromatography (HPLC) method described elsewhere.[9]

3. RESULTS AND DISCUSSION

3.1. Effects of IFNα and EGF on Cell Cycle Kinetics and EGF-R Expression on KB Cells

Exposure of KB cells to 1000 IU/ml IFNα for 48 h induced a significant decrease of the S phase cell fraction, proliferative index and of both Ki 67 and PCNA expression as evaluated with cytofluorimetric assay. 10 nM EGF did not cause significant changes of the S phase cell fraction or of the other cell proliferation markers (Fig. 1). Incubation of IFNα-treated KB cells with 10 nM EGF for 12 h restored the S phase fraction, proliferative index and expression of Ki 67 and PCNA to almost control values (Fig. 1). Therefore, the growth inhibition induced by IFNα is completely reverted by EGF in human epidermoid cancer KB cells.

3.2. Effects of IFNα and EGF on Hypusine Synthesis in KB Cells

Since IFNα and EGF alter the proliferative status of tumour cells we evaluated their effects on hypusine synthesis after labelling KB cells with [^{3}H]-putrescine. Exposure of KB cells to 10 nM EGF for 10 min or for 12 h did not significantly affect hypusine synthesis (Fig. 2). However, IFNα caused hypusine synthesis to decrease by about 75%; hypusine formation progressively returned almost to control values when the cells were incubated with EGF for 12 h (Fig. 2). Therefore, the antiproliferative effects induced by IFNα were associated with decreased hypusine synthesis, which, in turn, was restored by the addition of EGF to IFNα-treated KB cells.

Agents that suppress hypusine synthesis induce reversible arrest at the G_1-S boundary of the cell cycle.[32–34] On the other hand, hypusine synthesis increases after mitogen treatment of human peripheral blood lymphocytes.[34] Moreover, both polyamine[35] and hypusine levels (together with the enzymes that regulate their metabolism)[9,11,18,36,37] are correlated to normal and malignant growth. This coincides with our finding that modifications of the cell cycle induced by IFNα are paralleled by reduction of hypusine synthesis.

3.3. Effects of IFNα on Cytotoxicity of TP40

Manipulations of protein synthesis machinery is a key feature of strategies designed to inhibit tumour cell growth. Attempts are being made to construct agents

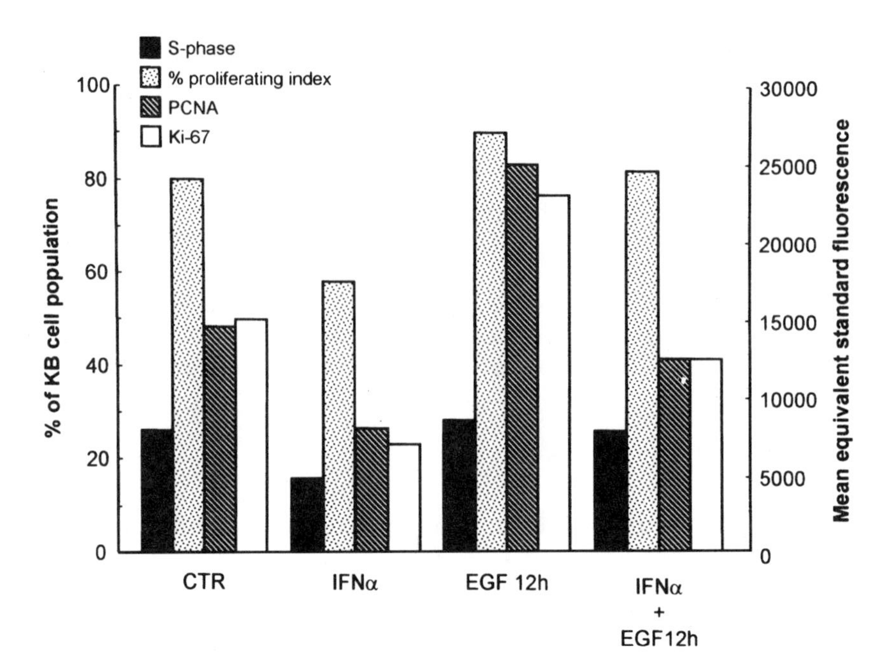

Figure 1. Effects of *IFNα* and EGF on KB cell cycle distribution. The percent distribution of KB cells in S-phase of cell cycle was evaluated by FACS analysis after nuclear labelling with propidium iodide. For the evaluation of PCNA and Ki-67 expression, the cells were labelled with anti-Ki67 or anti-PCNA MAbs and, after incubation with an FITC anti-mouse antibody, they were analysed by FACS. The levels of Ki67 and PCNA are expressed as mean equivalent standard fluorescence calculated comparing the fluorescence intensity of the antigen versus that of a standard fluorescent microsphere. The percent proliferation index was calculated as described under "Materials and Methods." S-phase (■); % proliferation index (▨); Ki-67 (□); PCNA (▧).

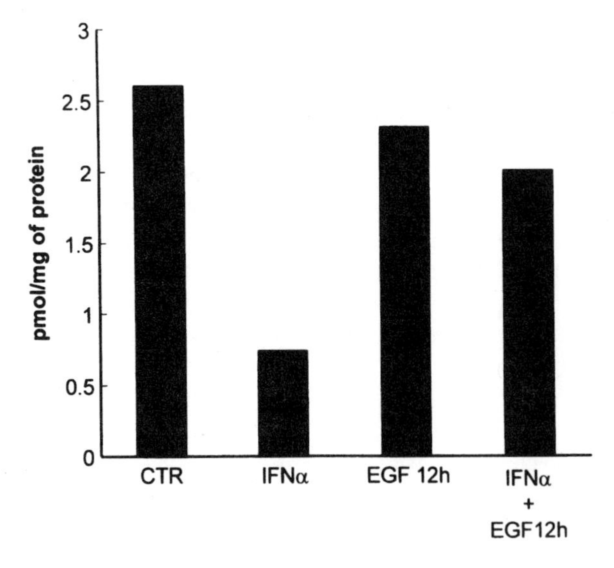

Figure 2. Effects of *IFNα* and EGF on hypusine synthesis in KB cells. Hypusine levels were determined after labelling of KB cells with [terminal methylenes ^{3}H] spermidine 3 HCl and the subsequent HPLC analysis of hydrolyzed cell protein extract as described under "Materials and Methods." The results are expressed in bars as the mean of the data from three different experiments performed in triplicate ± standard deviation (*SD*).

that induce tumour cytotoxicity by inhibiting the translational process. A case in point is the fusion protein TP40 which is a genetically engineered construct consisting of a recombinant toxin derived from *Pseudomonas aeruginosa* and of the targeting component, transforming growth factor-α, one of the natural ligands of EGF-R, which, in turn, is over-expressed on tumour cells.[29,30,38] Thus TP40 preferentially recognises cancer cells through EGF-R binding. The cytotoxic action of the targeted toxin is exerted through the physiological internalisation process of the bound EGF-R.[29] After internalisation the recombinant toxin inhibits protein synthesis through ADP-ribosylation of the EF2. Because IFNα upregulates EGF-R expression on the cell surface and reduces hypusine synthesis, which is necessary for eIF-5A activation, we asked if IFNα could increase the cytotoxicity of TP40 which targets EGF-R and inhibits another step of protein synthesis.

The concentration of TP40 that inhibited protein synthesis by 50% (ID:50) in untreated KB cells was 4.9 ng/ml as evaluated with a [^{3}H]L-leucine incorporation assay (Fig. 3). When KB cells were exposed to 1000 IU/ml IFNα the ID:50 of TP40 was 0.18 ng/ml. Moreover, the effects of TP40 were antagonised by the addition of saturating concentrations of EGF. This observation demonstrates that cytotoxicity is mediated by specific binding of the fusion protein to EGF-R (Fig. 3). The latter effect could be attributed both to IFNα-induced EGF-R upregulation and to IFNα- and TP40-induced inhibition of multiple steps of protein synthesis.

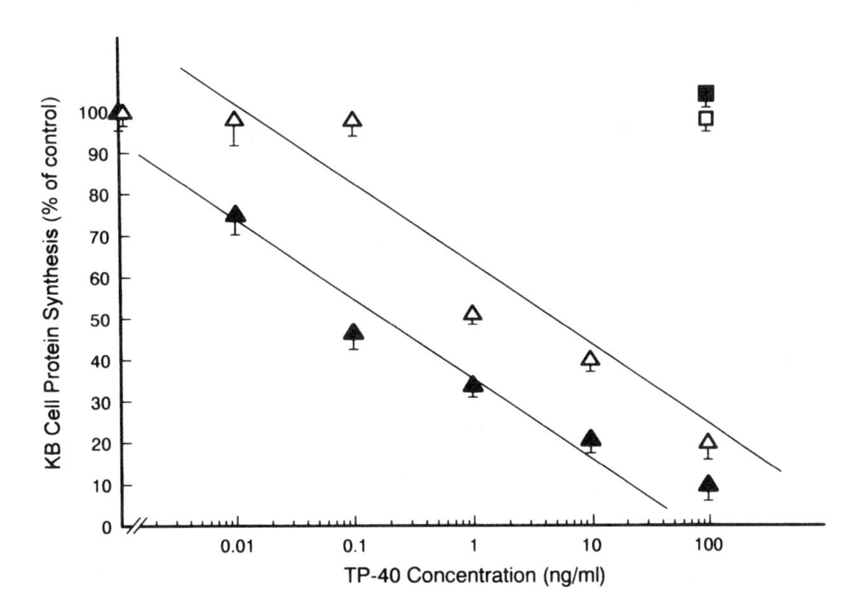

Figure 3. Effects of *IFNα* on cytotoxicity induced by TP40 in KB cells. KB cells were grown without (△) and with (▲) 1000 IU/ml *IFNα* for 48 h. Control cells (□) or *IFNα*-treated cells (■) incubated with 100 ng/ml TP40 and 100 nM EGF. KB cells were cultured with and without *IFNα* and/or EGF, then incubated with various concentrations of TP40 and finally labelled with [^{3}H] L-Leucine. Proteins were extracted from cells and processed as described under "Materials and Methods." Protein-associated radioactivity was determined with a β-counter. The percent inhibition of protein synthesis was calculated from the data of each experimental point versus the protein synthesis of KB cells not incubated with TP40. *Points*, average of results from three different experiments performed in triplicate. *Bars*, standard errors.

3.4. Ara-C Modulates TRF-R Expression and Intracellular Non Haem Iron Levels on AML Blasts and on HL-60 and U-937 Cell Lines

We have examined fresh leukaemic cells from 16 patients with AML and we have evaluated the effects of ara-C on the expression of TRF-R on these cells. We have found that treatment with 100 nM ara-C for 48 hours increased cell surface expression of the TRF-R on 13/16 (81%) leukaemic cultures (Fig. 4A). On the leukaemic cells from 7/16 (44%) patients we have recorded a 2–3-fold TRF-R increase while in the remaining cultures (37%) TRF-R expression was about 1-fold enhanced by ara-C (Fig. 1C). Similar results were obtained on HL-60 and U-937 after 48 hrs exposure to 100 nM ara-C (Fig. 4A).

We have evaluated the modifications of intracellular non haem iron content induced by ara-C on AML blasts from 5 patients. We have selected leukaemic cells in which ara-C increased TRF-R expression. We have found that ara-C decreased the intracellular non haem iron levels in all leukaemic blast cell lines 24 hours after plating (Fig. 4B). Twenty-four hours later, when a 2- or 3-fold decrease of the non haem metal levels occurred in the unexposed cells also, the intracellular non haem iron content was

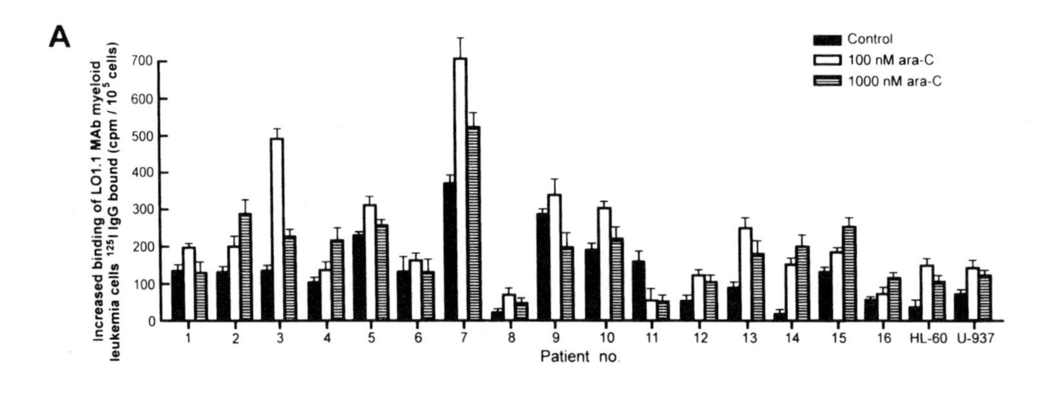

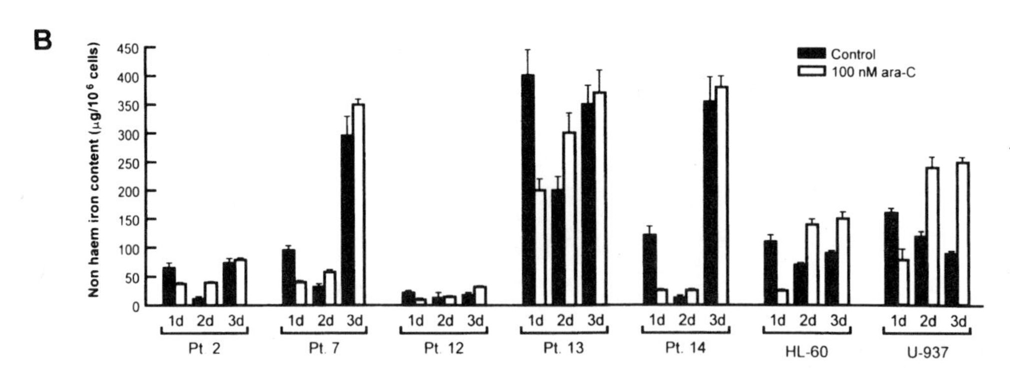

Figure 4. *Ara-C* effects on *TRF-R* expression and non haem iron content on AML cells. (A) Binding of 1.0 mg/ml anti-*TRF-R* MAb to untreated (■) or 48 h 100 nM (□) or 1000 nM *ara-C*-treated (▤) cells derived from 16 AML patients and HL-60 and U-937 cell lines, as described in "Materials and Methods". Points, average of triplicate determinations. Bars, SD. (B) AML cells were grown in 100 mm dishes. The cells were exposed for different times to 100 nM *ara-C* and then intracellular non haem iron content was spectophotometrically determined as described in "Materials and Methods." Non haem iron levels were expressed as µg/10^6 cells in untreated (■) or 100 nM (□) *ara-C*-treated AML cells.

reconstituted almost to normal levels in the ara-C exposed cells (Fig. 4B). Addition of fresh growth medium caused a strong increase of intracellular non haem iron in both untreated and ara-C-treated cells. Similar results were obtained in HL-60 and U-937 cells (Fig. 4B). Seventy-two hours after plating, following resuspension in fresh growth medium, recovery of non haem iron content was detected in the control cells while the iron storages remained stable or were further increased in ara-C-treated cells (Fig. 4B). Recovery of intracellular non haem iron in the ara-C treated cells was paralleled by the upregulation of TRF-R. The timing of these events suggests that the enhanced expression of TRF-R is induced by iron depletion in order to overcome the lack of an essential metabolite caused by ara-C in the tumour cells. It is known that a low intracellular iron content activates the synthesis of TRF-R at the transcriptional level through the stimulation of specific transactivating factors named iron responsive elements (IREs).[39] In fact, an early depletion of intracellular non haem iron was induced by ara-C on the leukaemic cells and recovery of iron content occurred when TRF-R was upregulated following a longer exposure to the drug. These findings suggest that the non haem iron depletive effects of ara-C-in AML cells are counteracted by changes in the surface expression of TRF-R.

3.5. Ara-C and Desferioxamine Cooperate on HL-60 and U-937 Apoptosis

On the bases of our observations on the non haem iron-depleting effect of ara-C, we have evaluated the effects of the combination between ara-C and desferioxamine on the growth and on the onset of programmed cell death of HL-60 and U-937 cells. We found that, after 96 hrs of cell culture, apoptosis was detectable in the 25% of HL-60 and 5% of U-937 control cells while 10 nM ara-C caused apoptosis in the 74% of HL-60 and 30% of U-937 cell population (Fig. 5). 0.05 μg/ml and 0.5 μg/ml desferioxamine induced apoptosis in the 49% and 60% of HL-60 and 7% and 5% of U-937 cells, respectively (Fig. 5B), while the combination of desferioxamine 0.05–0.5 μg/ml with 10 nM ara-C always induced programmed cell death in about 100% of HL-60 and 80% of U-937 cells (Fig. 5).

Combined iron depletive therapy has been proposed as a therapeutic approach to haematologic malignancies and human tumours. New pharmacological agents that block iron intake have been recently identified.[40] Moreover, desferioxamine is a powerful inhibitor of hypusine synthesis and again the interference with the post-translational modifications of eIF-5A potentiates the anti-tumour activity of an anti-neoplastic agent.[20] In fact, the combination of desferioxamine with ara-C at cytostatic concentrations, as suggested by our *in vitro* results, could be indeed of therapeutical interest considering that iron chelators have few side effects and might not produce cumulative toxicity when given together to ara-C.

4. CONCLUSIONS

In conclusion, we have studied the inhibition of tumour cell growth induced by IFNα and the ability of EGF to antagonise the effects of IFNα by modulating hypusine synthesis. The concurrent inhibition of eIF-5A induced by IFNα and of EF2 induced by TP40 could be relevant in determining the anti-cancer therapy (42). Moreover, we describe that the effects of ara-C on growth inhibition and apoptosis of human

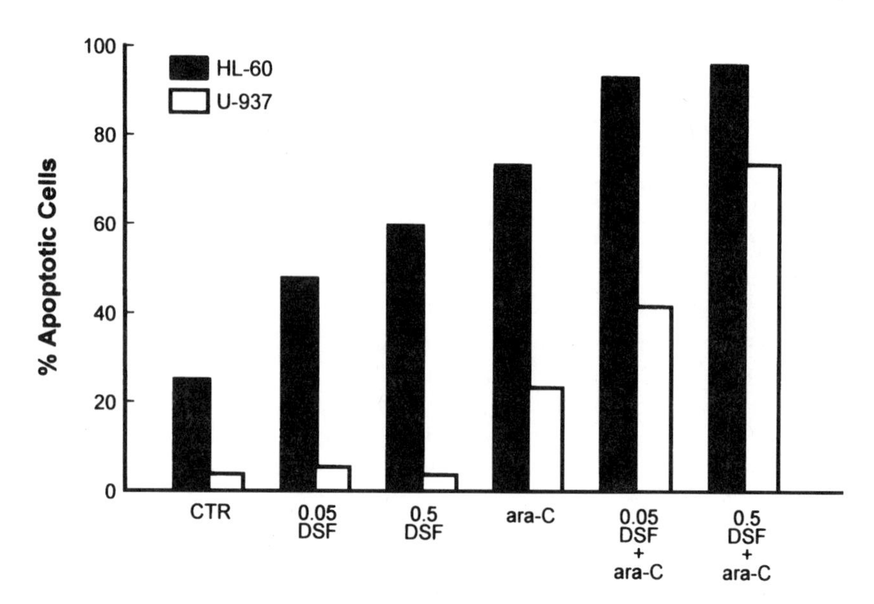

Figure 5. Effects of *ara-C* and desferioxamine on HL-60 and U-937 apoptosis. HL-60 and U-937 cells were grown in complete medium and in humidified atmosphere. Apoptosis was demonstrated and quantitized by FACS analysis labelling with propidium iodide 96h after seeding of HL-60 (■) and U-937 (□) cells. Unexposed controls (CTR); (B) 0.05 μg/ml *Desferioxamine* (0.05 DSF); 0.5 μg/ml DSF (0.5 DSF); 10 nM *ara-C* (*ara-C*); 0.05 μg/ml DSF + 1 nM *ara-C* (0.05 DSF + *ara-C*); 0.5 μg/ml DSF + 1 nM *ara-C* (0.5 DSF + *ara-C*).

myeloid leukaemia cells in culture are potentiated by the iron chelator desferioxamine. Again the inhibition of the post-translational modification of eIF-5A potentiates the antitumour activity of an anticancer agent.

REFERENCES

1. Milovic, V., Deubner, C., Zeuzem, S., Piiper, A., Caspary, W.F., and Stein, J. (1995) EGF stimulates polyamine uptake in Caco-2 cells. Biochem. Biophys. Res. Commun. **206**, 962–968.
2. Wojciechowski, K., Trzeciak, L., Konturek, S.J., and Ostrowski (1995) Inhibition of acid secretory response and induction of ornithine decarboxylase and its mRNA by TGF alpha and EGF in isolated rat gastric glands. J. Regul. Pept. **56**, 1–8.
3. Panagiotidis, C., Artandi, S., Calame, K., and Silverstein, S.J. (1995) Polyamines alter sequence-specific DNA-protein interactions. Nucleic Acid Res. **23**, 1800–1809.
4. Cooper, H.L., Park, M.H., Folk, J.E., Safer, B., and Braverman, R. (1983) Identification of the hypusine-containing protein hy+ as translation initiation factor eIF-4D. Proc. Natl. Acad. Sci. USA **80**, 1854–1857.
5. Park, M.H., Wolff, E.C., and Folk, J.E. (1993) Hypusine: its post-translational formation in eukaryotic initiation factor 5A and its potential role in cellular regulation. BioFactors **4**, 95–104.
6. Park, M.H. and Wolff, E.C. (1988) Cell-free synthesis of deoxyhypusine. Separation of protein substrate and enzyme and identification of 1,3-diaminopropane as a product of spermidine cleavage. J. Biol. Chem. **263**, 15264–15269.
7. Wolff, E.C., Park, M.H., and Folk, J.E. (1990) Cleavage of spermidine as the first step in deoxyhypusine synthesis. The role of NAD. J. Biol. Chem. **265**, 4793–4799.
8. Park, M.H., Chung, S.I., Cooper, H.L., and Folk, J.E. (1984) The mammalian hypusine-containing protein, eukaryotic initiation factor 4D. J. Biol. Chem. **259**, 4563–4565.
9. Beninati, S., Abbruzzese A., and Folk J.E. (1990) High-performance liquid chromatographic method for determination of hypusine and deoxyhypusine. Anal. Biochem. **184**, 16–20.

10. Park, M.H., Cooper, H.L., and Folk, J.E. (1982) The biosynthesis of protein-bound hypusine (N^{ε}-(4-amino-2-hydroxybutyl)lysine): lysine as the amino acid precursor and the intermediate role of deoxhypusine (N^{ε}-(4-amino-butyl)lysine). J. Biol. Chem. **257**, 7217–7222.
11. Abbruzzese, A., Park, M.H., and Folk, J.E. (1986) Deoxhypusine hydroxylase from rat testis. Partial purification and characterization. J. Biol. Chem. **261**, 3085–3089.
12. Abbruzzese, A., Park, M.H., and Folk, J.E. (1985) Deoxyhypusine hydroxylase: distribution and partial purification from rat testis. Fed. Proc. **44**, 1487.
13. Beninati, S., Nicolini, L., Jakus, J., Passeggio, A., and Abbruzzese, A. (1995) Identification of a substrate site for transglutaminases on the human protein synthesis initiation factor 5A.Biochem. J. **305**, 725–728.
14. Beninati, S., Gentile, V., Caraglia, M., Lentini, A., Tagliaferri, P., and Abbruzzese, A. (1998) Tissue transglutaminase expression affects hypusine metabolism in balb-C 3T3 cells. FEBS Letters **437**, 34–38.
15. Hershey, J.W.B. (1991) Translational control in mammalian cells. Annu. Rev. Biochem. **61**, 717–755.
16. Stiuso, P., Colonna, G., Ragone, R., Caraglia, M., Hershey, J.W.B., Beninati, S., and Abbruzzese A. (1999) Structural organization of the human eukaryotic initiation factor 5a precursor and its site-directed variant lys 50→arg. Amino Acids **16**, 91–106.
17. Park, M.H., Wolff, E.C., Smit-McBride, Z., Hershey, J.W.B., and Folk, J.E. (1991) Comparison of the activities of variant forms of eIF-4D: the requirement for hypusine or deoxyhypusine. J. Biol. Chem. **266**, 7988–7994.
18. Cooper, H.L., Park, M.H., and Folk, J.E. (1982) Posttranslational formation of hypusine in a single major protein occurs generally in growing cells and is associated with activation of lymphocyte growth. Cell **29**, 791–797.
19. Lentini, A., Mattioli, P., Nicolini, L., Pietrini, A., Abbruzzese, A., and Beninati, S. (1997) Anti-invasive effects of theophylline on experimental B16-F10 melanoma lung metastasis. Cancer J. **10**, 274–278.
20. Abbruzzese, A. (1988) Developmental pattern for deoxyhypusine hydroxylase in rat brain. J. Neurochem. **50**, 695–699.
21. Abbruzzese, A., Park, M.H., Beninati, S., and Folk, J.E. (1989) Inhibition of deoxyhypusine hydroxylase by polyamines and by a deoxyhypusine peptide. Biochem. Biophys. Acta **997**, 248–255.
22. Abbruzzese, A., Hanauske-Abel, H.M., Park, M.H., Henke, S., and Folk, J.E. (1991) The active site of deoxyhypusyl hydroxylase: use of catecholpeptides and their component chelator and peptide moieties as molecular probes. Biochem. Biophys. Acta **1077**, 159–166.
23. Jakus, J., Wolff, E.C., Park, M.H., and Folk, J.E. (1993) Features of the spermidine-binding site of deoxyhypusine synthase as derived from inhibition studies. J. Biol. Chem. **268**, 13151–13159.
24. Park, M.H., Wolff, E.C., and Folk, J.E. (1993) Is hypusine essential for eukaryotic cell proliferation? Trends Biochem. Sci. **18**, 475–479.
25. Budillon, A., Tagliaferri, P., Caraglia, M., Torrisi, M.R., Normanno, N., Iacobelli, S., Palmieri, G., Stoppelli, M.P., Frati, L., and Bianco, A.R. (1991) Upregulation of epidermal growth factor receptor induced by alpha-interferon in human epidermoid cancer cells. Cancer Res. **51**, 1294–1299.
26. Caraglia, M., Leardi, A., Corradino, S., Ciardiello, F., Budillon, A., Guarrasi, R., Bianco, A.R., and Tagliaferri, P. (1995) "α-Interferon potentiates epidermal growth factor receptor-mediated effects on human epidermoid carcinoma KB cells. Int. J. Cancer, **61**, 342–347.
27. Caraglia, M., Tagliaferri, P., Correale, P., Genua, G., Pinto, A., Del Vecchio, S., Esposito, G., and Bianco, A.R. (1993) Cytosine arabinoside increases the binding of [^{125}I]-labelled epidermal growth factor and [^{125}I]-transferrin and enhances the in vitro targeting of human tumor cells with anti-(growth factor receptor) mAb. Cancer Immunol. Immunother. **37**, 150–156.
28. Caraglia, M., Leardi, A., Improta, S., Perin, V., Ricciardi, B., Arra, C., Ferraro, P., Fabbrocini, A., Pinto, A., Bianco, A.R., and Tagliaferri, P. (1997) Transient exposure to cytarabine increases peptide growth factor receptor expression and tumorigenicity of melanoma cells. Anticancer Res. **17**, 2369–2376.
29. Tagliaferri, P., Caraglia, M., Muraro, R., Budillon, A., Pinto, A., and Bianco, A.R. (1994) Pharmacological modulation of peptide growth factor receptor expression on tumor cells as a basis for cancer therapy. Anti-Cancer Drugs **5**, 379–393.
30. Pai, L.H., Gallo, M.G., Fitzgerald, D.J., and Pastan, I. (1991) Antitumor activity of a transforming growth factor alpha-Pseudomonas exotoxin fusion protein (TGF-alpha-PE40). Cancer Res. **51**, 2808–2812.
31. Caraglia, M., Libroia, A.M., Corradino, S., Coppola, V., Guarrasi, R., Barile, C., Genua, G., Bianco, A.R., and Tagliaferri, P. (1994) "α-Interferon induces depletion of intracellular iron content and upregulation of functional transferrin receptors on human epidermoid cancer KB cells. Biochem. Biophy. Res. Commun. **203**, 281–288.

32. Park, M.H. (1987) Regulation of biosynthesis of hypusine in Chinese hamster ovary cells. Evidence for eIF-4D precursor polypeptides. J. Biol. Chem. **262**, 12730–12734.
33. Lalande, M. and Hanausske-Abel, H.M. (1990) A new compound which reversibly arrests T lymphocyte cell cycle near the G1/S boundary. Exp. Cell. Res. **188**, 117–121.
34. Park, M.H., Cooper, H.L., and Folk, J.E. (1981) Identification of hypusine, an unusual amino acid, in a protein from human lymphocytes and of spermidine as its biosynthetic precursor. Proc. Natl. Acad. Sci. USA **78**, 2869–2873.
35. Russell, D.H. and Durie, B.G. (1978) Polyamines as Biochemical Markers of Normal and Malignant Growth, 1–178, Raven Press, NY.
36. Lentini, A., Kleinman, H.K., Mattioli, P., Autuori, V., Nicolini, L., Pietrini, A., Abbruzzese, A., Cardinali, M., and Beninati, S. (1998) Inhibition of melanoma pulmonary metastasis by methylxanthines due to decreased invasion and proliferation. Melanoma Res. **8**, 131–137.
37. Beninati, S., Abbruzzese, A., and Cardinale, M. (1993) Differences in the posttranslational modification of proteins by polyamines between weakly and highly metastatic B16 melanoma cells. Int. J. Cancer **53**, 792–797.
38. Siegall, C.B., Xu, Y.H., Chaudary, V.K., Adhya, S., Fitzgerald, D., and Pastan I. (1989) Cytotoxic activities of a fusion protein comprised of TGF alpha and Pseudomonas exotoxin. FASEB J. **3**, 2647–2652.
39. Theil, E.C. (1990) Regulation of ferritin and transferrin receptor in mRNAs. J. Biol. Chem. **265**, 4771–4774.
40. Estrov, Z., Tawa, A., Wang, X-H., Dubé, I.D., Sulh, H., Cohen, A., Gelfand, E.W., and Freedman, M.H. (1987) In vitro and in vivo effects of deferoxamine in neonatal acute leukemia. Blood **69**, 757–761.
41. Leardi, A., Caraglia, M., Selleri, C., Pepe, S., Pizzi, C., Notaro, R., Fabbrocini, A., Musicò, M., Abbruzzese, A., Bianco, A.R., and Tagliaferri, P. (1998) desferioxamine increases iron depletion and apoptosis induced by ara-c on human myeloid leukemic cells. Br. J. Haematol. **102**, 746–752.
42. Caraglia, M., Passeggio, A., Beninati, S., Leardi, A., Nicolini, L., Improta, S., Pinto, A., Bianco, A.R., Tagliaferri, P., and Abbruzzese, A. (1997) Interferon α2 recombinant and epidermal growth factor modulate proliferation and hypusine synthesis in human epidermoid cancer KB cells. Biochem. J. **324**, 737–741.

DIET, FIBERS, AND COLON CANCER

Jean Faivre and Claude Bonithon-Kopp

Registre des Cancers Digestifs (Equipe Associèe INSERM-DGS, CRI 95-05)
Facultè de Mèdecine, 7 Boulevard Jeanne d'Arc, 21033 Dijon Cèdex, France

Epidemiological studies of colorectal cancer, particularly studies of time trends, of migrants, and of religious groups, indicate that environmental factors are of overwhelming importance in determining the incidence of the disease in the population. Of the possible environmental factors important in colorectal cancer carcinogenesis, diet has been the focus of attention. Recent advances have shown that genetic constitution is also important in determining susceptibility to the disease. This implies that colorectal cancer can be caused by an interaction between dietary factors and genetic predisposition. Among the dietary components that have been linked to decreased risk of colorectal cancer is the high intake of dietary fibre. The hypothesis of a fibre-depleted aetiology for colorectal cancer was particularly put forward by Burkitt.[1] This was based on the careful clinical observations of the difference in patterns of diseases between Western and traditional African societies. He noted the rarity of colorectal cancer in African countries compared to the west and found that African diets were higher in dietary fibre and lower in refined carbohydrates than west European diets. Results from laboratory animal studies and from analytical epidemiological studies provide evidence that a high fibre intake decreases the risk of colorectal cancer. However, some studies have failed to support the theory of a fibre-depleted diet as a causative factor in the aetiology of colorectal cancer. The objective of this review is to present the evidence for and against the protective role of dietary fibre against colorectal cancer.

1. TERMINOLOGY

The first attempt to analyse fibre was to assay crude fibre. It is defined as a component which is insoluble in hot dilute acid or alkali and which is composed of cellulose and lignins. Later on it was shown that many of the physiological effects of fibre on the colon were related to the extent of fermentation during colonic transit. Crude

Advances in Nutrition and Cancer 2, edited by Zappia *et al.*
Kluwer Academic / Plenum Publishers, New York, 1999.

fibre which is excreted unmetabolised in stools is not the important factor. Dietary fibre was defined as the group of substances in plant cell wall material that is indigestible by human small intestine enzymes.[2] The components of dietary fibre include all cellulose, hemicellulose, pectins, lignins, waxes, and cutins. Dietary fibre can be classified into two major categories: watersoluble fibre that are largely fermented by the intestinal microflora, resulting in almost complete digestion (hemicellulose, pectine, gums, mucilages) and water insoluble fibre which are much less fermentable (cellulose, lignins). The fermentation process results in an increase of bacterial mass and in the formation of short chain fatty acid, in particular butyric. acid.

Resistant starches which escape digestion in the small bowel and enter the colon must be added to dietary fibre. These polysaccharides physiologically behave similarly to hemicellulose. This is why the term non starch polysaccharides or complex carbohydrates has been proposed. This undigested starch can be in very variable amounts. It is well known that the proportion of starch components which remain undigested may vary considerably according to the way foods are prepared. The problem of definitions can explain why epidemiological findings are far from unanimous.

2. LABORATORY ANIMAL STUDIES

Many reports have been concerned with the influence of dietary fibre in chemically initiated experimental carcinogenesis, mainly in rats. Some investigations have examined the effect of a single source of fibre. Of 14 experimental protocols presented within eight papers[3–10] on the effect of dietary pectin on large-bowel tumorigenesis, three displayed evidence of tumour inhibition. In three other experiments tumour enhancement was demonstrated, while eight other protocols showed no effect of pectin. Concerning cellulose, nine papers have been published concerning 19 experimental protocols.[6,8,11–15] Two experiments suggested an increased risk with a high consumption of cellulose, nine no effect, and eight a decreased risk. But if only experiments in which cellulose was given during the promoting phase, are taken into account there is no reduction in tumour incidence. These data imply that cellulose may protect against tumorigenesis only in the initiating period. Mucilaginous substances, like Fybogel (ispaghul fibre), appear to have a protective effect.[15] Interestingly, the protective effect of Fybogel on carcinogenesis was observed during the promoting phase. Guar gum,[4,5,7,8] alfalfa,[10,12] carrageenan,[10] and cutin[8] seemed to have no effect on large-bowel carcinogenesis or in some cases appeared even to increase the risk.

Concerning bran, the effect varies with the bran source. Corn bran[16–18], rice bran[5], soybean bran,[6,16] and oat bran[11] do not seem to be protective. Wheat bran seems more interesting;[4,7,9,10,12,16,19–25] a protective effect was found in all studies where the diet contained 2%–6% fat. The results vary in rats given a high-fat diet. Available data also suggest that wheat bran might be more effective during the promoting phase than during the initiating phase.[26]

So it is apparent from animal studies of dietary fibre and colon cancer that attention should be focused on differentiating the types of fibres. They can have an enhancing effect, a protective effect, or no effect on large-bowel carcinogenesis depending on the nature, amount, form, and the period of administration. Mucilaginous substances and wheat bran appear to be of particular interest. But interpretation of the data is difficult because they have been provided by various experimental protocols in which dietary fibres were administrated to rats belonging to various strains, treated by carcinogens differing in nature and amount, with differences in the composition of diets, sometimes not even corrected for isocaloricity. Furthermore, different tumorigenic parameters have been used as end points. There is an urgent need for animal model studies to be standardised to the greatest feasible extent. The relevance of these data to human cancer has to be determined by intervention studies involving dietary fibre supplementation.

3. ANALYTICAL EPIDEMIOLOGICAL STUDIES

The results of case-control studies are not uniform. There are 12 case-control studies which support a protective effect of dietary fibre,[27–38] 9 studies which found no significant effect,[39–47] and two studies which indicate an increased risk associated with a high fibre intake.[48–49] A meta-analysis of 13 studies[50] found an inverse association between fibre intake and colorectal cancer with a relative risk of 0.53 for the highest quintile compared to the lowest quintile of intake. However when this analysis was restricted to studies that used validated dietary questionnaires and incorporated qualitative data into estimates of nutrient intakes, the relative risk estimates for colorectal associated with consumption of more dietary fibre were closer to 1. Similar findings have also been reported for a meta-analysis of 16 case-control studies with an odds ratio of 0.6 for highest versus lowest intake of fibre.[51]

The first three cohort studies do not report data on fibre and the incidence of colorectal cancer. Among the most recent studies seven cohort studies find no significant association with dietary fibre[54–58] while two suggest a protective effect.[59–60]

There is considerable evidence that a high proportion of colorectal cancers arise in adenomas. Taking into account epidemiological and histopathological data, Hill et al.[61] have suggested that large bowel cancer could be the result of a multistage process. According to this hypothesis the factors acting at each step of the adenoma-carcinoma: appearance of the adenoma, growth towards a large adenoma and adenoma transformation in carcinoma may be at least partly different. A very important reason for the difficulties in interpreting available studies on diet and colorectal cancer is that little attention has been given to the different steps of colorectal carcinogenesis.

Furthermore most studies which address the question of relationship between diet and colorectal adenomas do not consider adenoma size. Several case-control studies have found protective association with total fibre,[62–64] fibre from cereals in subjets with small adenomas[65] or fibre from vegetables and fruits.[66] In two studies the association was seen only in women[61,66] and in other studies daily intake of fibre was not associated with a lower risk of adenoma.[47,64] Two prospective studies are available. One of them conducted in US men did not find a significant association between the dietary intake of total, cereal or vegetable fibre and colorectal adenomas, although a slight reduction in risk was observed with increasing intake of fruit fibre.[68] In the other one conducted in US women no significant association between fibre intake and the risk of colorectal adenoma was found.[58] Evidence from case-control studies and cohort studies for the hypothesis that fibre intake is related to the risk of colorectal cancer or adenoma is inconclusive.

The reasons why there is not complete agreement on the effect of dietary fibre on colorectal carcinogenesis are multiple. Some limitations are inherent to dietary studies. The quality of a study depends largely on the quality of the dietary questionnaire, the degree of precision that is asked for and the ability of the dietician to obtain precise and unbiased answers. When the quality of the interview is similar in cases and controls avoiding any differential bias, there is a certain degree of misclassification of dietary intake which may results in an under estimation of the association. The quality of the food composition table is of major importance. It should be as detailed and complete as possible. A particularly important aspect of the problem is fibre. It has been observed that the fibre content may vary considerably, even for foods called by the same name. The application of values from food composition tables may lead to large errors. Fibre cannot be defined as simple chemical substance. It represents a complex mixture of different types of polysaccharides with different physiological aspects in the colon and rectum. Contrasting results from case-control and cohort studies could be explained by major variations in the definition of fibre and by the fact that food composition tables do not distinguish between the different types of fibre. It is hoped that the data on the detailed composition of foods regarding their types of fibre will soon be available and will enable a re-examination of the protective effect of dietary fibre. The problem is further complicated by nutrients which have a fibre-like effect such as resistant starch. Depending on the way the food is cooked, part of its starch content will arrive undigested in the large bowel and will be fermented by bacteria. Apparent discrepancies between case-control and cohort studies can been explained by the fact that cohort studies on cancer mortality investigate factors with initial and intermediate steps of the adenoma-carcinoma sequence, whereas case-control studies on cancer investigate preferentially factors associated with the latest steps.

4. INTERVENTION STUDIES

Analytical studies are not sufficient to serve as basis for firm dietary advice. Faced with this situation it is attractive to test the most attractive hypotheses within the framework of chemopreventive studies. In such studies a possible preventive agent is administered to individuals in order to reduce their probability of developing the disease.

Five chemopreventive studies, aimed at evaluating the possibility of primary prevention of colorectal cancer with fibre supplements have been or are being carried out in the world. A study with the large bowel as the end point requires thousands of subjects and at least a 10-year follow-up period. In order to reduce sample size and length of follow-up a population with adenomas (adenoma recurrence or adenoma growth) has been selected as the target population.

Wheat bran supplementation is proposed in four studies. Its effect at the dose of 22.5 g/day, together with vitamin C (4 g/day) and vitamin E (400 mg/day) has been evaluated in patients with familial polyposis and with the rectum left in place.[69] The ratio between the number of adenomas at the time of the follow-up examination and the number of adenomas at base line was the main trial outcome. Altogether 58 subjects were randomised and followed up by rigid proctosigmoidoscopy every 6 months. The overall compliance to treatment was 74%. Intent to treat analysis suggested a limited effect of the treatment in the group receiving wheat bran, vitamin C, and vitamin E compared with the group receiving vitamins alone or a placebo. There were significant differences only at 33 and 39 months.

In the Toronto study the intervention group received counselling on diet high in fibre (50 g/day) and low in fat (20% of energy from fat). The control group continued with its usual diet.[70] Overall 201 subjects were randomised and the compliance to treatment rate was 71%. Compliance to the final colonoscopy was 82%. After 12 months of counselling fibre consumption was 35 g in the intervention group and 16 g in the control-group and fat consumption was 25% and 33% of energy respectively. After two years the relative risk (RR) for adenoma recurrence was 1.2. There was a reduced risk of adenoma recurrence in women (RR = 0.7) and an increased risk in men (RR = 1.7), which did not reach the significant level because of the small sample size. Thus the issue of a gender effect related to adenoma recurrence remains a definite question to be addressed by much larger studies.

The effect of wheat bran (25 g/day) on adenoma recurrence was considered in the Australian study.[71] The effect of a low fat diet (<25% of calories from fat) and of β carotene (20 mg/day) were also tested using a 2 × 2 factorial design. It has the advantage of allowing the effect of the combination of two treatments to be evaluated and of giving more power to the study compared with a parallel scheme with the same number of participants. In this study 411 subjects were randomised. The compliance rate to treatment was 74% and compliance to colonoscopy was 95% at two years and 74% at 4 years. There was no evidence that any intervention reduced the recurrence rate of adenomas at two or four years. A significant reduction in the incidence of large adenomas (> 1 cm) was found in the low fat diet group. The effect was observed when the low fat diet was combined with wheat bran.

A multicentre European study performed within the European Cancer Prevention Organisation (ECP) has been assessing a mucilaginous substance in the form of Ispashula husk (3.8 g/day).[72] This dose was recommended by the manufacturer to obtain stool bulking. In this study a second group received calcium (2 g/day) and a third group a placebo. The duration of the study was 3 years and 665 subjects were randomised. The final colonoscopy was performed in 87% of the subjects. The compliance for supplements was obtained in 77% of the subjects. The results suggest a deleterious effect of fibre on the recurrence of colorectal adenoma (OR adjusted = 1.63 95% CI 1.01–2.69) and are compatible with a modest beneficied effect of calcium (OR adjusted = 0.66 95% CI 0.38–1.17).[73]

In another study in Arizona the assessment of the effect of wheat bran (30 g/day) on adenoma recurrence is on-going.[73] Overall 1400 subjects were randomised in this study. No results are available from this study yet.

It is not yet possible to draw any definitive conclusion on the effect of fibre supplementation in colorectal carcinogenesis. Available results do not support a protective effect of wheat bran or mucilaginous substances on adenoma recurrence. The results concerning adenoma growth are not clear.

5. CONCLUSION

Overall, the epidemiology of fibre and colorectal cancer is somewhat inconsistent, perhaps because of the heterogeneous nature of fibre and differences in the way in which fibre is measured. Moreover it is possible that the effects of fibre must be considered in the context of total diet and of the interactions with various other dietary components. Available data from intervention trials do not support a protective effect of dietary fibre on adenoma recurrence. The role in the latest steps of the adenoma-carcinoma sequence has yet to be specified.

ACKNOWLEDGMENTS

This study performed within the ECP colon group was supported by the Europe Against Cancer Program and the French Ministry of Health (PHRC).

REFERENCES

1. Burkitt D.P. (1971) Epidemiology of cancer of the colon and rectum. Cancer 28:3–13.
2. Trowell H. (1974) Definition of fibre. Lancet i:503.
3. Bauer H.G., Asp N.G., Oste R., Dehlqvist A., and Fredlund P.E. (1979) Effect of dietary fiber on the induction of colorectal tumors and fecal β-glucuronidase activity in the rat. Cancer Res 39:3752–6.
4. Bauer H.G., Asp N.G., Dahlqvist A., Fredlund P.E., Nyman M., and Oste R. (1981) Effect of two kinds of pectin and guar gum on 1,2-dimethylhydrazine initiation of colon tumors and on fecal β-glucuronidase activity in the rat. Cancer Res 41:2518–23.
5. Casteleden W.M. (1977) Prolonged survival and decrease in intestinal tumors in dimethyhydrazine-treated rats fed a chemically defined diet. Br J Cancer 35:491–5.
6. Freeman H.J., Spiller G.A., and Kim Y.S. (1978) A double-blind study on the effect of purified cellulose dietary fiber on 1,2-dimethylhydrazine-induced rat colonic neoplasia. Cancer Res 38: 2912–17.
7. Jacobs L.R. (1983) Enhancement of rat colon carcinogenesis by wheat bran consumption during the stage of 1,2-dimethylhydrazine administration. Cancer Res 43:4057–61.
8. Klurfeld D.M. (1987) The role of dietary fiber in gastrointestinal diseases. J Am Diet Assoc 87:1172–7.
9. Reddy B.S., Mori H., and Nicolais N. (1981) Effect of dietary wheat bran an dehydrated citrus fiber on azoxymethane-induced intestinal carcinogenesis in Fischer 344 rats. J Natl cancer Inst 66:553–7.
10. Watanabe K., Reddy B.S., Weisburger J.H., and Krichevsky D. (1979) Effect of dietary alfalfa, pectin, and wheat bran on azoxymethane or methylnitrosourea-induced colon carcinogenesis in F344 rats. J Natl Cancer Inst 63:141–5.
11. Jacobs L.R. and Lupton J.R. (1986) Relationship between colonic luminal pH, cell proliferation and colon carcinogenesis in 1,2-dimethylhydrazine treated rats fed high fiber diets. Cancer Res 46: 1727–34.
12. Nigro N.D., Bull A.W., Klopfer B.A., Pak M.S., and Campbell R.L. (1979) Effect of dietary fiber on azoxymethane-induced intestinal carcinogenesis in rats. J Natl Cancer Inst 62:1097–102.
13. Prizont R. (1987) Absence of large bowel tumors in rats injected with 1,2-dimethylhydrazine and fed high dietary cellulose. Dig Dis Sci 32:1418–21.
14. Ward J.M., Yamamoto R.S., and Weisburger J.H. (1973) Cellulose dietary bulk and azoxymethane-induced intestinal cancer. J Natl Cancer Inst 51:713–15.
15. Wilpart M. and Roberfroid M. (1987) Intestinal carcinogenesis and dietary fibers: the influence of cellulose or fybogel chronically given after exposure to DMH. Nutr Cancer 10:39–51.
16. Barnes D.S., Clapp N.K., Scott D.A., Oberst D.L., and Berry S.S. (1983) Effect of wheat, rice, corn, and soybean bran on 1,2-dimethylhydrazine induced large bowel tumorigenesis in F344rats. Nutr Cancer 5:1–9.
17. Freeman H.J., Spiller G.A., and Kim Y.S. (1980) A double-blind study on the effect o differing purified cellulose and pectin fiber diets on 1,2-dimethylhydrazine-induced rat colonic neoplasia. Cancer Res 40:2661–5.
18. Reddy B.S., Maeura Y., and Wayman M. (1983) Effect of dietary corn bran and autohydrolyzed lignin on 3,2-dimethyl-4-aminobiphenyl-induced intestinal carcinogenesi in male F344 rats. J Natl Cancer Inst 71:419–23.
19. Abraham R., Barbolt T.A., and Rodgers J.B. (1980) Inhibition by bran of the colonic cocarcinogenicity of ble salts in rat given dimethyhydrazine. Exp Mol Pathol 33:133–43.
20. Barbolt T.A. and Abraham R. (1978) The effect of bran on dimethylhydrazine-induced colon carcinogenesis in the rat. Proc Soc Exp Biol Med 157:656–9.
21. Barbolt R.A. and Abraham R. (1980) Dose response, sex difference and the effect of bran in dimethylhydrazine induced intestinal tumorigenesis in rats. Toxicol Appl Pharmacol 55:417–22.
22. Cruse J.P., Lewin M.R., and Clark C.G. (1978) Failure of bran to protect against experimental colon cancer in rats. Lancet 2:1278–80.

23. Fleiszer D., Murray D., Macfarlan J., and Brown R. (1978) Protective effect of dietary fibre against chemically-induced bowel tumours in rats. Lancet 2:552–3.
24. Wilson R.B., Hutcheson D.P., and Widemanf V. (1977) Dimethylhydrazine induced colon tumors in rats fed diets containing beef fat or corn oil with and without wheat bran. Am J Clin Nutr 30:176–81.
25. Reddy B.S. and Mori H. (1981) Effect of dietary wheat bran and dehydrated citrus fiber on 3,2-dimethyl-4-aminobiphenyl-induced intestinal carcinogenesis in F344 rats. Carcinogenesis 2:21–5.
26. Nauss K.M., Locnishar M., Sondergaard D., and Newberne P.M. (1984) Lack of effect of dietary fat on N-nitrosomethyl urea induced colon tumorigenesis in rats. Carcinogenesis 5:255–60.
27. Modan B., Barell V., and Lubin F. (1975) Low-fiber intake as an etiologic factor in cancer of the colon. J Natl Cancer Inst 55:15–18.
28. Kune S., Kune G.M., and Watson F. (1987) Case-control study of dietary etiological factors: the Melbourne Colorectal Cancer Study. Nutr Cancer 9:43–56.
29. Tuyns A.J., Haelterman M., and Kaaks R. (1987) Colorectal cancer and the intake of nutrients: oligosaccharides are a risk factor, fats are not: a case-control study in Belgium. Nutr Cancer 11:189–204.
30. West D.W., Slattery M.L., Robison L.M., Schuman K.L., Ford M.H., Mahoney A.W., et al. (1989) Dietary intake and colon cancer: sex- and anatomic site-specific associations. Am J Epidemiol 128:989–99.
31. Benito E., Stiggelbout A., Bosch F.X., Obrador A., Kaldor J., Mullet M., et al. (1991) Nutritional factors in colorectal cancer risk: a case-control study in Majorca. Int J Cancer 49:161–7.
32. Whittemore A.S., Wu-Williams A.H., Lee M., Zheng S., Gallagher R.P., Jiao D.A. et al. (1990) Diet, physical activity, and colorectal cancer among Chinese in North America and China. J Natl Cancer Inst 82:915–26.
33. Freudenheim J., Graham S., Marshall J.R., Haughey B.P., and Wilkinson G. (1990) A case-control study of diet and rectal cancer in Western New-York. Am J Epidemiol 131:612–24.
34. Young T. and Wolf D.A. (1988) Case-control study of proximal and distal colon cancer and diet in Winconsin. Int J Cancer 42:167–75.
35. Bidoli E., Franceschi S., Talamini R., Barra S., and La Vecchia C. (1992) Food consumption and cancer of the colon and rectum in North-Eastern Italy. Int J Cancer 50:223–9.
36. Iscovich J.M., LíAbbe K.A., Castelleto R., Calzona A., Bernedo A., Chopita N.A. et al. (1992) Colon cancer in Argentina. I. Risk from intake of dietary items. Int J Cancer 51:851–7.
37. Meyer T. and White E. (1993) Alcohol and nutrients in relation to colon cancer in middle-aged adults. Am J Epidemiol 138:225–36.
38. Arbman G., Axelson D., Ericsson-Begodsui A.B., Fredriksson M., Nilsson E., and Sjodahl R. (1992) Cereal fiber, calcium, and colorectal cancer. Cancer 69:2042–8.
39. Dales L.G., Friedman G.D., Ury H.K., Grossman S., and Williams S.R., (1979) A case-control study of relationship of diet and other traits to colorectal cancer in American blacks. Am J Epidemiol 109:132–44.
40. Miller A.B., Howe G.R., Jain M., Craib K.J.P., and Harrison L. (1983) Food items and food groups as risk factors in a case-control study of diet and colorectal cancer. Int J Cancer 32:155–61.
41. Macquart-Moulin G., Riboli E., CornÈe J., Charnay B., Berthezene P., and Day N. (1986) Case-control study on colorectal cancer and diet in Marseilles. Int J Cancer 38:183–91.
42. Peters R.K., Garabrant D.H., and Yu M.C. (1989) A case-control study of occupational and dietary factors in colorectal cancer in youn men by subsite. Cancer Res 49:5459–68.
43. Lee H.P., Gourley L., Duffy S.W., Esteve J., Lee J., and Day N. (1989) Colorectal cancer and diet in an Asian population—case-control study among Singapore Chinese. Int J Cancer 43:1007–16.
44. Pickle L.W., Greene M.H., and Ziegler R.G. (1984) Colorectal cancer in rural Nebraska. Cancer Res 44:363–9.
45. Bristol J.B., Emmett P.M., and Heaton K.W. (1985) Sugar, fat, and the risk of colorectal cancer. Br Med J 291:1467–70.
46. Gerhardsson de Verdier M., Hagman U., Steineck G., Rieger A., and Norell S.E. (1990) Diet, body mass, and colorectal cancer: a case-referent study. Int J Cancer 46:832–8.
47. Faivre J., Boutron M.C., Senesse P., Couillault C., Belghiti C., and Meny B. (1997) Environmental and familial risk factors in relation to the colorectal adenoma-carcinoma sequence: results of a case-control study in Burgundy (France). Eur J Cancer Prev 6:127–31.
48. Martinez I., Torres R., and Frias Z. (1981) Factors associated with adenocarcinomas of the large bowel in Puerto Rico. Rev Latinoam Oncol Clin 13:13–20.
49. Potter J.D. and McMichael A.J. (1986) Diet and cancer of the colon and rectum: a case-control study. J Natl Cancer Inst 76:557–69.

50. Howe G.R., Benito E., Castelleto R., CornÈe J., EstËve J., Gallagher R.P. et al. (1992) Dietary intake of fiber and decreased risk of cancers of the colon and rectum: evidence from the combined analysis of 13 case-control studies. Cancer Causes Control 84:1887–96.
51. Trock B., Lanza E., and Greenwald P. (1990) Dietary fiber, vegetables, and colon cancer, critical review and meta-analyses of the epidemiological evidence. J Natl Cancer Inst 82:650–61.
52. Morgan J.W., Frazer G.F., Philips R.L., and Andress M.H. (1988) Dietary factors and colon cancer incidence among seventh day adventists. Am J Epidemiol 128:918 (abstract).
53. Goldbohm R.A., Van-den-Brandt P.A., Van-t-Veer P., Brants H.A., Dorant E., Sturmans F. et al. (1994) A prospective cohort study on the relation between meat consumption and the risk of colon cancer. Cancer Res 54:718–23.
54. Giovannucci E., Rimm E.B., Stampfer M.J., Colditz G.A., Ascherio A., and Willett W.C. (1994) Intake of fat, meat, and fiber in relation to risk of colon cancer in men. Cancer Res 54:2390–7.
55. Gaard M., Tretli S., and Loken E.B. (1996) Dietary factos and risk of colon cancer: a prospective study of 50,535 young Norwegian men and women. Eur J Cancer Prev 5:445–54.
56. Willet W.C., Stampfer M.J., Colditz G.A.; Rosner B.A., and Speizer F.E. (1990) Relation of meat fat and fiber intake to the risk of colon cancer in a prospective study among women. N Engl J Med 323:1664–72.
57. Steinmetz K.A., Kushi L.H., Bostick R.M., Folsom A.R., and Potter J.D. (1994) Vegetables, fruits, and colon cancer in the Iowa womenís health study. Am J Epidemiol 139:1–15.
58. Fuchs C.S., Giovanucci E.L., Colditz G.A., Hunter D.J., Stampfer M.J., Rosner B., Speizer F.E., and Willett W.C. (199) Dietary fiber and the risk of colorectal cancer and adenoma in women. N Engl J Med 340:169–76.
59. Heibrun L., Nomura A., Hankin J., and Stermmermann G.N. (1989) Diet and colorectal cancer with special reference to fiber intake. Int J Cancer 44:1–6.
60. Thun M.J., Calle E.E., Namboodiri M.M., Flanders W.D., Coates R.J., Byers T., et al. (1992) Riks factors for fatal colon cancer in a large prospective study. J Natl Cancer Inst 84:1491–500.
61. Hill M.J., Morson B.C., and Bussey H.J.R. (1978) Aetiology of adenoma-carcinoma sequence in the large bowel. Lancet i:245–7.
62. Hoff G., Moen E., Trygg K., Frolich W., Sauar J., Vatn M., Gjones E., and Larsen S. (1986) Epidemiology of polyps in the rectum and sigmoid colon. Evaluation of nutritional factors. Scand J Gastroenterol 21:199–204.
63. Kune G.A., Kune S., Read A., Mac Gowan K., Penfold C., and Watson L.F. (1991) Colorectal polyps, diet, alcohol, and family history of colorectal cancer: a case-control study. Nutr Cancer 16:25–30.
64. Neugut A.I., Garbowski G.C., and Leew C. (1993) Dietary risk factors for the incidence and recurrence of colorectal adenomatous polyps. A case-control study. Ann Intern Med 118:91–5.
65. Sandler R.S., Lyles C.M., Peipins L.A., Mc Auliffe C.A., Woosley J.T., and Kupper L.L. (1993) Diet and risk of colorectal adenomas: macronutrients cholesterol and fiber. J Natl Cancer Inst 85:884–91.
66. Macquart-Moulin G., Riboli E., Cornee J., Kaas R., and Berthezene P. (1987) Colorectal polyps and diet: a case-control study in Marseilles. Int J Cancer 40:179–88.
67. Little J., Logan R.F.A., Hawtin P.G., Hardcastle J.D., and Turner I.D. (1993) Colorectal adenomas and diet: a case-control study of subjects participating in the Nottingham faecal occult blood screening programme. Br J Cancer 67:177–84.
68. Giovanucci E., Stampfer M.J., and Colditz G. (1992) Relationship of diet to risk of colorectal cancer adenoma in men. J Natl Cancer Inst 84:91–8.
69. De Cosse J.J., Miller H.H., and Lesser M.L. (1989) Effect of wheat fiber and vitamin C and E on rectal polyps in patients with familial adenomatous polyposis. J Natl Cancer Inst 81:1290–7.
70. Mc Keown-Eyssen G., Holloway C., Jazmaji V., Britht-See E., Dion P., and Bruce W.R. (1988) A randomized trial of vitamins C and E in the prevention of recurrence of colorectal polyps. Cancer Res 48:4701–5.
71. Mac Lennan R., Macrae F.A., Bain C., Battistutta D., Chapuis P., and Gratten H. (1995) Randomized trial of intake of fat, fiber, and beta carotene to prevent colorectal adenomas. J Natl Cancer Inst 87:1760–6.
72. Faivre J., Couillault C., Kronborg O., Rath U., Giacosa A., De Oliveira H., Obrador T., OíMorain C., and E.C.P. Colon Group (1997) Chemoprevention of metachronous adenomas of the large bowel: design and interim results of a randomized trial of calcium and fibre. Eur J Cancer Prev 6:132–8.
73. Vargas P.A. and Alberts D.S. (1992) Primary prevention of colorectal cancer through dietary modification. Cancer 70:1229–35.

18

PHYTOCHEMICALS AS MODULATORS OF CANCER RISK

H. Leon Bradlow, Nitin T. Telang, Daniel W. Sepkovic, and
Michael P. Osborne

Strang Cancer Research Laboratory
New York, NY 10021

INTRODUCTION

The role of diet as a modulator of cancer initiation and promotion has been suggested on epidemiological grounds showing changes in cancer incidence as people migrate from low cancer incidence (breast and prostate) countries to high incidence countries.[1] By the third generation the descendants of these migrants have acculturated and now have the same cancer incidence as natives of the host country. Conversely immigrants from these same countries where stomach and esophageal cancer has a high incidence show a drop by the third generation when they move into a country with a low incidence of these cancers.[2]

Because hormone-related cancers are high incidence cancers in the US and Europe we sought to explore the role of dietary agents present in foods consumed heavily in low breast and prostate incidence countries but to a lesser extent in Western countries. These include such compounds as indole-3-carbinol found in cruciferous vegetables (broccoli, bok-choi, brussels sprouts, cabbage, and cauliflower), epigallocatechin gallate (green tea), curcumin (turmeric), and genistein (soy products).

Because hormone-related cancers depend in whole or part on the presence of estradiol we have explored the role of a variety of cancer-promoting agents such as oncogenes, viruses, and carcinogens as modulators of estrogen.[3–6] All of these consistently act to decrease C-2 hydroxylation and increase C-16a-hydroxylation of estradiol (Table 1a,b,c).

The ability of dietary agents to oppose the effect of these tumorigenic agents on estrogen metabolism as well as direct non-hormonal effects of these compounds on cell regulation and apoptosis have been the subject of intensive study. Certain of these dietary agents also appear to act as antioxidants blocking oxidative damage to DNA.

Advances in Nutrition and Cancer 2, edited by Zappia *et al.*
Kluwer Academic / Plenum Publishers, New York, 1999.

Table 1a. Influence of the Human Papilloma Virus on 16α-Hydroxylation of Estradiol by Human Tissues

	% 16α-hydroxylation before HPV infection	After HPV infection	Fold increase
Foreskin	0.085	1.7	20
Cervix	2.7	16.2	6
Larynx	1.9	3.8	2
Cell line fron cervical CA HPV-16		8.2	
Average per 100 µg tissue			

Table 1b. Alteration of Estrogen Metabolism in Human Mammary Epithelial Cells by Oncogenes and Carcinogens

Cell line	E2 metabolism (2/16α-OHE1)	P
184-B5	4.2 ± 0.3	
184-B5/BP	0.5 ± 0.1	<0.03
184-B5/HER	0.5 ± 0.1	<0.03
MDA-MB-231	0.4 ± 0.1	<0.03

Values are mean =/– SD, n = 4/cell line.
Ratio is 2-OHE1/16α-OHE1 formation.

Table 1c. Effect of Chemical Carcinogens on the Metabolism of 17β-Estradiol (E_2) in Human Mammary Epithelial 184-B5 Cells

	E_2 metabolite[b] (relative abundance /10^7 cells)		
Treatment[a]	2-OHE_1	16α-OHE_1	C2/C16α ratio
DMSO	342 ± 73[c]	53 ± 13[f]	6.45
DMBA	52 ± 3[d]	65 ± 5[g]	0.80
BP	45 ± 5[e]	78 ± 8[h]	0.58

[a]cell cultures incubated with 0.1% DMSO, 39 µM DMBA, or 39 µM BP for 24 hours and with 10^{-8} M E_2 for the subsequent 48 hours. The culture medium processed for the product isolation and identification by the GC-MS assay.
[b]Values are mean ± SD, n = 4.
[c–d,c–e]P = 0.001.
[f–g]n.s.
[f–h]P = 0.01.

METHODS

A. Estradiol Metabolism

The action of these dietary agents on estrogen metabolism has been tested in vitro in cell culture systems, and in vivo in animal models and in human volunteers. In cell culture systems the compound to be tested and estradiol are added to the culture dishes after a 48-hour pre-incubation. After a 48-hour incubation period the medium is removed and extracted by solid phase extraction. The extract is derivitized and the metabolites are analyzed by GC-MS using single-ion monitoring to both identify and

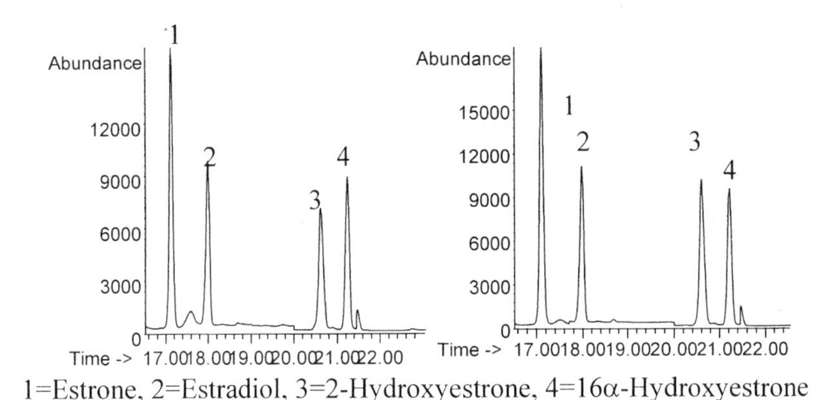

Figure 1. Results of GC-MS analysis of estrogens. Buccal epithelial cells from a normal individual untreated and treated with I3C.

quantitate the metabolites formed (Fig. 1). Deuterated estrone and estradiol were also used to help confirm the identity of the metabolites in some human studies (Fig. 2).[7] In some early studies estradiol specifically labeled with tritium at metabolically reactive sites was used. In these experiments the extent of reaction was determined by measuring the amount of tritium present in the media or body fluids as water at the end of the incubation.[8] Human metabolic studies were carried out using this same specifically labeled estradiol[9] or subsequently by measuring the levels of 2-hydrooxyestrone (2-OHE1) and 16a-hydroxyestrone (16a-OHE1) present in the urine using a specific ELISA assay.[10] The ability of dietary agents to oppose the effects of tumorigenic agents on estrogen metabolism was studied by the simultaneous addition of dietary agents and carcinogens to the culture medium.

B. Direct Effects

Measurements of direct effects of these compounds were carried out by measuring cell proliferation, changes in the distribution of cells within the different phases of the cell cycle,[11] as well as by measurements of cyclins and kinases regulating cell cycle changes.[12] Apoptosis was measured from laddering, nuclear condensation, and changes in the relative amounts of Bcl2 and Bax.

C. Antitumor Activity

When the antitumor activity of these compounds was assayed in vitro the cytostatic and cytotoxic activity of the compounds was determined from total cell count and fraction of cells in different phases of the cell cycle. In animal studies the compounds were incorporated into the diet or given by gavage and the indices measured were the time until the first tumor was observed and the percentage of animals with tumors at the end of the study period.[13–15]

RESULTS

The effects of the various compounds are discussed separately since their activity and mechanism of action vary considerably.

```
File        : C:\HPCHEM\1\DATA\NYU2C3.D
Operator    :
Acquired    : 18 Jan 1996  13:43    using AcqMethod EESIM33.M
Instrument  :   MS_5971
Sample Name:
Misc Info   :
Vial Number: 3
```

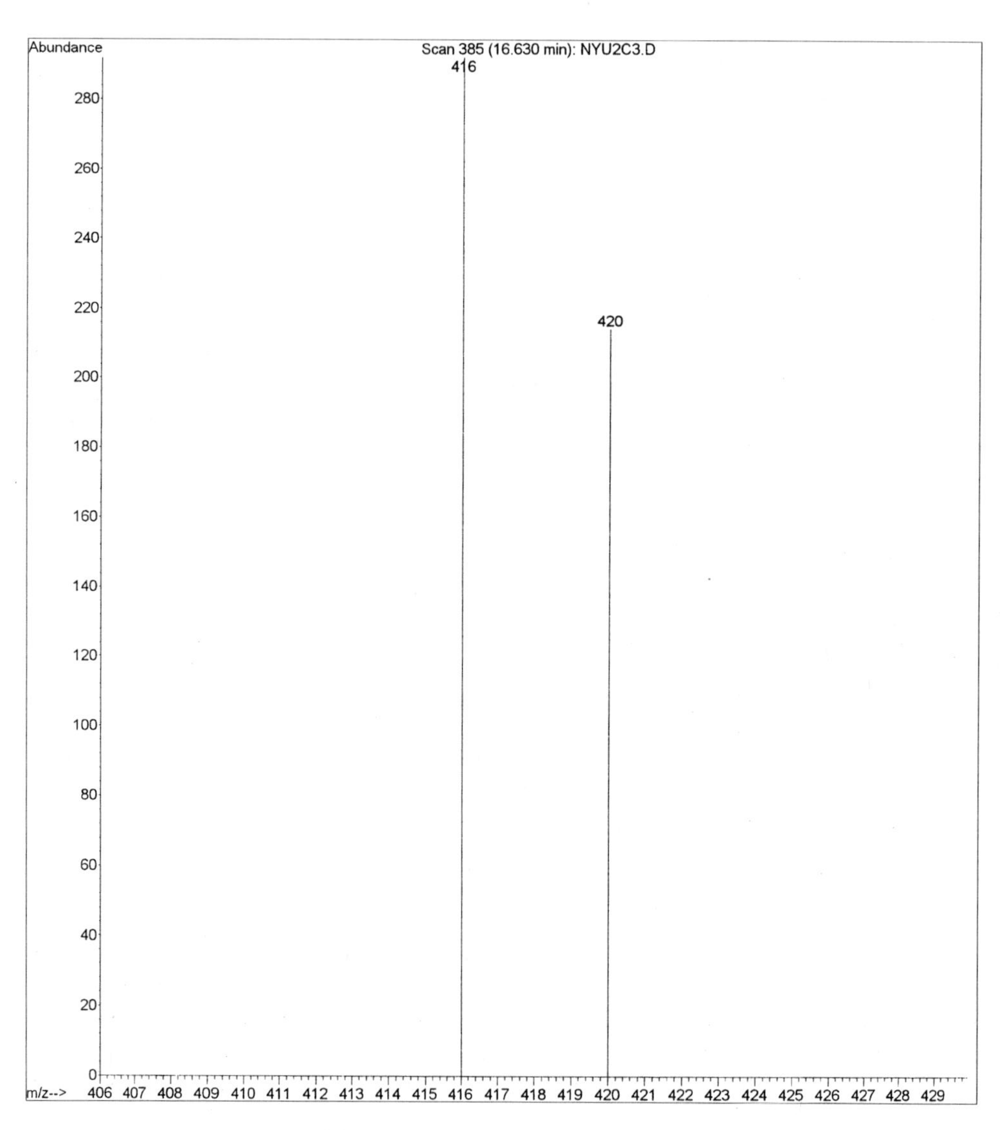

Figure 2. Use of deuterated estrogens.

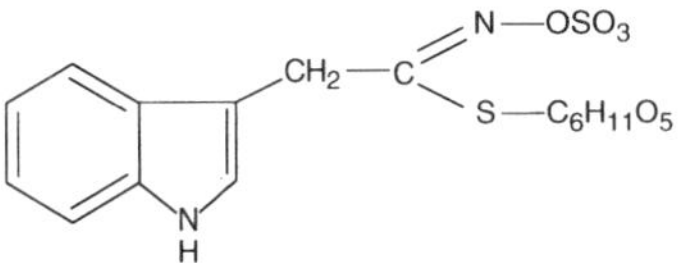

Indole-3-glucosinolate

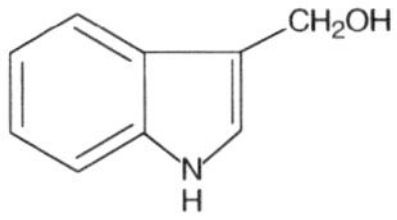

Indole-3-carbinol (I3C)

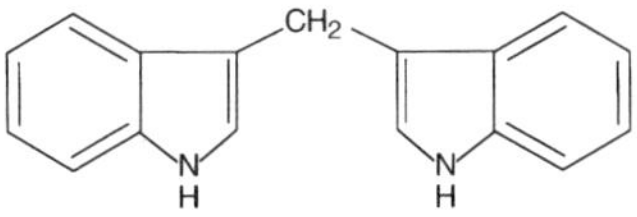

Diindolylmethane

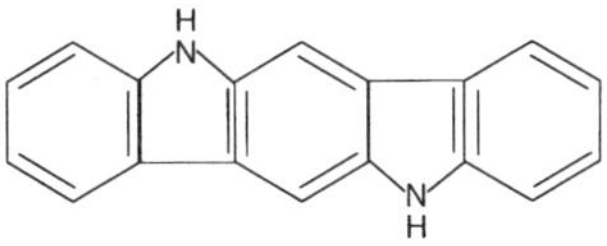

Indole (3, 2, b) carbazole (ICZ)

Figure 3. Structures of I3C, its precursor, and its metabolites.

A. Indole-3-Carbinol

Indole-3-carbinol (I3C) is a compound present in cruciferous vegetables (broccoli, brussels sprouts, cabbage, cauliflower, and bok-choi) as the glucosinolate (Fig. 3). This compound is cleaved by the enzyme myrosinase present in these plants to yield I3C. Early observations showed that it is capable of altering estrogen metabolism in the direction of 2-hydroxylation by inducing the enzyme P4501A2.[16] Among related indole compounds in cell culture I3C is the most potent at inducing 2-hydroxylation of estrogens.[17] In animal models I3C is not active when given IP or IV but only when given orally. This suggests that I3C is in reality a pro-drug which is converted by stomach acid to the dimer diindolylmethane (DIM) and indolylcarbazole (ICZ) which are potent compounds when given IP or IV (Fig. 4). The seeming paradox that I3C is active in cell culture in the absence of acid was resolved when it was shown that I3C is quantitatively converted to DIM in cell culture during a 24 hr incubation period. Studies in rats showed that a single dose of I3C given by gavage resulted in an increase in 2-hydroxylation which peaked at 12 hours and then decreased back to baseline by 24 hours.[18]

The increase in 2-hydroxylation was also observed in human subjects in both acute and long term studies.[19,20] No difference between men and women was observed in the response. In long term studies the response was sustained throughout the treatment period.

Animal studies showed that starting I3C treatment at 5–6 wks of age markedly decreased MMTV-induced mammary tumor development,[13,14] Subsequent studies by Grubbs showed that initiation of I3C treatment after tumor initiation with DMBA blocked the formation of mammary tumors almost totally. When NMU was used as the initiating agent I3C treatment resulted in a marked decrease in mammary tumors but not as dramatic as after DMBA15. Kojima demonstrated a similar decrease in endometrial tumors in the Donryku rat fed I3C.[21]

Cell culture and animal studies have demonstrated that the HPV virus inhibits 2-hydroxylation and stimulates 16α-hydroxylation.[22,23] Conversely it was found that 16a-OHE1 promoted increased expression of HPV and 2-OHE1 decreased expression of the virus and blocked formation of papillomatous cysts in nude mice.[24] Based on these observations it was shown that feeding I3C to patients with laryngeal papillomas resulted in a partial or complete remission in 70% of such patients if they showed an increase in 2-hydroxylation.[25,26] Studies by Dashwood and colleagues showed that feeding I3C protected against IQ-DNA adduct formation and aberrant crypt formation in the rat colon.[27,28] Although the colon is not usually considered to be a hormone-responsive organ, it has been shown to possess estrogen receptors.

Patients carrying a defective variant of P4501A1, called MSP-1, have a decreased rate of 2-hydroxylation and do not respond to induction with I3C.[29,30] Patients with this defect showed an increased risk for breast cancer.

Recent studies by Auborn and colleagues have demonstrated some previously unreported activities for I3C.[31] When cervical keratinocytes were incubated in the presence of I3C a marked drop in the level of the mRNA for P4501B1 was observed as well as a 400% decrease in the formation of 4-OHE1, the hydroxylation product of this enzyme. Because the 4-hydroxyestrogens are known carcinogens this is an important protective effect of I3C. Exercise has also been shown to alter the extent of 2- and 4-hydroxylation of estrogens with a sharp increase in 2-hydroxylation being observed in women engaged in vigorous exercise.[32,33] Hudson and colleagues have demonstrated that I3C inhibits the expression of enzymes such as ornithine decarboxylase as well as the growth of colon cancer cell lines, further demonstrating the diverse activities of I3C.[34]

Studies by Dashwood and colleagues[35] reported that I3C might act as a tumor promoter in fish treated with aflatoxin but only when given after aflatoxin but not when I3C treatment was started before the aflatoxin treatment. Studies in the neonatal mouse[36] treated with DEN revealed no promotinal effect for I3C. Other studies from this laboratory in the F-344 rat initiated with DEN also showed that I3C had no promotional effect in this model.

B. Sulforaphanes

Talalay and colleagues have isolated an interesting anti-cancer drug from broccoli, which they called sulforaphane (Fig. 4). It occurs in the plant as the glucosinolate derivative. The complex is cleaved by myrosinase, an enzyme liberated when

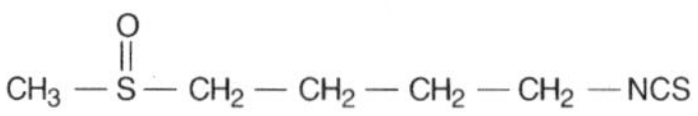

$CH_3 - S(=O) - CH_2 - CH_2 - CH_2 - CH_2 - NCS$

Sulforaphane

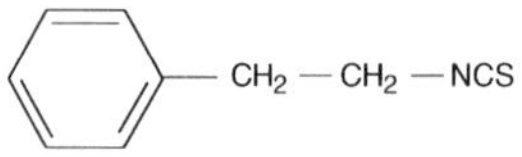

$C_6H_5 - CH_2 - CH_2 - NCS$

Phenethyl isothiocyanate

Figure 4. Structure of sulforaphane.

the plant is eaten. The concentration of this compound was reported to vary widely in different strains of broccoli and under different growing conditions by as much as fifteen-fold. The compound was shown to be a member of the isothiocyanate family of anticancer drugs which are widely distributed in the plant kingdom. Its primary activity was shown to be as a potent phase II enzyme inducer increasing the level of enzymes such as quinone reductase, NADPH reductase, glutathione sulfotransferases, and glucuronyl transferases in cell culture systems.[37] These enzymes are capable of conjugating activated carcinogens and converting them to water-soluble compounds which are inactive and can be readily cleared by the kidney. There is also some evidence that in addition to their phase II inducing ability they can also inactivate P450–2E4, a potent hydroxylating enzyme capable of activating potential carcinogens. No activity against P450–1A2 which carries out 2-hydroxylation of estrogens has been reported.

Sulforaphene, and its sulfide and sulfone analogs along with synthetic acetylnorbornyl isothiocyanates were studied in Sprague-Dawley rats pretreated with DMBA. A dose-dependent 30–60% reduction in mammary tumors was seen in the animals treated with sulfurathene. A lesser decrease was observed when the norbornyl isothiocyanates were used.[38] Cross-breeding experiments between cultivars and wild strains have developed a new strain which is particularly rich in sulforaphene in seedling plants and reportedly low in other glucosinolates.[39] The isothiocyanates as a class have also been shown to be active against a variety of other tumors.[40,41] Phenethyl isothiocyanate has been most extensively studied.

The isothiocyanates are not reported to have any direct effects on cells in terms of altering the cell cycle or increasing apoptosis. It has recently been reported that the isothiocyanates are metabolized by conjugation with glutathione to yield dithiocarbamate derivatives which are then further rearranged to yield acetyl cysteine derivatives.[42] The activity of these further metabolites is not known. Because the rate of this metabolic conversion is not uniform, it is not clear what therapeutic dose will be needed to maintain a therapeutic response in human subjects.

C. Epigallocatechin Gallate

Epigallocatechin gallate (Fig. 5) a compound found in green tea and to a lesser extent in black tea, has been extensively studied for its activity as an antitumor agent in a remarkably diverse variety of tumors. It has been shown to be active as a topical agent against skin tumors induced by carcinogens and promoted by TPA.[43] It has also showed dramatic activity in a variety of lung tumor models including NNK tumor induction.[44] When tested against a variety of gastrointestinal carcinogens including

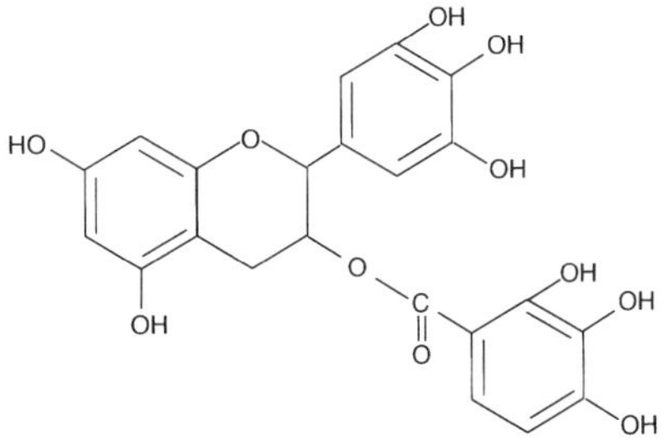

Figure 5. Structure of epigallocatechin gallate.

ENNG, MNNG, and azoxymethane it showed substantial inhibitory activity.[45,46] It has also been shown to be active in vivo against prostate cancer and breast cancer and in vitro against the corresponding cell lines.[47,48]

When studied in breast cancer cell lines EGCG showed a significant ability to alter estrogen metabolism in a protective direction, increasing the ratio of C-2/C-16α metabolites.[49] However it is primarily considered to be active as an antioxidative agents.

Despite this extensive literature, except in China and Japan where green tea is consumed in large amounts, EGCG as the compound or as green tea has failed to be considered as an antitumor agent in the Western world despite its substantial potency and absence of toxicity. The lack of a corporate champion may well be one of the problems. When compared with I3C and genestein for its ability to block proliferation and induce apoptosis EGCG was intermediate in potency, more potent than I3C but less potent than genistein (Table 1).[12]

D. Limonene

Limonene is a diterpene found in citrus peel along with its mono-oxygenated derivative, perillyl alcohol. It has been extensively studied by Gould and colleagues for its antitumor activity.[50] On a molar basis limonene is less potent than some of the other phytochemicals under study and in animal studies limonene must be used at doses up to 5% of the diet. Limonene is believed to act as an inhibitor of farnesylation of the ras oncogene.[51–58] In addition it is thought to interfere with the first steps in isoprenylation from hydroxymethylglutarate. No human trials with limonene have been reported.

E. Curcumin

Curcumin, a close relative of caffeic acid, is the principal ingredient of the Indian spice turmeric (Fig. 6). Pure curcumin is a dimethoxy compound while commercial curcumin contains 10–20% of the monodesmethyl compound. Both the pure material and the commercial grade material are fully potent as antitumor agents against a variety of tumors.

Animal studies have shown that it acts to inhibit DMBA-induced mammary tumors at a dose of 100–200 mg/kg body weight.[59] Formation of DNA adducts was also decreased.[60] Rao[61] and others[62] have shown that it is also potent at blocking colon cancer.

Applied topically it also acts to block topical tumors induced by DMBA and phorbol esters.[63] Menon[64] and others have shown that it will also block lung metastases induced by B16F10 melanoma cells. Comparative studies showed that it is more active than catechins in this model. Zheng and colleagues report that the urinary excretion of isoflavanoids is inversely correlated with the risk for breast cancer.[65]

It has also been demonstrated to be active against BP-induced forestomach tumors.[66] When combined with catechins there was a synergistic response. Curcumin

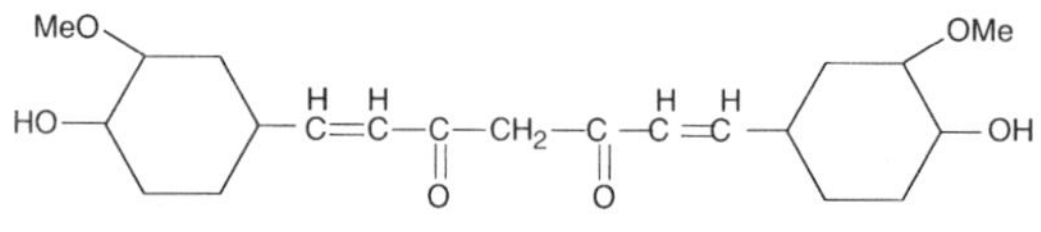

Figure 6. Structure of curcumin.

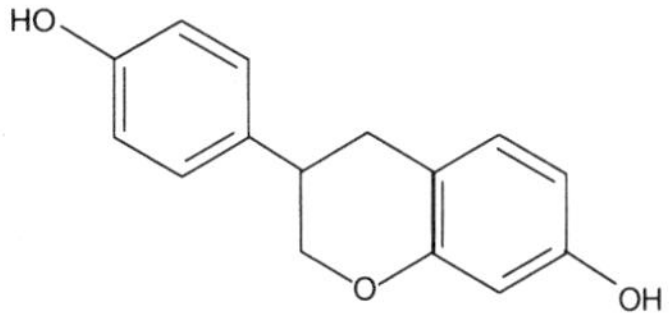

Equol

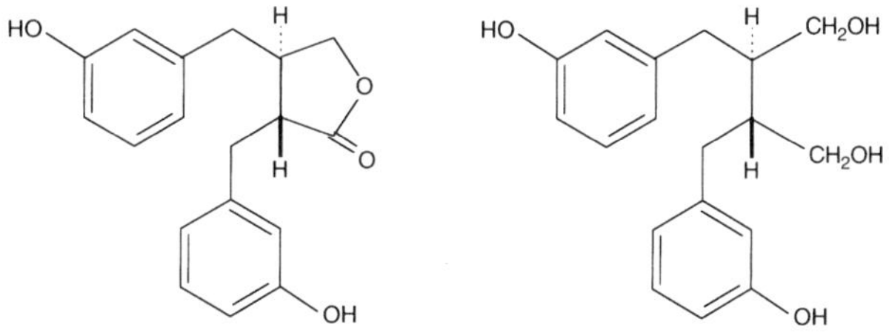

Enterolactone Enterodiol

Figure 7. Structure of the lignan metabolites enterolactone and enterodiol.

has also been demonstrated to block lipoidal peroxidation and superoxide formation, both known DNA-damaging agents.[67]

These results support a widespread antitumor activity for curcumin and its close metabolites.

F. Lignans and Isoflavones

These two classes of compounds have frequently been studied together because of their origin in dietary components. The principal active lignans are enterolactone and enterodiol (Fig. 7), which are derived by the action of intestinal bacteria on the parent lignans (metaresinol and secoisolariciresinol) which are found in such foods as flaxseed, a particularly rich source.[68] The relative yield of the two lignan products varies considerably between individuals and between the sexes.[69]

The isoflavones (genistein, daidzein, and equol) (Fig. 8) are found in a variety of foods but especially in soy products.[70] They are also extensively metabolized in the body

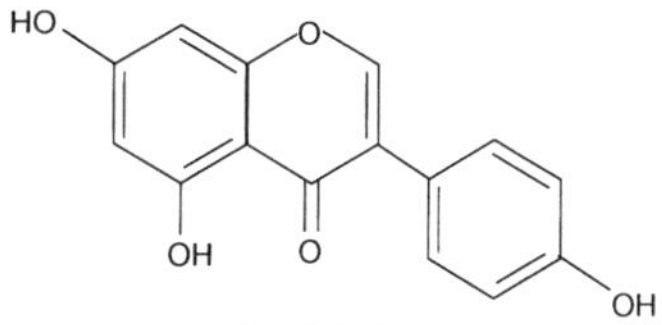

Genistein

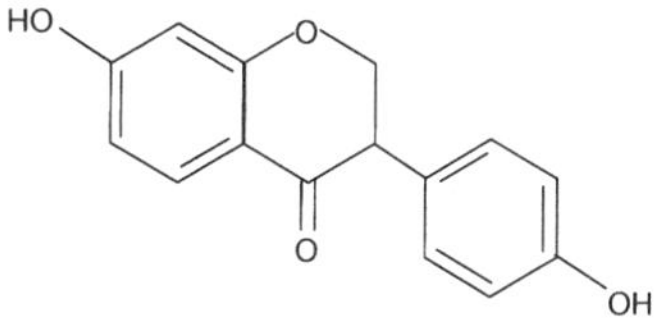

Daidzein

Figure 8. Structure of the isoflavones and their metabolites.

to a variety of compounds; whether the total activity is due to the parent compounds or the metabolites has still not been resolved. All of these compounds are excreted as glucosiduronate or sulfate conjugates.[71]

Epidemiological studies suggest that populations consuming large amounts of these dietary items have lower rates of prostate and breast cancer.[72] Measurement of lignans in the prostatic fluid of Asian and Western men confirm differences in the levels of these compounds directly in the prostate in these two populations.[73]

The action of these compounds involves several different hormonal roles. One is the ability of these compounds to inhibit aromatase activity in peripheral tissues.[74] This is particularly important in postmenopausal women where peripheral metabolism is the principal source of circulating estrogen. Enterolactone was the most potent of these compounds while enterodiol was less potent than the isoflavones which are of intermediate activity. The lignans and isoflavones also act to displace E2 and E1 from SHBG, the principal protein carrier of estrogens in the blood.[75] They are thought to interfere with the transport of the estrogens to target sites. In addition, blocking the binding of estrogens to SHBG can interfere with the role of the complex as a second messenger binding to a cell surface receptor and inducing intracellular cAM.[76] A third is their ability to inhibit 5α-steroid reductase and 17β- steroid reductase. Although both lignans and isoflavones are active at inhibiting 17β-SDH there is considerable variability in the potency of these compounds at inhibiting 5α-reductase with different responses in the inhibition of the Type I and Type II reductases.[77] Because of the importance of this reaction on the activation of testosterone by reduction to dihydrotestosterone this inhibition is of considerable interest.

A number of studies have shown that feeding a supplement of a lignan precursor like flaxseed results in a substantial increase in circulating enterolactone and enterodiol levels in the circulation and excretion in the urine, up to 10–20 fold greater than the baseline level.[78] Telang[79] has showed that genistein is potent in inhibiting cell progression and holds cells at G0/G1. The response to isoflavonoids is not always simple. A recent report showed that the isoflavone tangeretin, when given alone in cell culture was inhibitory to the growth of mammary tumor cells and acted synergistically with tamoxifen. However in vivo, the opposite effect was observed. Substantial reduction of the inhibitory effect of tamoxifen was observed when tangeretin was fed along with the antihormone.[80]

These results suggest a feasibility study to see if supplementary feeding of isooflavones and lignans results in a significant level of cancer prevention.

G. Hydroxyphenyl Cinnamic Acid Derivatives

This group of dihydroxyphenyl cinnamic acid derivatives, originally found in bee propolis, have been studied at varying levels of interest for the different major members of this class of antitumor agents. This group includes caffeic acid and its phenylethyl ester (Fig. 9). Studies in the Min-mouse show a substantial decrease in colonic tumors in mice treated with the latter compound.[81]

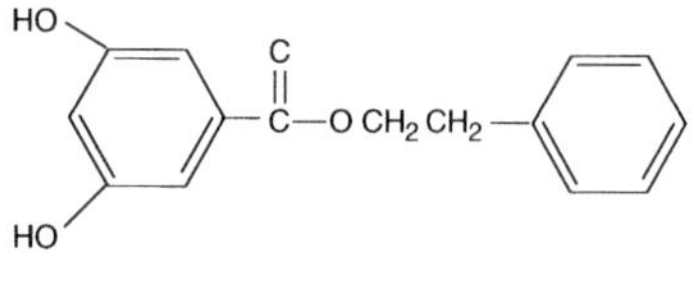

Figure 9. Caffeic acid phenylthyl ester.

SUMMARY

These results, describing antitumor activity of some of the phytochemicals that have been actively studied, suggest that dietary changes could play a role in decreasing the incidence of a variety of tumors. I3C and the other compounds discussed may well be only prototypes for other as yet unexplored phytochemicals present in the diet.

There have been no attempts to explore the possibilities of synergistic action among the various phytochemicals, I3C, limonene, curcumin, epigallocatechin gallate, sulforaphene, or genistein. Mixtures of these compounds might well show potency at lower doses for each of the compounds and show even greater promise than that already demonstrated.

REFERENCES

1. Graham, S. (1983) Toward a dietary control of cancer. Epidemiol Rev 5:38–50.
2. Wynder E.L., Fujita Y., and Harris R.E. (1991) Comparative epidemiology of cancer between the United States and Japan. A second look. Cancer 67:746–763.
3. Telang N.T., Narayanen R., Bradlow H.L., and Osborne, M.P. (1991) Coordinated expression of intermediate biomarkers for tumorigenic transformation in ras transfected mouse mammary epithelial cells. Breast Cancer Res Treat 18:155–163.
4. Bradlow H.L., Hershcopf R.J., Martucci C.P., Fishman J., and Osborne M.P. (1985) Estradiol 16a-hydroxylation in the mouse correlates with mammary tumor virus incidence and the presence of murine mammary tumor virus: A possible model for the hormonal etiology of breast cancer in humans. Proc Natl Acad Sci USA 82:6295–6299.
5. Telang N.T., Axelrod D., Bradlow H.L., and Osborne M.P. (1990) Metabolic biotransformation of estradiol in human mammary explant cultures. In: Biochemistry of Breast Cyst Fluid: Correlation with Breast Cancer Risk, eds. A. Angeli, H.L. Bradlow, F.I. Chasalow, and L. Dogliotti, Ann NY Acad Sci 586:70–78.
6. Auborn K.J., Woodworth C., DiPaolo J.A., and Bradlow H.L. (1991) The interaction between HPV Infection and estrogen metabolism in cervical carcinogenesis Int J Cancer 49: 867–869.
7. Raju U., Levitz M., Sepkovic D.W., Bradlow H.L., Dixon M., and Miller W.R. (1997) estradiol (E2) and estrone (E1) metabolislm in human breast cysts J Soc Gyn Invest 85a:#T185.
8. Telang N.T., Bradlow H.L., Kurihara H., and Osborne M.P. (1989) In vitro biotransformation of estradiol by explant cultures of murine mammary tissues. Breast Cancer Res Treat 13:173–181.
9. Schneider J., Kinne D., Fracchia A., Pierce V., Anderson K.E., Bradlow H.L., and Fishman, J. (1982) Abnormal oxidative metabolism of estradiol in women with breast cancer. Proc Natl Acad Sci USA, 79:3047–3050.
10. Bradlow H.L., Sepkovic D.W., Klug T., and Osborne M.P. (1998) Application of an improved ELISA assay to the analysis of urinary estrogen metabolites Steroids 63:406–413.
11. Telang N.T., Katdare M., Bradlow H.L., Osborne M.P., and Fishman J. (1997) Inhibition of proliferation and modulation of estradiol metabolism: Novel mechanisms for breast cancer prevention by the phytochemical indole-3-carbinol. Proc Soc Exp Biol Med 216:246–251.
12. Telang N.T., Katdare M., Bradlow H.L., and Osborne M.P. (1999) Cell cycle regulation, apoptosis, and estradiol biotransformation: Novel endpoint biomarkers for human breast cancer prevention. J Clin Ligand Assay in press.
13. Bradlow H., Michnovicz J.J., Telang N.T., and Osborne M.P. (1991) Effect of dietary indole-3-carbinol on estradiol metabolism and spontaneous mammary tumors in mice. Carcinogenesis 12:1571–1574.
14. Malloy V.L., Bradlow H.L., and Orentreich N. (1998) Interaction between a semisynthetic diet and indole-3-carbinol on mammary tumor incidence in Balb/cfC3H Mice Anticancer Res 17: 4333–4338.
15. Grubbs C., Steele V.E., Casebolt T., et al. (1994) Chemoprevention of Chemically Induced Mammary Carcinogenesis by Indole-3-Carbinol AACR Meeting Abst #1305.
16. Tiwari R.K., Bradlow H.L., Telang N.T., and Osborne M.P. (1994) Selective Responses of human breast cancer cells to indole-3-carbinol, a chemo-preventive agent. J Natl Cancer Inst 86:126–131.

17. Niwa T., Swaneck G., and Bradlow, H.L. (1994) Alterations in estradiol metabolism in MCF-7 cells induced by treatment with indole-3-carbinol and related com-pounds. Steroids 59:523–527.
18. Bradlow H.L., Telang N.T., Sepkovic D.W., and Osborne M.P. (1997) 2-Hydroxyestrone: The ÒgoodÓ estrogen International Symposium on DHEA Transformation into Androgens and Estrogens in Target Tissues: Intracrinology, Quebec. J Endocrinol 150:S259–S265.
19. Michnovicz J.J., and Bradlow H.L. (1990) Induction of estradiol metabolism by dietary indole-3-carbinol in humans. J Natl Cancer Inst 50:947–950.
20. Bradlow, H.L., Michnovicz J.J., Wong G.Y.C., Halper M.P., Miller D., and Osborne M.P. (1994) Long term responses of women to indole-3-carbinol or a high fiber diet. Cancer Epidemiol Biomarkers Prevention 3:591–595.
21. Kojima T., Tanaka T., and Mori M. (1994) Chemoprevention of spontaneous endo-metrial cancer in female Donryu rats by dietary indole-3-carbinol. Cancer Res. 54:1446–1449.
22. Auborn K., Abramso A., Bradlow H.L., Sepkovic D.W., and Mulloolly V. (1999) Estrogen metabolism and laryngeal papillomatosis: A pilot study on dietary prevention Anticancer Resh. 18: 4569–4574.
23. Auborn K., Abramson A., Bradlow H.L., Sepkovitz D.W., and Mulloly V. (July 1995) Cruciferous vegetables as adjunct therapy for laryngeal papillomatosis, a pilot study Workshop on Respiratory Papillomatosis Quebec, Canada July.
24. Newfield K.M., Goldsmith A., Bradlow H.L., and Auborn K. (1993) Estrogen metabolism and human papillomavirus-induced tumors of the larynx: Chemo-Prophylaxis with indole-3-carbinol Anticancer Res 13:337–342.
25. Rosen C.A., Thompson J.W., Woodson G.E., Hengesteg A.P., and Bradlow H.L. (1998) Preliminary results of the use of indole-3-carbinol for ecurrent respiratory papillomatosis. Otolaryngol Head Neck Surg 118:810–815.
26. Coll D.A., Rosen C.A., Auborn K., Potsic W.P., and Bradlow H.L. (1997) Treatment of recurrent respiratory papillomatosis with indole-3-carbinol Am J Otolaryngol 18:283–285.
27. Xu M., Schut H.A., Bjeldanes L.F., Williams D.E., Bailey G.S., and Dashwood R.H. (1997) Inhibition of 2-amino-3-methylimidazo[4,5-f]quinoline-DNA adducts by indole-3-carbinol: dose-response studies in the rat colon Carcinogenesis 18:2149–2153.
28. Xu M., Bailey A.C., Hernandez J.F., Taoka C.R., Schut H.A., and Dashwood R.H. (1997) Protection by green tea, black tea, and indole-3-carbinol against 2-amino-3-methylimidazo[4,5-f] quinoline-induced DNA adducts and colonic aberrant crypts in the F344 rat. Carcinogenesis 17:1429–1434.
29. Taioli E., Bradlow H.L., Sepkovic D.W., Garbers S., Trachman J., and Gartes S. (1996) Cyp-1A1 genotype, estradiol metabolism, and breast cancer in African-Americans AACR Spring # 1697 in Press in Cancer Prevention and Detection.
30. Taioli E., Bradlow H.L., Garbers S.V., Sepkovic D.W., Osborne M.P., Trachman J., Ganguly S., and Garte S.J. (1999) Cyp1A1 genotype, estradiol, metabolism, and breast cancer in African-Americans Cancer Prevention and Detection in press.
31. Yuan F., Chen D., Liu K., Sepkovic D.W., Bradlow H.l., and Auborn K. (1999) Anti-estrogenic effect of indole-3-carbinol in cervical cancer cells: Implications for estrogen-related carcinogenesis Submitted.
32. De Cree C., Ball P., Seidlitz B., Van Kranenburg G., Geurten P., and Keizer H.A. (1998) Responsiveness of plasma 2- and 4-hydroxycatecholestrogens to training and to graduate submaximal and maximal exercise in an untrained woman. Interuni-versity project on reproductive endocrinology in women and exercise Int J Sports Med 19:20–25.
33. Snow R., Barbieri R., and Frisch R. (1991) Estrogen 2-hydroxylase oxidation and menstrual function among elite oarswomen J Clin Endocrinol Metab. 69:369–376.
34. Manson M.M., Hudson E.A., Ball H.W.L., Barrett M.C., Clark H.L., Judah D.J., Verschoyle R.D., and Neal G.E. (1998) Chemoprevention of afllatoxin B1-induced carcinogenesis by indole-3-carbinol in rat liver—predicting the outcome using early biomarkers Carcinogenesis in press.
35. Dashwood R.H. (1998) Indole-3-carbinol: anticarcinogen or tumor promoter in brassica vegetables? Chem Biol Interact 110:1–5.
36. Oganesian A., Hendricks J.D., and Williams D.E. (1997) Long term dietary indole-3-carbinol inhibits diethylnitrosamine-initiated hepatocarcinogenesis in the infant mouse model. Cancer Lett 116:87–94.
37. Zhang Y. and Talalay P. (1994) Anticarcinogenic activities of organic isothiocyanates: chemistry and mechanisms Cancer Res 54:1976s–1981s.

38. Zhang Y., Kensler TW., Cho C-G., Posner G.H., and Talalay P. (1994) Anticarcinogenic activities of sulforaphene and structurally related synthetic norbornyl isothiocyanates Proc Natl Acad Sci USA 91:3147–3150.
39. Faukner K., Mithen R., and Williamson G. (1998) Selective increase of the potential anticarcinogen 4-methylsulphinylbutyl glucosinolate in broccoli. Carcinogenesis 19:605–609.
40. Posner G.H., Cho C.G., Green J.V., Zhang Y., and Talalay P. (1994) Design and synthesis of bifunctional isothiocyanate analogs of sulforaphene: correlation between structure and potency as inducers of anticarcinogenic detoxification enzymes. J Med Chem 37:170–176.
41. Pereira M.A., Grubbs C.J., Barnes L.H., Li H., Olson G.R., Eto I., Juliana M., Whitaker L.M., Kelloff G.J., Steele V.A., and Lubet R.A. (1996) Effects of the phytochemicals, curcumin and quercitin, upon azoxymethane-induced colon cancer and 7,12-dimethylbenz[a]anthracene-induced mammary cancer in rats. Carcinogenesis 17:305–311.
42. Shapiro T.A., Fahey J.W., Wase K.L., Stephenson K.K., and Talalay P. (1998) Human metabolism and excretion of cancer chemopreventive glucosinolates and isothiocyanates of cruciferous vegetables Cancer Epi Biomarkers and Preven 7:1091–1100.
43. Katiyar S.K., Mohan R.R., Agarwal R., and Mukhtar H. (1997) Protection against induction of mouse skin papillomas with low and high risk of conversion to malignancy by green tea polyphenols. Carcinogenesis 18:497–502.
44. Shi S.T., Wang Z.Y., Smith T.J., Hong J.Y., Chen W.F., Ho C.T., and Yang C.S. (1994) Effects of green tea and black tea on 4-(methylnitrosamino)-1-(3-pyridyl)-1-butanone bioactivation, DNA methylation, and lung tumorigenesis in A/J mice. Cancer Res 54:4641–4647.
45. Yamane T., Knakatani H., Kikuoka N., Matsumoto H., Iwata Y., Kitao Y., Oya K., and Takahashi T. (1996) Inhibitory effects and toxicity of green tea polyphenols for gastrointestinal carcinogenesis. Cancer 77:1662–1667.
46. Yamane T., Hagiwara N., Taleishi M., Akachi S., Kim M., Okuzumi J., Kitao Y., Inagake M., Kuwata K., and Takahasi T. (1991) Inhibition of azoxymethane-induced colon carcinogenesis in rat by green tea polyphenol fraction. Jpn J Cancer Res 82:1336–1339.
47. Hirose M., Mizoguchi Y., Yaono M., Tanak H., Yamaguchi T., and Shirai I. (1997) Effects of green tea catechins on the progression or late promotion stage of mammary gland carcinogenesis in female Sprague-Dawley rats pre-treated with 7,12-dimethylbenz(a)anthracene Cancer Lett 112:141–147.
48. Kujiki H., Yoshizawa S., Horiuchi T., Suganuma M., Yatsunami J., Nishiwaki S., Okabe S., Nishiwaki-Matsushima R., Okuda T., and Sugimura T. (1992) Anticarcinogenic effects of epigallocatechin gallate. Prev Med 21:503–509.
49. Michnovicz J.J. and Bradlow H.L. (1994) Dietary cytochrome P450 modifiers in the control of estrogen metabolism. In: Food Phytochemicals for Cancer Prevention I. eds. M.T. Huang, T. Osawa, C.T. Ho and R.T. Rosen, Published by the American Chemical Society, Washington, DC pp. 282–293.
50. Ariazi E.A., and Gould M.N. (1996) Identifying differential gene expression in monoterpene-treated mammary carcinomas using subtractive display. J Biol Chem 271:29286–29294.
51. Karlson J., Borg-Karlson A.K., Unelius R., Shoshan M.C., Wilking N., Ringborg U., and Linder S. (1996) Inhibition of tumor cell growth by monoterpenes in vitro: evidence of a Ras-independent mechanism of action. Anticancer Drugs 7:422–429.
52. Crowell P.L., Siar A.A., and Burke Y.D. (1995) Antitumorigenic effects of limonene and perillyl alcohol against pancreatic and breast cancer. Adv Exp Med Biol 401:131–136.
52. Broitman S.A., Wilkinson J. 4th, Cerda S., and Branch S.K. (1996) Effects of monoterpenes and mevinolin on murine colon tumor CT-26 in vitro and its hepatic ÒmetastasesÓ in vivo. Adv Exp Med Biol 401:111–130.
53. Reddy B.S., Wang C.X., Samaha H., Lubet R., Steele V.E., Keloff G.J., and Rao C.V. (1997) Chemoprevention of colon carcinogenesis by dietary perillyl alcohol. Cancer Res 57:420–425.
54. Haag J.D. and Gould M.N. (1994) Mammary carcinoma regression induced by perillyl alcohol, a hydroxylated analog of limonene Cancer Chemother Pharmacol 34:477–483.
55. Gould M.N. G.C.J., Shang R., Wang B., Kennan W.S., and Haag J.D. (1994) Limonene chemoprevention of mammary carcinoma induction following direct in situ transfer of v-Ha-ras. Cancer Res 54:3540–3543.
56. Karlson J, Borg-Karlson AK, Unelius R, Shoshan MC, Wilking N, Ringborg U, Linder S. (1996) Inhibition of tumor cell growth by monoterpenes in vitro: evidence of a Ras-independent mechanism of action. Anticancer Drugs 7:422–429.

57. Ariazi E.A. and Gould M.N. (1996) identifying differential gene expression in monoterpene-treated mammary carcinoma using subtractive display J. Biol Chem 271:29286–29294.
58. Whysner J. and Williams G.M. (1996) D-limonene mechanistic data and risk assessment: absolute species-specific cytotoxicity, enhanced cell proliferation, and tumor promotion. Pharmacol Ther 71:127–136.
59. Conney A.H., Lou Y.R., Xie J.G., Osawa T., Newmark H.L., Liu Y., Chang R.L., and Huang M.T. (1997) Some perspectives on dietary inhibition of carcinogenesis: Studies with curcumin and tea. Proc Soc Exptl Biol Med 216:234–245.
60. Mehta K., Pantazis P., McQueen T., and Aggaral B.B. (1997) Antiproliferative effect of curcumin (diferuloylmethane) against human breast tumor cell lines. Anticancer Drugs 8:470–481.
61. Rao C.V., Rivenson A., Simi B., and Reddy B.S. (1995) Chemoprevention of colon cancer by dietary curcumin Ann NY Acad Sci 768:201–204.
62. Kawamori T., Lubet R., Steele V.E., Kelloff G.J., Kaskey R.B., Rao C.V., and Reddy B.S. (1999) Chemopreventive effect of curcumin, a naturally occurring anti-inflammatory agent, during the promotion/progression stages of colon cancer. Cancer Res 59:597–601.
63. Huang M.T., Ma W., Yen P., Xie J.G., Han J., Frenkel K., Grunberger D., and Conney A.H. (1997) Inhibitory effects of topical application of low doses of curcumin on 12-O-tetradecanoylphorbol-13-acetate-induced tumors promotion and oxidized DNA bases in mouse epidermis. Carcinogenesis 18:83–88.
64. Menon L.G., Kuttan R., and Kuttan G. (1995) Inhibition of lung metastasis in mice induced by B16F10 melanoma cells by polyphenolic compounds. Cancer Lett 95:221–225.
65. Singh Hvh S.V., Hu X., Srivastava S.K., Singh M., Xia H., Orchard J.L., and Zaren H.A. (1998) Mechanism of inhibition of benzo[a]pyrene-induced forestomach cancer in mice by dietary curcumin. Carcinogenesis 19:1357–1360.
66. Sreejayan N., and Rao M.N. (1996) Free radical scavenging activity of curcuminoids, Arzneimittelforschung 46:169–171.
67. Richard S.E., Orcheson L.J., Seidl M.M., Luyengi L., Fong H.H., and Thompson L.L. (1996) Dose-dependent production of mammalian lignans in rats and in vitro from the purified precursor secoisolariciresinol diglycoside in flaxseed. J Nutr 126:2012–2019.
68. Zheng, W., Dai Q., Custer L.J., Shu X-O., Wen W-Q., Jin F., and Franke A.A. (1999) Urinary excretion of isoflavonoids and the risk of breast cancer. Cancer Epidem Biomarkers Preven 8:36–40.
69. Morton M.S., Matos-Ferreira A., Abranches-Monteiro L., Correia R., Blacklock N., Vhan P.S., Cheng C., Lloyd S., Chieh-ping W., and Griffith K. (1997) Measurement and metaboism of isoflavonoids and lignans in the human male. Cancer Lett. 19:145–151.
70. Joannou G.E., Kelly G.E., Reeder A.Y., Waring M., and Nelson C.A. (1995) urinary profile study of dietary phytoestrogens. The identification and mode of metabolism of new isoflavonoids J Steroid Biochem Mol Biol 54:167–184.
71. Kurzer M.S., Lampe J.W., Martini M.C., and Adlercreutz H. (1995) Fecal lignan and isoflavonoid excretion in premenopausal women consuming flaxseed powder. Cancer Epidemiol Biomarkers Prev 4:353–358.
72. Adlercreutz H., Fotsis T., Kurzer M.S., Wahala K., Makela T., and Hase T. (1995) Isotope dilution gas chromatographic-mass spectrometric method for the determination of unconjugated lignans and isoflavonoids in hman feces, with preliminary results in omnivorous and vegetarian women. Anal Biochem 225:101–108.
73. Bambagiotti-Alberti M., Cooron S.A., Ghiara C., Moreti G., and Raffaelli A. (1994) Investigation of mammalian lignan precursors in flax seed: first evidence of secoisolariciresinol diglucoside in two isomeric forms by liquid chromato-graphic/mass spectrometry. Rapid Commun. Mass Spectrom 6:929–932.
74. Adlercreutz H., van der Wildt J., Kinzel J., Attalla H., Wahala K., Makela T., Hase T., and Fotsis T. (1995) Lignan and isoflavanoid conjugates in human urine. J Steroid Biochem Mol Biol 52:97–103.
75. Wang C., Makela T., Hase T., Adlercreutz H., and Kurzer M.S. (1994) Lignans and flavanoids inhibit aromatase enzyme in human preadipocytes J Steroid Biochem Mol Biol 50:205–212.
76. Martin M.E., Haourigui M., Pelissero C., Benassayag C., and Nunez E.A. (1996) Interactions between phytoestrogens and human sex steroid binding protein. Life Sci 58:429–436.
77. Fortunati N., Fissore F., Comba A., Becchis M., Catalano M.G., Fazzari A., Berta L., and Frairia R. (1996) Sex steroid-binding protein and its membrane receptor in estrogen-dependent breast cancer: biological and pathophysiological impact. Horm Res 45:202–206.

78. Morton M.S., Wilcox G., Wahlqvist M.L., and Griffiths K. (1994) Determination of lignans and isoflavanoids in human female plasma following dietary supplementation. J Endocrinol 142:251–9.
79. Lampe J.W., Martini M.C., Kurzer M.S., Adlercreutz H., and Slavin J.L. (1994) Urinary lignan and isoflavanoid excretion in premenopausal women consuming flaxseed powder. Am J Clin Nutr 60:122–128.
80. Bracke M.E., Depypere H.T., Boterberg T., Van Marck V.L., Vennekens K.M., Vanluchene E., Nuytunck M., Serreyn R., and Marcek M.M. (1999) Influence of tangeretin on tamoxifenÕs therapeutic benefit in mammary cancer J Natl Cancer Inst 91:354–359.
81. Carothers A.M., Barrett C.C., Mahmoud J.C., Bolinski R.T., Churchill M.R., Isaacs J., Grunberger D., and Bertagnolli M.M. (1999) Caffeic acid phenethyl ester (CAPE) prevents tumors in Min/+ mice and regulation of P53 function in vitro, in Cancer Prevention: Novel ed by H Leon Bradlow, Jack Fishman, and Michael P. Osborne Ann NY Acad Sci in press.

LOW DOSE EXPOSURE TO CARCINOGENS AND METABOLIC GENE POLYMORPHISMS

Emanuela Taioli and Seymour Garte[1]

Policlinico IRCCS, University of Milan
Via F. Sforza 28, 20122 Milan
Italy
[1] EOHSI, UMDNJ-Robert Wood Johnson Medical School
Piscataway NJ 08855

1. INTRODUCTION

Metabolic gene polymorphisms encode for enzymes which are involved in both metabolism and conjugation of environmental as well endogenous compounds. Some of the products of the metabolic process are carcinogens. Two main categories of metabolic genes are known: Phase I genes, which include CYP1A1, CYP2E1, CYP2D6, and Phase II genes, such as GSTM1, GSTT1, NAT2. Polymorphisms have been described in these genes, with different frequencies according to ethnicity and geographic area.[1] Several case-control studies have been conducted to study the association between metabolic gene polymorphisms and cancer of various sites, with special focus on lung and bladder cancer. It has been suggested that these genes play a role in cancer risk only when they interact with environmental exposure, since the substrates of their gene products are xenobiotic chemicals or their metabolites.[2] This form of gene-environment interaction (GEI) has been described as "Type 2" GEI by Khoury[3] and Ottman.[4] According to this model, the presence or absence of the genetic risk factor is irrelevant for disease causation, if there is no exposure to an environmental agent. When the dose of environmental exposure (such as smoking) is analyzed with respect to metabolic susceptibility gene polymorphisms, two patterns are seen. In one case, a low exposure-gene (LEG) effect is observed, in which a decreasing degree of interaction occurs as a function of increasing exposure dose. A high exposure-gene (HEG) effect is observed when there is an increased degree of interaction as a function of exposure dose.[2]

Advances in Nutrition and Cancer 2, edited by Zappia *et al.*
Kluwer Academic / Plenum Publishers, New York, 1999.

Table 1. Effect of GSTM1 deletion on lung cancer risk according to smoking dose

	Nonsmokers	<40 pack-years	≥40 pack-years
GST+	1.0 (Ref)	7.6 (3.5–16.6)	19.2 (8.9–41.1)
GST–	1.0 (0.3–3.6)	9.1 (4.1–20.0)	13.0 (5.9–28.9)
London, 1995			
GST+	1.0 (Ref)	2.6 (1.5–4.3)	3.6 (2.3–5.8)
GST–	1.0 (0.6–1.7)	3.7 (2.3–6.0)	6.5 (4.3–9.9)
Kihara, 1995			
		≤20 pack-years	>20 pack-years
GST+	1.0 (Ref)	1.6 (0.4–6.9)	2.9 (0.9–9.6)
GST–	0.8 (0.1–7.2)	0.8 (0.2–3.8)	3.3 (1.0–10.9)
Brockmoller, 1993			

M-H odds ratio for non-smokers: 1.0 (0.6–1.4).

2. REVIEW OF THE LITERATURE

2.1 Studies on Cancer Risk, Genetic Polymorphisms, and Carcinogen Dose

We have reviewed the published literature, and selected articles that met the following criteria: case-control studies with hospital or community controls on lung cancer and CYP1A1, GSTM1, which included a stratified analysis by smoking dose. Studies where controls had other diseases of the lung were excluded. Table 1 reports the results of studies on GSTM1 and lung cancer according to smoking dose. The crude odds ratios (OR) have been calculated from the published tables, and subjects who did not smoke and did not carry the homozygous GSTM1 deletion were set as the reference group. While the study from London,[5] conducted on Caucasians and African-Americans, shows a LEG effect, the studies from Kihara[6] in Japaneses and Brockmoller[7] in Caucasians show a HEG effect. In non smokers, the average OR for the GSTM1 deletion, calculated with the Mantel-Haenzsel (M-H) method, is 1.0 (95% CI:0.6–1.4). Unfortunately, the difference in smoking dose used by the various Authors does not allow for a summary OR across strata of smoking. Two studies on CYP1A1*2 (Msp1), one in Japanese,[8] the other in Brazilians,[9] indicate a LEG effect, although the data are not comparable, due to differences in the reference group (one has non smokers as a reference, the other has low level of smoking). In addition, Nakachi presents the wild type and the heterozygous together, and limits the analysis to squamous cell carcinoma.

The studies on CYP1A1*4 are controversial: the data from London[10] and Taioli[11,12] indicate a LEG effect, while the data from Kelsey[13] suggest a HEG effect. The comparison among studies is difficult, since different reference groups and different smoking cut offs have been used by the various authors.

Two studies have analyzed the combined genotype GSTM1 and CYP1A1 as risk factors for lung cancer in relation to smoking dose. As shown in Table 2, the study from Nakachi[14] on Japanese with squamous cell carcinoma shows that the CYP1A1*2 polymorphism alone increases the risk of lung cancer at low smoking dose, but the effect

Table 2. Effect of CYP1A1 polymorphisms on lung cancer risk according to smoking dose

CYP1A1*2			
	<3 × 10^5 lifetime cig	3–4 × 10^5 lifetime cig	>4 × 10^5 lifetime cig
wt+hetero	1.0 (Ref)	6.6 (2.0–21.4)	11.7 (4.0–34.5)
homozygous	7.3 (1.7–32.5)	13.3 (2.6–66.3)	13.2 (2.2–79.9)
Nakachi, 1991			
	Nonsmokers	<40 pack-years	≥40 pack-years
wt	1.0 (Ref)	4.2 (1.4–12.4)	18.1 (6.9–47.5)
hetero+homo	1.3 (0.3–6.9)	3.7 (1.1–12.3)	12.6 (4.5–35.2)
Sugimura, 1993			
CYP1A1*4			
	Nonsmokers	≤35 pack-years	>35 pack-years
wt	1.0 (Ref)	6.2 (2.5–15.4)	10.8 (4.3–27.1)
hetero+homo	—	8.2 (2.6–25.7)	26.1 (8.0–85.5)
London, 1995			
	≤15 pack-years	>15 pack-years	
wt	1.0 (Ref)	12.5 (4.3–36.8)	
hetero+homo	8.5 (2.3–32.1)	6.4 (0.7–55.2)	
Taioli, 1995 and 1998			
	≤24 pack-years	25–49 pack-years	>50 pack-years
wt	1.0 (Ref)	8.0 (3.0–21.3)	14.7 (5.6–38.4)
hetero+homo	2.0 (0.3–11.8)	5.0 (1.2–21.4)	4.0 (0.8–19.4)
Kelsey, 1994			

is less evident at high doses for the homozygous variant. The additional presence of the GSTM1 homozygous deletion increases the risk of lung cancer at both smoking doses, but especially at high doses. A similar effect was observed for CYP1A1*3 polymorphism, while the additional presence of the GSTM1 deletion increases the risk of cancer much more at low doses, consistent with a LEG effect. At high smoking dose, the risk associated with the double polymorphism seems to reach a plateau.

Another study in the Japanese population (see Table 3), limited to squamous and small cell carcinoma[15] showed no effect of CYP1A1*2 on cancer risk at both smoking doses, and an increased risk when GSTM1 is also present, especially at high smoking doses.

2.2 Effect of Different Carcinogen Doses on Other Endpoints

Few studies have assessed the effect of different doses of carcinogens on intermediate endpoints in relation to metabolic polymorphisms. Table 4 shows the results of exposure to different doses of 1,2-epoxy-3-butene (MEB;[16,17]) or diepoxybutane (DEB;[18]) on sister chromatid exchange in lymphocytes from subjects carrying homozygous deletions of GSTT1 or GSTM1. This set of experiments show that the relative induction of SCE by carcinogens is higher in the presence of deletion of either GSTM1 or GSTT1, with more effect at higher doses, as expected from a HEG effect.

In contrast, a study[19] on benzo-a-pyrene urinary levels in workers according to CYP1A1*2 genotype and smoking dose shows that the urinary levels are higher in subjects carrying the CYP1A1 polymorphism at low smoking doses, but the effects of the polymorphism almost disappears at higher smoking doses. Similar examples on carcinogen-adduct levels according to cotinine urinary levels and other metabolic gene polymorphisms in genes such as NAT2 and CYP2E1 have been published.[20]

Table 3. Effect of both CYP1A1 polymorphisms and GSTM1 deletion on lung cancer risk according to smoking dose

GSTM1+	CYP1A1*2	≤31.1 × 10^4 lifetime cig	>32.1 × 10^4 lifetime cig
	Wt	1.0 (Ref)	9.0 (2.0–40.5)
	Hetero	0.4 (0.1–7.3)	13.0 (3.2–53.5)
	Homo	8.3 (1.5–46.4)	10.0 (1.0–99.0)
GSTM1−	Wt	2.6 (0.6–12.0)	11.8 (2.9–48.6)
	Hetero	2.2 (0.4–11.5)	16.3 (4.0–66.8)
	Homo	16.0 (3.4–74.6)	20.0 (3.2–124.0)
GSTM1+	CYP1A1*3		
	Wt	1.0 (Ref)	13.7 (3.5–53.3)
	Hetero	3.9 (0.8–19.9)	19.1 (4.0–92.2)
	Homo	—	13.7 (0.7–277.1)
GSTM1−	Wt	4.0 (0.9–16.7)	18.2 (4.8–68.6)
	Hetero	3.0 (0.5–16.4)	21.9 (4.7–100.8)
	Homo	41.0 (8.0–209.1)	27.3 (2.8–268.3)
Nakachi, 1993			
GSTM1+	CYP1A1*2	<40 pack-years	≥40 pack-years
	Wt	1.0 (Ref)	1.6 (0.5–5.2)
	Hetero	1.2 (0.3–4.5)	1.8 (0.6–5.6)
	Homo	0.3 (0.1–4.5)	0.6 (0.1–5.1)
GSTM1−	Wt	0.4 (0.1–2.60)	1.7 (0.6–5.2)
	Hetero	1.0 (0.2–4.0)	3.1 (1.0–9.0)
	Homo	1.6 (0.3–8.0)	4.1 (1.2–14.4)
Kihara, 1995			

Table 4. Effect of metabolic gene polymorphisms on sister chromatid exchange (SCE) and 1-hydroxy pyrene (HP) urinary levels according to exposure to different doses of carcinogens

SCE			
GSTT1	125 mM MEB	250 mM MEB	
+	4.6	8.9	
−	5.5	12.5	
Bernardini, 1998			
GSTM1	50 mM MEB	250 mM MEB	
+	2.9	22.3	
−	4.2	29.3	
Uuskula, 1995			
GSTM1+	2 mM DEB	5 mMDEB	
GSTT1+	33.0	69.5	
GSTT1−	58.6	112.2	
GSTM1−			
GSTT1+	33.4	70.0	
GSTT1−	59.5	116.1	
Norppa, 1995			
HP			
CYP1A1*2	Non Smokers	≤15 cig/day	>15 cig/day
Wt	7.0	11.5	19.0
Hetero	8.5	20.5	22.0
Merlo, 1998			

3. ANALYSIS OF DOSE EFFECTS IN TYPE 2 GENE ENVIRONMENT INTERACTION

We have previously described[2] a mathematical approach to the analysis of case control data that fit a Type 2 GEI model. In this approach, the coefficient of the genetic risk factor in a multiple regression model (b_g in Eq. 1) is defined as equal to 0.

$$F(OD) = a + b_e E + b_g G + b_{eg} EG \tag{1}$$

where OD is the odds of disease, E is the environmental exposure, and G is the genetic risk factor. Thus equation 1 becomes equation 2:

$$F(OD) = a + b_e E + b_{eg} EG. \tag{2}$$

which can be written in a different way:

$$F(OD) = a + b_e(1 + \alpha G)E \tag{3}$$

where $\alpha = b_{eg}/b_e$.

If the value of α is positive (>0), then the effect of the GRF is to modify the effect of the exposure by increasing the odds of disease. If, the genetic factor is protective, then the α term would be negative(<0). If the GRF has no effect on the risk from the exposure, then $\alpha = 0$.

If the slope of a plot of α vs. exposure dose is positive, then the gene environment interaction follows a direct (HEG) dose effect. If the slope is negative, then an inverse (LEG) dose effect is operative.

Since the definition of a HEG effect is an increase in the value of the α term with increased exposure, while the LEG effect results from an inverse relation between α and exposure level, it would be useful to examine in more detail the mathematical relationship between α and exposure dose. This can be done if we make the assumption, generally agreed to be valid, that for diseases at very low incidence, the probability of an individual getting the disease is roughly equal to the incidence. This assumption allows us to substitute incidence for probability in the formulas for the relevant odds ratios. Thus the probability that a case was exposed is substituted by the incidence among exposed individuals (Y). The odds ratio for disease in exposed individuals is then

$$\Psi_e = \frac{Y(1 - Y_0)}{Y_0(1 - Y)} \tag{4}$$

where Y_0 is the incidence rate of the non exposed group, and Y is the incidence of the exposed group.

We know that the incidence Y is a function of dose. If we now define another incidence G as a function of dose in the presence of a genetic risk factor, then the odds ratio for the group with both exposure and the genetic risk factor is

$$\Psi_{eg} = \frac{G(1 - Y_0)}{Y_0(1 - G)} \tag{5}$$

As described above (and with more detail in ref 2), the modified logistic regression equation for odds of the disease (OD) for type 2 GEI given in equation 2 contains 2 coefficients b_e and b_{eg} that are equal to $b_e = \mathrm{Ln}\Psi_e$ and $b_{eg} = \mathrm{Ln}(\Psi_{eg}/\Psi_e)$, where Ψ_e and Ψ_{eg} are the odds ratios for exposure in the absence and in the presence of the genetic risk factor respectively. Since α is defined as the ratio of the coefficients b_{eg}/b_e then

$$\alpha = \frac{\mathrm{Ln}\Psi_{eg} - \mathrm{Ln}\Psi_{e}}{\mathrm{Ln}\Psi_{e}} \tag{6}$$

Therefore we can derive the following equation for α in terms of several parameters that are functions of dose.

$$\alpha = \frac{[\mathrm{Ln}G + \mathrm{Ln}(1-Y)] - [\mathrm{Ln}(1-G) + \mathrm{Ln}Y]}{[\mathrm{Ln}Y + \mathrm{Ln}(1-Y_0)] - [\mathrm{Ln}Y_0 + \mathrm{Ln}(1-Y)]} \tag{7}$$

Although we know that $Y = f(D)$, and $G = g(D)$, and assuming a positive gene environmental interaction, that $G = g(D) > Y = f(D)$, we do not know what these functions are, and therefore any further analysis of α as a function of dose cannot be done with precision. However the change in α with respect to dose can be shown to depend on the nature of the functions Y and G. For example, if both of these functions are simple linear functions of dose, (such as $Y = A_1D + B$, and $G = A_2D + B$, with A_1, A_2, and B being constants and $A_2 > A_1$), then it can be proven that the sign of $d\alpha/dD$ is always negative. This implies that for diseases which follow a linear dose response relationship with respect to the involvement of a particular genetic risk factor, the type 2 gene environment interaction will always produce a LEG effect. For more complex functions of Y and G, the situation can be more complex; for some functions, the sign of $d\alpha/dD$ may depend on the dose and/or the relation between Y and G, and for others, a HEG effect may always be seen. This means that it may be possible, knowing the detailed behavior of α as a function of dose, to make some inferences regarding the form of the dose response relationship relevant to the genetic risk factor in question, and therefore to shed some light on possible mechanistic scenarios.

4. DISCUSSION

Metabolic genes act by modifying the risk associated with environmental exposure to carcinogens by changing the amount of metabolized carcinogens available at the target tissue. Polymorphisms in metabolic genes are known, but for some of them the functional significance has not been studied. It is hard to establish the independent role of each of them on cancer risk, since each metabolic gene encodes for enzymes that contribute to the metabolism of several different carcinogen products. In addition, certain enzyme systems increase the amount of carcinogen metabolites, certain others decrease it, and the balance among these different enzymes determines the quantity of carcinogen that reach DNA. However, their association with certain types of cancer, such as lung or bladder, is well documented, confirmed independently by several groups, and biologically plausible. When an attempt is made to look at the risk associated with a metabolic gene polymorphism according to different levels of exposure to environmental carcinogens, the results are less clear. This is partly due to the small sample size of almost all the published studies, to different selection of controls, and to different categorizations of the levels of exposure. For example, very rarely are the distribution of genotypes in cases and controls non exposed to the environmental factor (for example, non smokers) is reported. For lung cancer, this is a natural consequence of the low frequency of the disease among non smokers. In this case, the reference category cannot be defined. When this is possible (Table 1), the odds ratio of cancer associated with the presence of the GSTM1 deletion in subjects who are non smokers is 1.0 (95% CI: 0.6–1.4), confirming the hypothesis of a type 2 GEI.

Further research in this area is needed, to confirm that subjects who carry a certain metabolic polymorphism, if not exposed to the carcinogen that is metabolized by that specific gene product, do not have an increased risk of cancer. It would be also interesting to know what is the risk of cancer in subjects who carry a metabolic polymorphism and are exposed to very low dose of carcinogen, such as non smokers exposed to passive smoking. Unfortunately, no data have been published on this topic, and the categorizations used in the articles reported in Tables 1–3 are too wide to allow any comment on low levels of exposure. Another aspect that has not been studied is the separate effect of dose and length of exposure to carcinogens in subjects carrying a metabolic polymorphism. So far, only measurements of integrated dose, such as pack-years of smoking, have been used.

The study of GEI according to exposure levels has the potential to substantially enhance our understanding of the role of genetic factors such as metabolic gene polymorphisms in cancer etiology, as well as to elucidate important quantitative mechanisms involved in human environmental carcinogenesis.

REFERENCES

1. Garte S.J., Trachman J., Crofts F., Taioli E., Toniolo P., Buxbaum J., and Bayo S. CYP1A1 genotypes in African-Americans and Caucasians. Human Heredity 46, 121–127, 1996.
2. Taioli E., Zocchetti C., and Garte S.J. Models of interaction between metabolic genes and environmental exposure in cancer susceptibility. Environm. Health Persp. 106, 67–70, 1998.
3. Khoury M.J. and James L.M. Population and familial relative risks of disease associated with environmental factors in the presence of gene-environment interaction. Am J Epidemiol. 137, 1241–50, 1993.
4. Ottman R. An epidemiologic approach to gene-environment interaction. Genet. Epidemiol. 7, 177–185, 1990.
5. London S.J., Daly A.K., Cooper J., Navidi W.C., Carpenter C.L., and Idle J.R. Polymorphism of glutathione S-transferase M1 and lung cancer risk among African-Americans and Caucasians in Los Angeles county, California. JNCI, 87, 1246–1253, 1995.
6. Kihara M., Noda K., and Kihara M. Distribution of GSTM1 null genotype in relation to gender, age, and smoking status in Japanese lung cancer patients. Pharmacogenetics 5, S74–S79, 1995.
7. Brockmoller J., Kerb R., Drakoulis N., Nitz M., and Roots I. Genotype of glutatione S-transferase class μ isoenzymes μ and Ψ in lung cancer patients and controls. Cancer Research, 53, 1004–1011, 1993.
8. Nakachi K., Imai K., Hayashi S., Wtanabe J., and Kawajiri K. Genetic susceptibility to squamous cell carcinoma of the lung in relation to cigarette smoking dose. Cancer Research, 51, 5177–5180, 1991.
9. Sugimura H., Suzuki I., Hamada G.S., Iwase T., Takahashi T., Nagura K., Iwata H., Watanabe S., Kino I., and Tsugane S. Cytochrome P-450 1A1 genotype in lung cancer patients and controls in Rio de Janeiro, Brazil. Cancer Epidem Biom Prev, 3, 145–148, 1994.
10. London S.J., Daly A.K., and Fairbrother K.S. Lung cancer risk in African-Americans in relation to a race-specific CYP1A1 polymorphism. Cancer Research, 55, 6035–6037, 1995.
11. Taioli E., Crofts F., Trachman J., Demopoulos R., and Garte SJ. A specific African-American polymorphism is associated with adenocarcinoma of the lung. Cancer Research, 55, 472–73, 1995.
12. Taioli E., Ford J., Trachman J., Li Y., Demopoulos R. and Garte S. Lung cancer risk and CYP1A1 genotype in African-Americans. Carcinogenesis, 19: 813–18, 1998.
13. Kelsey K.T., Wiencke J.K., and Spitz M.R. A race-specific genetic polymorphismin the CYP1A1 gene is not associated with lung cancer in African-Americans. Carcinogenesis 15, 1121–1124, 1994.
14. Nakachi K., Imai K., Hayashi S., and Kawajiri K. Polymorphisms of the CYP1A1 and glutathione S-transferase genes associated with susceptibility to lung cancer in relation to cigarette dose in a Japanese population. Cancer Research, 53, 2994–2999, 1993.
15. Kihara M., Kihara M., and Noda K. Risk of smoking for squamous and small cell carcinomas of the lung modulated by combinations of CYP1A1 and GSTM1 gene polymorphisms in a Japanese population. Carcinogenesis, 16, 2331–2336, 1995.
16. Bernardini S., Hirvonen A., Pelin K., and Norppa H. Induction of sister chromatid exchange by 1,2-

epoxy-3-butene in cultured human lymphocytes: influence of GSTT1 genotype. Carcinogenesis, 19, 377–80, 1998.

17. Uuskula M., Jarventaus H., Hirvonen A., Orsa M., and Norppa H. Influence of GSTM1 genotype on sister chromatid exchange induction by styrene-7,8-oxide and 1,2-epoxy-3-butene in cultured human lymphocytes. Carcinogenesis, 16, 947–50, 1995.
18. Norppa H., Hirvonen A., Jarventaus H., Uuskula M., Tasa G., Ojajarvi A., and Sorsa M. Role of GSTT1 and GSTM1 genotypes in determining individual sensitivity to sister chromatid exchange induction by diepoxybutane in cultured human lymphocytes. Carcinogenesis, 16, 1261–64, 1995.
19. Merlo F., Andreassen A., Weston A., Pan C.-F., Haugen A., Valerio F., Reggiardo G., Fontana V., Garte S., Puntoni R., and Abbondandolo A. Urinary excretion of 1-hydroxypyrene as a marker for exposure to urban air levels of plycyclic aromatic hydrocarbons. Cancer Epidemiol Biomark Prevent, 7, 147–155, 1998.
20. Vineis P. Molecular epidemiology: low-dose carcinogens and genetic susceptibility. Int. J. Cancer, 71, 1–3, 1997.

20

CARCINOGEN-DNA ADDUCTS AS TOOLS IN RISK ASSESSMENT

Luisa Airoldi,* Roberta Pastorelli, Cinzia Magagnotti, and Roberto Fanelli

Department of Environmental Health Sciences
Istituto di Ricerche Farmacologiche Mario Negri
Via Eritrea 62, 20157 Milan, Italy

1. INTRODUCTION

Effective strategies for preventing human cancer resulting from exposure to chemical carcinogens of environmental, occupational or life-style sources, require the identification of the carcinogens to which people are exposed and the evaluation of the degree of exposure. The use of biological markers to detect such exposure and the early events in the process of carcinogenesis has increased greatly in recent years and their use is expected to be of help for the assessment of future disease risk.[1–3]

Many carcinogens are known to react with macromolecules, either directly or after metabolic activation, to form adducts.[4] DNA or protein adducts are currently used as biochemical endpoints to evaluate human exposure to carcinogens.[5,6] Adducts to proteins such as albumin or hemoglobin, though not involved in carcinogenesis, are useful surrogate for DNA adducts.[6] The formation of DNA adducts is thought to be the relevant step with respect to chemical carcinogenesis. If not repaired, the persistence of DNA lesions may lead to permanent genetic changes in critical genes, an event that can lead to the initiation and progression of cancer.[7] The presence of adducts imply that a carcinogen has been absorbed, distributed in different tissues and metabolically activated and has escaped detoxification and DNA repair mechanisms. Thus the amount of specific DNA adducts provides a good indication of exposure to chemical carcinogens, is expression of the genetic damage resulting from exposure, and accounts for some of the factors affecting individual susceptibility to cancer such

* Corresponding author

Advances in Nutrition and Cancer 2, edited by Zappia *et al.*
Kluwer Academic / Plenum Publishers, New York, 1999.

as genetic polymorphisms of enzymes involved in the activation/detoxification of carcinogens.[5–8]

The relevance of DNA adducts to carcinogenesis is highlighted by several observations. For instance, correlation between the capacity of different compounds of forming DNA adducts and their carcinogenic potency in vivo or their ability to transform cells in vitro have been reported.[9–11] In some instances DNA adduct levels are higher in target than in non-target tissues.[12] Moreover, the presence of DNA adducts can affect the regulation of oncogenes and tumor suppressor genes.[13] Adducts have been reported to be formed in the *P53* gene of bronchial epithelial cells treated with benzo(a)pyrene diolepoxide at position that are mutational hotspots in human lung cancer, providing a direct etiological link between exposure to a defined chemical and human cancer.[14]

There are several potential sites for adduct formation in DNA, but reactive species binding is not random.[15,16] Adduct formation depends on the electrophilic reactivity of the species involved, on stereochemical factors, and nucleophilicity of the DNA sites. Guanine is the base more frequently modified by chemical carcinogens, each one giving adducts at specific sites. For instance, the activation of small alkylating agents such as N-nitroso compounds leads to the formation of methylating or ethylating species that bind predominantly at the N7 position of guanine, though the O^6-alkylation also occurs, the latter being more relevant with respect to the mutagenesis and carcinogenesis of these compounds.[17–19] The exocyclic N^2 position of guanine is the preferred binding site of polycyclic aromatic hydrocarbons;[20,21] aromatic amines and heterocyclic aromatic amines bind to the C8 position,[22] whereas aflatoxins bind to the N7 position.[23,24]

2. METHODS FOR CARCINOGEN-DNA ADDUCT ANALYSIS

Humans are exposed to mixtures of carcinogens present in the environment at very low levels, so the methodologies used to detect DNA adducts have to be highly sensitive and easily applied to routine analyses. Specificity is also required to allow identification of the chemicals to which people are exposed. Finally, DNA must be obtained non-invasively.

The best indication of DNA damage is given by the measurement of the adducts in target DNA, however, since DNA in target organs is not readily accessible in humans, white blood cells or lymphocytes from peripheral blood are the most common DNA sources.[8,25] Exfoliating cells from bladder and oral cavity, placentas or autopsy tissues represent additional DNA sources.[26–30] Urinary excretion of carcinogen adducted DNA bases may also give indication of exposure.[31,32]

Many different analytical techniques have been developed to this purpose, the most frequently applied include ^{32}P-postlabelling, immunoassays, HPLC with fluorescence detection, and mass_spectrometry. Table 1 shows the sensitivity and the amount of DNA required for the methods most frequently used for the analysis of adducts.

2.1. ^{32}P-Postlabeling Assay

The ^{32}P-postlabelling assay is the most sensitive procedure currently available allowing the detection of 1 adduct in 10^{10} normal nucleotides. The method requires only 1–10 μg DNA and can be applied to several different classes of chemicals.[33–35]

Table 1. Sensitivity of methods for carcinogen-DNA adducts analysis

Method	Sensitivity (adducts/ normal nucleotides)	DNA/Assay (μg)
^{32}P-Postlabelling	$1/10^{10}$	1–10
Immunoassays	$1/10^{8}$	25–100
Synchronous fluorescence spectrometry	$3–10/10^{8}$	~100
Fluorescence line narrowing spectrometry	$1/10^{7}$	~100
Mass spectrometry	$1/10^{9}$	~100

The assay involves firstly an enzymatic hydrolysis of DNA to deoxinucleotides 3′-monophosphate, using micrococcal endonuclease and spleen phosphodiesterase. This step is followed by an enrichment of adducted nucleotides by elimination of normal nucleotides, accomplished by butanol extraction or nuclease P1 treatment.[33,36] Butanol enrichment is a solvent-solvent partitioning procedure in which bulky aromatic and lipophilic adducted nucleotides are extracted preferentially in an organic phase from an acidic aqueous medium. Nuclease P1 enrichment takes advantage of the preferential 3′-dephosphorylation of normal nucleotides, while adducted nucleotide are more resistant to this enzymatic activity. In subsequent ^{32}P-labelling step, dephosphorylated normal nuclueotides are not substrates for 5′-phosphorylation. Although most polycyclic aromatic hydrocarbon-DNA adducts are nuclease P1 resistant, arylamine adducts are lost.[34]

Conversion of 3′-monophosphate adducted nucleotide are converted to the corresponding 5′-^{32}P-labelled 3′,5′-diphosphates by transfer of γ ^{32}P-ATP using T4 polynucleotide kinase. Labelled diphosphates are then separated by multidirectional anion-exchange TLC. The adduct spots are located and quantitated on the TLC plates by autoradiography. HPLC is also used for the separation of ^{32}P-labelled adducts.[37]

^{32}P-Postlabelling does not require prior knowledge of adduct structure and can be applied to the measurement of unidentified DNA adducts. The method does not give information on the chemical structure of the adducts, but tentative identification can be determined by co-chromatography with chemically synthesized DNA adducts.

Due to its versatility, this method is the most frequently used to detect human exposure to a variety of compounds including environmental chemical carcinogens such as polycyclic aromatic hydrocarbons, aromatic amines, and small alkylating agents.[35]

2.2. Immunoassays

Immunological techniques more frequently used to detect carcinogen-DNA adducts include radioimmunoassays, solid-phase enzyme-linked immunosorbent assay or ultrasensitive enzymatic radioimmunoassay.[38–42] The sensitivity of these methods is generally high enough to detect 1 adduct per 10^6–10^8 normal nucleotides and they require 10–100 μg DNA. The specificity of the antibody is important, since cross-reactivity with unmodified nucleotides or similar adducts may give confounding results.

In competitive radioimmunoassays, a radiolabeled adduct is incubated with the specific antibody, the antibody-antigen complex formed is precipitated and quantified by radioactivity measurement. In experimental samples the adduct levels are determined by the degree to which they compete with the labelled antigen to form the antibody-antigen complex thus decreasing the radioactivity in the precipitate. Quantitation is made by comparison with standard curves constructed by adding known amounts of unlabeled adduct to the labelled antigen and measuring the decrease in radioactivity in the precipitate.

In enzyme-linked immunosorbent assay higher sensitivity is obtained by linking a suitable enzyme to the antibody-antigen complex.[41] Since the enzyme can convert the specific substrate to a product that can be measured by spectrophotometry, fluorescence_spectroscopy or radiometry, the signal is greatly amplified.[41]

Enzyme-linked immunosorbent assays can be either direct or competitive. Typically, in a competitive assay wells in microtiter plates are coated with the antigen (i.e. DNA adducted in vitro). Unbound coating antigen is washed away. The antibody is mixed with experimental samples or with various dilutions of modified DNA standard before being added to the wells. After washing, a second antibody recognizing the first one and labelled with an enzyme is added. After washing again, the specific substrate for the enzyme is added and the enzymatic activity is measured.

Highly specific antibodies have been raised against DNA adducts resulting from exposure to different carcinogens including polycyclic aromatic hydrocarbons, aromatic amines, micotoxins, and nitrosamines and used in immunoassays.[38]

In addition to the immunoassays described, immobilized antibodies have been used in immunoaffinity chromatography for adduct purification prior to analysis with other techniques such as ^{32}P-postlabelling, HPLC or gas chromatography-mass spectrometry.[43–45] Immunohistochemical analyses of DNA adducts have also been described.[28]

2.3. Fluorescence Spectroscopy

A number of molecules, when exposed to high intensity light, enter into an excited state and instead of returning to the ground state by loosing energy through vibration and rotation within the molecule, as most molecules do, they loose energy by emitting a photon: fluorescence is the result of such photon emission. Therefore, this technique can be applied only to detect DNA adducts from those carcinogens that fluoresce, such as polycyclic aromatic hydrocarbons and aflatoxins.[46–48]

However, the sensitivity of traditional fluorescence technique is not sufficient to detect adduct formation from exposure to low levels of environmental carcinogens. Improvements have been obtained by using synchronous fluorescence spectroscopy. Typically, fluorescence spectra are obtained by exciting the samples at a fixed wavelength and monitoring emission at different wavelengths or by monitoring emission at a fixed wavelength and varying excitation wavelengths. Synchronous spectra are obtained by varying both excitation and emission frequencies simultaneously to maintain a constant wavelength difference between them. By selecting the appropriate wavelength interval the spectrum obtained can became a single narrow emission band. Sensitivity has been reported to range between 3 and 10 adducts per 10^8 normal nucleotides, using at least 100 μg DNA. Synchronous fluorescence spectroscopy has been applied to measure benzo(a)pyrene adducts in lymphocytes from occupationally exposed workers or aflatoxin B_1-guanine adducts excreted in urine in African people.[46–48]

Higher specificity is obtained by the use of fluorescence line narrowing spectrometry. A narrow-line laser is used for the excitation of samples cooled at 4.2°K: only the subset of molecules with a transition at the laser frequency absorbs the narrow light and only this selected subset of molecules fluoresces, consequently a fluorescence line narrowed spectrum is obtained.[49] This approach can identify different polycyclic aromatic hydrocarbon-DNA adducts in a mixture eliminating the need of separating adducts before analysis. However, the sensitivity of this technique (about 5 adducts per 10^6 normal nucleotides) is too low for human biomonitoring.

2.4. Mass Spectrometry

When coupled to gas chromatography for the separation of complex mixture in biological samples, this technique is highly sensitive and since it gives structural information of the analyte, it provides the specificity the other methods used for carcinogen-DNA adduct analysis do not provide.

A mass spectrometer is essentially constituted of an ion source, a mass analyzer, a detector, and a data system. The ion source is where ions are produced. Electron impact is the classical method of ionization, in which a beam of electrons interacts with the analyte molecules in the gas phase by extracting an electron and producing positively charged ions. The ions are separated within the mass analyzer (e.g. a magnetic sector or a quadrupole) according to their mass/charge (m/z) ratio and reach the detector, usually a multiplier, and generate a mass spectrum that is characteristic for each chemical compound. In general, this type of ionization produces molecular ions of low intensity and many fragment ions, thus limiting the sensitivity of the technique. This is why less energetic processes such as chemical ionization, are preferred. In the chemical ionization mode a reagent gas (e.g. methane or ammonia) is introduced into the ion source, so that the initial ionization is usually by electron impact on the gas present in excess over the analyte molecules, producing ionized gas and thermal electrons. In the positive chemical ionization the sample is ionized by proton transfer with the formation of protonated or reagent gas-adducted molecular ions. In the negative ion chemical ionization analyte molecules with high electron affinity (e.g. fluorinated derivatives) form singly charged negative ions by interacting with thermal electrons. Mass spectra obtained in the chemical ionization mode show less fragmentation than in the electron impact mode. By monitoring only selected ions and using stable isotope labelled analogs of the analyte it is possible to reach the sensitivity and specificity needed for carcinogen-DNA adducts and detect 1—10 adducts per 10^9 normal nucleotide using 100 μg DNA.[45,50–53]

The development of atmospheric pressure ionization techniques has much improved the use of liquid chromatography-mass spectrometry, but its use for DNA adducts analysis is still limited.[54]

Despite the potential of mass spectrometry based methods for the quantitation of DNA adducts, the requirement of extensive cleanup of biological samples has limited their application in molecular epidemiology studies involving large populations.

3. DNA ADDUCTS IN RISK ASSESSMENT

Carcinogen-DNA adducts have been measured in humans in occupational and environmental settings. In most studies non-target sites have been considered (e.g. lymphocytes, white blood cells), since these have been shown to be good monitors

Table 2. Recent publications on DNA adduct measurement in humans

Carcinogen	Exposure	DNA Source	Method	Reference
PAH	occupational, environmental, smoking	white blood cells; lymphocytes; macrophages	^{32}P-Postlabelling	56, 59, 60, 61, 62
PAH	smoking	placenta; white blood cells	Immunoassay	30, 63
B(a)P	occupational	white blood cells; lung tissue	HPLC/FS	64
B(a)P	smoking	lymphocytes	GC-MS	45
4-ABP	smoking	bladder biopsies; lung biopsies;	GC-MS	65, 66

of internal dose. Table 2, which is not intended to give a comprehensive summary of the studies in this field, reports some recent examples (publication years 1997–1998).

The quantitation of DNA adducts has been proposed to identify groups of individuals at high cancer risk. To date only a few epidemiological studies have incorporated biological markers in risk assessment. In one such study polycyclic aromatic hydrocarbon-DNA adducts in peripheral leukocytes were significantly associated with lung cancer risk, the odds of having elevated adducts being more than 6 times greater in cases than in controls.[55] Another study showed that polycyclic aromatic hydrocarbon-DNA adducts in white blood cells were significantly associated with the risk of bladder cancer.[56] The best example is perhaps given by a large epidemiological study linking the exposure to dietary aflatoxins and hepatitis B seropositivity to the risk of hepatocellualr carcinoma in China.[57,58] Aflatoxin exposure was assessed by measuring urinary aflatoxin-N7-guanine adducts. Individuals who were positive for aflatoxin adducts in their urine had 3.4 times higher risk of developing liver cancer, individuals who had aflatoxin adducts and were also positive for hepatitis B antigen had a 60-fold increased relative risk of hepatocellular carcinoma compared to controls.[58] The same study revealed no statistically significant association between hepatocellular cancer risk and dietary aflatoxins consumption, thus underlining the relevance of biomarker measurements in epidemiological studies.

4. CONCLUSIONS

At present, most cancer risk estimates are based on studies on genetically homogeneous laboratory animals using doses that often are several order of magnitude higher than those to which humans are exposed. The extrapolation from these data to the low level of exposure in the general population may be misleading, because it does not take into account that interindividual variability may be relevant to cancer susceptibility. Incorporation of biomarkers into risk assessment would reduce uncertainties and give more accurate estimates of risk. Most of the methods available for measuring DNA adducts, such as ^{32}P-postlabelling or immunoassays meet the requirement of sensitivity needed to monitor human exposure to carcinogens at low environmental levels, but they give little structural information of the compounds to which people have been exposed. Mass spectrometry based methods instead are both highly sensitive and specific, though lenghty sample preparation prior to analysis has limited their application in molecular epidemiology studies. However, with the increasing use of powerful techniques such as antibody affinity chromatography for sample cleanup, gas chromatography or liquid chromatography coupled to mass spectrometry are expected to be more frequently used, since the specificity of the biomarker is essential in assessing the hazard associated with the exposure to a specific chemical.

5. SUMMARY

Genotoxic chemicals are known to react with DNA either directly or after metabolic activation to form adducts, a step thought to be relevant with respect to chemical carcinogenesis. Evaluation of cancer risk due to exposure to chemicals requires information about the internal dose which depends on individual variation in rates of metabolic activation and detoxification. The presence and the amount of specific DNA adducts provide a good indication of chemical exposure and genetic damage resulting from exposure to carcinogens and account for some of the factors affecting individual susceptibility to cancer. Analysis of DNA adducts requires that the sensitivity of the methods be sufficiently high to allow the detection of about 1 adduct/109 normal nucleotides. Most suitable methods are based on ^{32}P-postlabelling, immunoassays or physico-chemical techniques such as HPLC coupled to synchronous fluorescence spectroscopy or gas chromatography-mass spectrometry. These methods have been used to assess human exposure to a variety of chemical carcinogens including polycyclic aromatic hydrocarbons, aromatic amines, heterocyclic aromatic amines or aflatoxins. In some instances, the use of DNA-adducts has given accurate estimates of risk.

ACKNOWLEDGMENTS

The editorial assistance of the staff of the G.A. Pfeiffer Memorial Library is gratefully acknowledged.

This work was supported by the European Commision contract BMH4-98-3243, by the Fondazione Lombardia per l'Ambiente and by the Associazione Italiana per le Ricerche sul Cancro.

REFERENCES

1. Perera, F.P. Molecular epidemiology: insights into cancer susceptibility, risk assessment, and prevention. J. Natl. Cancer Inst., 88:496–509, 1996.
2. La, D.K. and Swenberg, J.A. DNA adducts: biological markers of exposure and potential applications to risk assessment. Mutat. Res., 365:129–146, 1996.
3. Dale, C.M. and Garner, R.C. Measurement of DNA adducts in humans after complex mixture exposure. Food Chem. Toxicol., 34:905–919, 1996.
4. Miller, E.C. and Miller, J.A. Mechanisms of chemical carcinogenesis. Cancer, 47:1055–1064, 1981.
5. Wogan, G.N. Molecular epidemiology in cancer risk assessment and prevention: Recent progress and avenues for future research. Environ. Health Perspect., 98:167–178, 1992.
6. Skipper, P.L. and Tannenbaum, S.R. Protein adducts in the molecular dosimetry of chemical carcinogens. Carcinogenesis, 11:507–518, 1990.

7. Harris, C.C. Chemical and physical carcinogenesis: advances and perspectives for the 1990s. Cancer Res., 51:5023s–5044s, 1991.
8. Swenberg, J.A., Fedtke, N., Fennell, T.R., and Walker, V.E. Relationships between carcinogen exposure, DNA adducts and carcinogenesis. In: D.B. Clayson, I.C. Munro, P. Shubik, and J.A. Swenberg (eds.), Progress in Predictive Toxicology, pp. 161–184, Amsterdam: Elsevier. 1990.
9. Brookes, P. and Lawley, P.D. Evidence for the binding of polynuclear aromatic hydrocarbons to the nucleic acids of mouse skin: relation between carcinogenic power of the hydrocarbons and their binding to deoxyribonucleic acid. Nature, 202:781–784, 1964.
10. Lutz, W.K. Quantitative evaluation of DNA binding data for risk estimation and for classification of direct and indirect carcinogens. J. Cancer Res. Clin. Oncol., 112:85–91, 1986.
11. Poirier, M.C. The use of carcinogen-DNA adduct antisera for quantitation and localization of genomic damage in animal models and the human population. Environ. Mutagen., 6:879–887, 1984.
12. Wogan, G.N. and Gorelick, N.J. Chemical and biochemical dosimetry of exposure to genotoxic chemicals. Environ. Health Perspect., 62:5–18, 1985.
13. Nesnow, S., Ross, J.A., Mass, M.J., and Stoner, G.D. Mechanistic relationships between DNA adducts, oncogene mutations, and lung tumorigenesis in strain A mice. Exp. Lung Res., 24:395–405, 1998.
14. Denissenko, M.F., Pao, A., Tang, M., and Pfeifer, G.P. Preferential formation of benzo[a]pyrene adducts at lung cancer mutational hotspots in P53. Science, 274:430–432, 1996.
15. Gupta, R.C. Nonrandom binding of the carcinogen N-hydroxy-2-acetylaminofluorene to repetitive sequences of rat liver DNA in vivo. Proc. Natl. Acad. Sci. USA, 81:6943–6947, 1984.
16. Dipple, A. DNA adducts of chemical carcinogens. Carcinogenesis, 16:437–441, 1995.
17. Loveless, A. Possibile relevance of O-6 alkylation of deoxyguanosine to the mutagenicity and carcinogenicity of nitrosamines and nitrosamides. Nature, 223:206–207, 1969.
18. Singer, B. N-nitroso alkylating agents: formation and persistence of alkyl derivatives in mammalian nucleic acids as contributing factors in carcinogenesis. J. Natl. Cancer Inst., 62:1329–1339, 1979.
19. Saffhill, R., Margison, G.P., and O'Connor, P.J. Mechanisms of carcinogenesis induced by alkylating agents. Biochim. Biophys. Acta, 823:111–145, 1985.
20. Jerina, D.M., Chadha, A., Cheh, A.M., Schurdak, M.E., Wood, A.W., and Sayer, J.M. Covalent binding of bay-region diol epoxides to nucleic acids. Adv. Exp. Med. Biol., 283:533–553, 1991.
21. Jernstrom, B. and Graslund, A. Covalent binding of benzo[a]pyrene 7,8-dihydrodiol 9,10-epoxides to DNA: molecular structures, induced mutations, and biological consequences. Biophys. Chem., 49:185–199, 1994.
22. Humphreys, W.G., Kadlubar, F.F., and Guengerich, F.P. Mechanism of C8 alkylation of guanine residues by activated arylamines: evidence for initial adduct formation at the N7 position. Proc. Natl. Acad. Sci. USA, 89:8278–8282, 1992.
23. Essigmann, J.M., Croy, R.G., Nadzan, A.M., Busby, W.F., Jr., Reinhold, V.N., Buchi, G., and Wogan, G.N. Structural identification of the major DNA adduct formed by aflatoxin B1 in vitro. Proc. Natl. Acad. Sci. USA, 74:1870–1874, 1977.
24. Croy, R.G., Essigmann, J.M., Reinhold, V.N., and Wogan, G.N. Identification of the principal aflatoxin B1-DNA adduct formed in vivo in rat liver. Proc. Natl. Acad. Sci. USA, 75:1745–1749, 1978.
25. Phillips, D.H., Hemminki, K., Alhonen, A., Hewer, A., and Grover, P.L. Monitoring occupational exposure to carcinogens: detection by 32P-postlabelling of aromatic DNA adducts in white blood cells of iron foundry workers. Mutat. Res., 204:531–541, 1988.
26. Cuzick, J., Routledge, M.N., Jenkins, D., and Garner, R.C. DNA adducts in different tissues of smokers and non-smokers. Int. J. Cancer, 45:673–678, 1990.
27. Talaska, G., Schamer, M., Skipper, P., Tannenbaum, S., Caporaso, N., Unruh, L., Kadlubar, F.F., Bartsch, H., Malaveille, C., and Vineis, P. Detection of carcinogen-DNA adducts in exfoliated urothelial cells of cigarette smokers: association with smoking, hemoglobin adducts, and urinary mutagenicity. Cancer Epidemiol. Biomarkers. Prev., 1:61–66, 1991.
28. Hsu, T.M., Zhang, Y.J., and Santella, R.M. Immunoperoxidase quantitation of 4-aminobiphenyl- and polycyclic aromatic hydrocarbon-DNA adducts in exfoliated oral and urothelial cells of smokers and nonsmokers. Cancer Epidemiol. Biomarkers. Prev., 6:193–199, 1997.
29. Talaska, G., al-Juburi, A.Z., and Kadlubar, F.F. Smoking related carcinogen-DNA adducts in biopsy samples of human urinary bladder: identification of N-(deoxyguanosin-8-yl)-4-aminobiphenyl as a major adduct. Proc. Natl. Acad. Sci. USA, 88:5350–5354, 1991.
30. Arnould, J.P., Verhoest, P., Bach, V., Libert, J.P., and Belegaud, J. Detection of benzo[a]pyrene-DNA adducts in human placenta and umbilical cord blood. Hum. Exp. Toxicol., 16:716–721, 1997.
31. Shuker, D.E.G. and Farmer, P.B. Relevance of urinary DNA adducts as markers of carcinogen exposure. Chem. Res. Toxicol., 5:450–460, 1992.

32. Groopman, J.D., Donahue, P.R., Zhu, J.Q., Chen, J.S., and Wogan, G.N. Aflatoxin metabolism in humans: detection of metabolites and nucleic acid adducts in urine by affinity chromatography. Proc. Natl. Acad. Sci. USA, 82:6492–6496, 1985.
33. Gupta, R.C., Reddy, M.V., and Randerath, K. 32P-postlabeling analysis of non-radioactive aromatic carcinogen–DNA adducts. Carcinogenesis, 3:1081–1092, 1982.
34. Gupta, R.C. 32P-postlabelling analysis of bulky aromatic adducts. IARC Sci. Publ., 124:11–23, 1993.
35. Beach, A.C. and Gupta, R.C. Human biomonitoring and the 32P-postlabeling assay. Carcinogenesis, 13:1053–1074, 1992.
36. Reddy, M.V. and Randerath, K. Nuclease P1-mediated enhancement of sensitivity of 32P-postlabeling test for structurally diverse DNA adducts. Carcinogenesis, 7:1543–1551, 1986.
37. Levy, G.N. and Weber, W.W. High-performance liquid chromatographic analysis of 32P-postlabeled DNA-aromatic carcinogen adducts. Anal. Biochem., 174:381–392, 1988.
38. Poirier, M.C. Human exposure monitoring, dosimetry, and cancer risk assessment: the use of antisera specific for carcinogen-DNA adducts and carcinogen-modified DNA. Drug Metab. Rev., 26:87–109, 1994.
39. Strickland, P.T. and Boyle, J.M. Immunoassay of carcinogen-modified DNA. Prog. Nucleic Acid Res. Mol. Biol., 31:1–58, 1984.
40. Kriek, E., Den Engelse, L., Scherer, E., and Westra, J.G. Formation of DNA modifications by chemical carcinogens. Identification, localization, and quantification. Biochim. Biophys. Acta, 738:181–201, 1984.
41. Harris, C.C., Yolken, R.H., and Hsu, I-C. Enzyme immunoassays: applications in cancer research. In: H. Busch and L.C. Yeoman (eds.), Methods in Cancer Research, Vol. XX, pp. 213–243, New York: Academic Press. 1982.
42. Hsu, I.C., Poirier, M.C., Yuspa, S.H., Grunberger, D., Weinstein, I.B., Yolken, R.H., and Harris, C.C. Measurement of benzo(a)pyrene-DNA adducts by enzyme immunoassays and radioimmunoassay. Cancer Res., 41:1091–1095, 1981.
43. King, M.M., Cuzick, J., Jenkins, D., Routledge, M.N., and Garner, R.C. Immunoaffinity concentration of human lung DNA adducts using an anti-benzo[a]pyrene-diol-epoxide-DNA antibody. Analysis by 32P-postlabelling or ELISA. Mutat. Res., 292:113–122, 1993.
44. Airoldi, L., Magagnotti, C., Chiappetta, L., Bonfanti, M., Pastorelli, R., and Fanelli, R. Simultaneous immunoaffinity purification of O^6-methyl, O^6-ethyl-, O^6-propyl- and, O^6-butylguanine and their analysis by gas chromatography/mass spectrometry. Carcinogenesis, 16:2247–2250, 1995.
45. Pastorelli, R., Guanci, M., Cerri, A., Negri, E., La Vecchia, C., Fumagalli, F., Mezzetti, M., Cappelli, R., Panigalli, T., Fanelli, R., and Airoldi, L. Impact of inherited polymorphisms in glutathione S-transferase M1, microsomal epoxide hydrolase, cytochrome P450 enzymes on DNA, and blood protein adducts of benzo(a)pyrene-diolepoxide. Cancer Epidemiol. Biomarkers. Prev., 7:703–709, 1998.
46. Vahakangas, K., Haugen, A., and Harris, C.C. An applied synchronous fluorescence spectrophotometric assay to study benzo[a]pyrene-diolepoxide-DNA adducts. Carcinogenesis, 6:1109–1115, 1985.
47. Autrup, H., Bradley, K.A., Shamsuddin, A.K.M., Wakhisi, J., and Wasunna, A. Detection of putative adduct with fluorescence characteristics identical to 2,3-dihydro-2-(7′-guanyl)-3-hydroxyaflatoxin B1 in human urine collected in Murang'a district, Kenya. Carcinogenesis, 4:1193–1195, 1983.
48. Autrup, H., Wakhisi, J., Vahakangas, K., Wasunna, A., and Harris, C.C. Detection of 8,9-dihydro-(7′-guanyl)-9-hydroxyaflatoxin B_1 in human urine. Environ. Health Perspect., 62:105–108, 1985.
49. Sanders, M.J., Cooper, R.S., Jankowiak, R., Small, G.J., Heisig, V., and Jeffrey, A.M. Identification of polycyclic aromatic hydrocarbon metabolites and DNA adducts in mixtures using fluorescence line narrowing spectrometry. Anal. Chem., 58:816–820, 1986.
50. Strickland, P.T., Routledge, M.N., and Dipple, A. Methodologies for measuring carcinogen adducts in humans. Cancer Epidemiol. Biomarkers Prev., 2:607–619, 1993.
51. Farmer, P.B., Bailey, E., Naylor, S., Anderson, D., Brooks, A., Cushnir, J., Lamb, J.H., Sepai, O., and Tang, Y.S. Identification of endogenous electrophiles by means of mass spectrometric determination of protein and DNA adducts. Environ. Health Perspect., 99:19–24, 1993.
52. Lin, D., Lay, J.O.,Jr., Bryant, M.S., Malaveille, C., Friesen, M., Bartsch, H., Lang, N.P., and Kadlubar, F.F. Analysis of 4-aminobiphenyl-DNA adducts in human urinary bladder and lung by alkaline hydrolysis and negative ion gas chromatography-mass spectrometry. Environ. Health Perspect., 102 Suppl 6:11–16, 1994.
53. Friesen, M.D., Kaderlik, K., Lin, D., Garren, L., Bartsch, H., Lang, N.P., and Kadlubar, F.F. Analysis of DNA adducts of 2-amino-1-methyl-6-phenylimidazo[4,5-b]pyridine in rat and human tissues by alkaline hydrolysis and gas chromatography/electron capture mass spectrometry: Validation by comparison with ^{32}P-postlabeling. Chem. Res. Toxicol., 7:733–739, 1994.

54. Kambouris, S.J., Chaudhary, A.K., and Blair, I.A. Liquid chromatography/electrospray ionization tandem mass spectroscopy (LC/ESI MS/MS) analysis of 1,2-epoxybutene adducts of purine deoxynucleosides. Toxicology, 113:331–335, 1996.
55. Tang, D., Santella, R.M., Blackwood, A.M., Young, T.-L., Mayer, J., Jaretzki, A., Grantham, S., Tsai, W.-Y., and Perera, F.P. A molecular epidemiological case-control study of lung cancer. Cancer Epidemiol. Biomarkers Prev., 4:341–346, 1995.
56. Peluso, M., Airoldi, L., Armelle, M., Martone, T., Coda, R., Malaveille, C., Giacomelli, G., Terrone, C., Casetta, G., and Vineis, P. White blood cell DNA adducts, smoking, and NAT2 and GSTM1 genotypes in bladder cancer: a case-control study. Cancer Epidemiol. Biomarkers Prev., 7:341–346, 1998.
57. Ross, R.K., Yuan, J.-M., Yu, M.C., Wogan, G.N., Qian, G.-S., Tu, J.-T., Groopman, J.D., Gao, Y.-T., and Henderson, B.E. Urinary aflatoxin biomarkers and risk of hepatocellular carcinoma. Lancet, 339:943–946, 1992.
58. Qian, G.-S., Ross, R.K., Yu, M.C., Yuan, J.-M., GaoY.-T., Henderson, B.E., Wogan, G.N., and Groopman, J.D. A follow-up study of urinary markers of aflatoxin exposure and liver cancer risk in Shanghai, People's Republic of China. Cancer Epidemiol. Biomarkers Prev., 3:3–10, 1994.
59. Binkova, B., Topinka, J., Mrackova, G., Gajdosova, D., Vidova, P., Stavkova, Z., Pilcik, P.V., Rimar, V., Dobias, L., Farmer, P.B., and Sram, R.J. Coke oven workers study: the effect of exposure and GSTM1 and NAT2 genotypes on DNA adduct levels in white blood cells and lymphocytes as determined by 32P-postlabelling. Mutat. Res., 416:67–84, 1998.
60. Peluso, M., Merlo, F., Munnia, A., Valerio, F., Perrotta, A., Puntoni, R., and Parodi, S. 32P-postlabeling detection of aromatic adducts in the white blood cell DNA of nonsmoking police officers. Cancer Epidemiol. Biomarkers Prev., 7:3–11, 1998.
61. Van Schooten, F.J., Godschalk, R.W.L., Breedijk, A., Maas, L.M., Kriek, E., Sakai, H., Wigbout, G., Baas, P., Van'tVeer, L., and van Zandwijk, N. 32P-postlabelling of aromatic DNA adducts in white blood cells and alveolar macrophages of smokers: saturation at high exposures. Mutat. Res., 378:65–75, 1997.
62. Hemminki, K., Dickey, C., Karlsson, S., Bell, D., Hsu, Y., Tsai, W.-Y., Mooney, L.A., Savela, K., and Perera, F.P. Aromatic DNA adducts in foundry workers in relation to exposure, life style and CYP1A1 and glutathione transferase M1 genotype. Carcinogenesis, 18:345–350, 1997.
63. Mooney, L.A., Bell, D.A., Santella, R.M., Van Bennekum, A.M., Ottman, R., Paik, M., Blaner, W.S., Lucier, G.W., Covey, L., Young, T.-L., Cooper, T.B., and Glassman, A.H. Contribution of genetic and nutritional factors to DNA damage in heavy smokers. Carcinogenesis, 18:503–509, 1997.
64. Rojas, M., Alexandrov, K., Cascorbi, I., Brockmöller, J., Likhachev, A., Pozharisski, K., Bouvier, G., Auburtin, G., Mayer, L., Kopp-Schneider, A., Roots, I., and Bartsch, H. High benzo[a]pyrene diol-epoxide DNA adduct levels in lung and blood cells from individuals with combined CYP1A1 MspI/MspI-GSTM1 *0/*0 genotypes. Pharmacogenetics, 8:109–118, 1998.
65. Martone, T., Airoldi, L., Magagnotti, C., Coda, R., Randone, D., Malaveille, C., Avanzi, G., Merletti, F., Hautefeuille, A., and Vineis, P. 4-Aminobiphenyl-DNA adducts and p53 mutations in bladder cancer. Int. J. Cancer, 75:512–516, 1998.
66. Culp, S.J., Roberts, D.W., Talaska, G., Lang, N.P., Fu, P.P., Lay, J.O.Jr., Teitel, C.H., Snawder, J.E., Von Tungeln, L.S., and Kadlubar, F.F. Immunochemical, 32P-postlabeling, and GC/MS detection of 4-aminobiphenyl-DNA adducts in human peripheral lung in relation to metabolic activation pathways involving pulmonary N-oxidation, conjugation, and peroxidation. Mutat. Res., 378:97–112, 1997.

21

SIGNIFICANCE OF GENETIC POLYMORPHISMS IN CANCER SUSCEPTIBILITY

Eino Hietanen

Department of Clinical Physiology
Turku University Hospital
FIN-20520 Turku, Finland

INTRODUCTION

The variability of biotransformation enzyme activities is associated with various types of exposures and host factors, possibly originating from early childhood. Since many carcinogenic compounds require metabolic activation before being capable of reacting with cellular macromolecules, individual features of carcinogen metabolism may play an essential role in the development of environmental cancer.[1] As individual response to environmental mutagens and carcinogens vary there is no pure distinction between purely genetic or environmental cancers. Often there is no incompatibility between environmental and genetic origin of cancer as is the case e.g. with smoking where a chemical mixture induces cancer but individuals show different sensitivity to these agents causing a cancer. A complicating factor is the multistage etiology of carcinogenesis implying the involvement of many distinct events. However, it has become evident that the enzymes activating and inactivating exogenous carcinogens are involved.

Many compounds are converted to reactive electrophilic metabolites by the oxidative, mainly cytochrome P450-related enzymes (CYPs). Secondary metabolism, mainly involving epoxide hydrolase, glutathione S-transferases, UDPglucosyltransferases, sulfotransferases, and acetyltransferases and another subset of activating CYP isoforms, leads to the formation of the highly reactive metabolites that can bind to genomic DNA. Thus, the concerted action of these enzymes may be crucial in determining the final biological effect(s) of a carcinogen. The activity of these enzymes is modulated by many host factors and environmental factors. Host factors include both life-style factors, diseases, and genetic factors while environmental factors include exposures to various chemicals as well as dietary habits (Fig. 1).

Advances in Nutrition and Cancer 2, edited by Zappia *et al.*
Kluwer Academic / Plenum Publishers, New York, 1999.

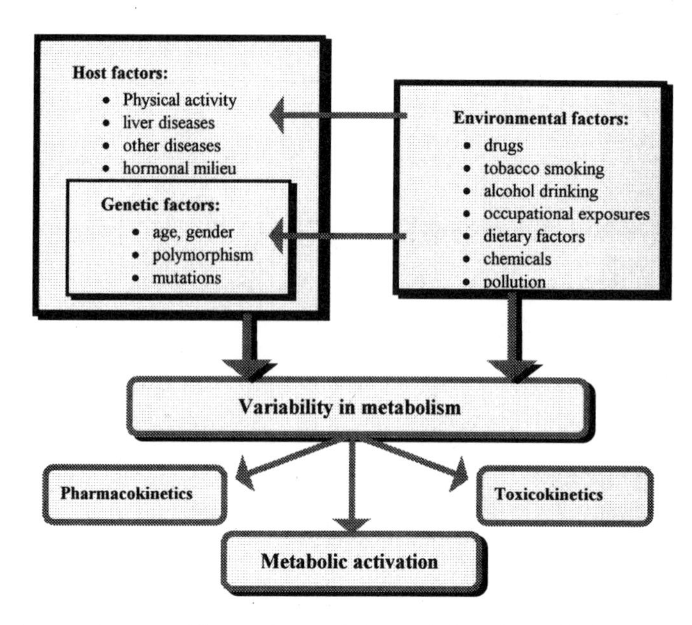

Figure 1. The host-environmental interactions and the regulation of xenobiotic-metabolizing enzymes.

PHENOTYPIC AND GENOTYPIC POLYMORPHISMS

The phenotypic metabolic variation has led to the discovery of genetic polymorphism of biotransformation involving enzymes participating in carcinogen metabolism. A growing number of genes encoding carcinogen-metabolising enzymes has subsequently been identified and cloned. Consequently, there is increasing knowledge of the allelic variants or genetic defects which give rise to the observed polymorphic variation. Development of rather simple new techniques, such as PCR based assays, has enabled identification of individual's genotype for a variety of metabolic polymorphisms with precision. Thus, recent knowledge of the genetic basis for individual metabolic variation has opened new possibilities for studies focusing on increased susceptibility to environmental cancer. The genetic polymorphism does not only increase the susceptibility but also sensitivity to cancer.[2,3] As an example a germ-line mutation in the tumour suppressor gene p53 predisposes to various childhood cancers and breast cancers in the absence of any known exposures.[4–6] On the other hand a mutation leading to a deficient superoxide dismutase activity as is in xeroderma pigmentosum, the sensitivity to solar radiation-induced cancers increase.[3]

A wide range of enzymes of toxicological importance have been shown to have different activity levels in the population which might also mean a different level in the sensitivity to cancers and many have been reported to be associated with enhanced risk for cancer.[7] The basis for genetic polymorphisms originates from the metabolic basis of the non-caussian distribution of the enzyme activity which has led to the detection of mutations leading to altered enzyme expression. Phenotypic metabolic variation has been related to genetic polymorphisms. A growing number of genes encoding carcinogen-metabolizing enzymes have been identified and cloned and knowledge is increasing on the allelic variants and gene defects giving rise to phenotypic variation.

Many of the polymorphic genes of carcinogen metabolism show also ethnic differences in gene structure and allelic distribution explaining some discrepant differences in data. Whether using the metabolic or genotypic markers for the detection of cancer risks the marker must fulfill certain criteria to be of use to detect persons with excessive cancer risk including a proper allelic distribution in the population and detected relationship to exposure.

In the case of some cytochrome P450 isozymes a polymorphic distribution may divide populations in terms of their susceptibility to e.g. cancer and may regulate pharmacokinetics of many drugs. The cytochrome P450 1A2 enzyme (CYP1A2) catalyses the metabolism of many aromatic and heterocyclic carcinogenic compounds present in the diet and tobacco smoke.[8] The metabolic distribution of cytochrome P4501A2 has shown a wide activity distribution and in some populations distribution of activities may suggest metabolic polymorphism.[9,10] Lately also genetic polymorphism has been suggested.[11] In our on-going study the possible metabolic polymorphism of CYP1A2 enzyme activity was studied in the homo- and heterozygotic twins who were discordant to their smoking habits. The CYP1A2 metabolic index was determined by assaying caffeine metabolites in urine.[12] CYP1A2 phenotyping of the Finnish twins is presented as a frequency distribution (Fig. 2). The metabolic ratio (MR) of CYP1A2 index in 49 non-smokers and in the 23 smokers were not normally distributed but showed a bimordial distribution. Probit transformation of CYP1A2 index in the Finnish twin population resulted in non-linear probit plot (Fig. 3). This probit transformation further supports that this group did not have a normal distribution of CYP1A2 but showed a phenotypic polymorphism suggesting that there might exist also genotype polymorphism as studies in mice a strain have suggested.[13]

CYP1A1 AND *CYP1A2* POLYMORPHISMS AND CANCER SUSCEPTIBILITY

Several studies have indicated an association of the genetic polymorphism of *CYP1A1* and cancer. A co-segregation of the CYP1A1 phenotype and polymorphism of the MspI restriction site in the *CYP1A1* gene have been reported but this discovery has been challenged later.[14,15] Recent studies have indicated that variant alleles at the MspI site in exon 7 could result in a more active CYP1A1 enzyme.[16,17] A significant correlation in a Japanese population between susceptibility to lung cancer and homozygosity for the rare MspI allele has been reported by Kawajiri et al.[18] and Nakachi et al.[19] A point mutation resulting in an amino acid substitution (Ile to Val) in the heme binding region of the CYP1A1 protein has been found by Hayashi et al.[20] This genotype results in an altered enzyme activity and was shown to be associated with squamous cell and small cell types of lung cancer. There are significant ethnic differences in the frequency of CYP1A1 alleles, and both the m2 and the Val alleles appear to be rare in Caucasians.[21–23] Japanese studies have proposed that especially squamous cell cancer risk is increased even at low cigarette smoking doses when there is a susceptible (m2m2) genotype (Table 1).[19,24,25]

As CYP1A2 catalyses the metabolism of many dietary aromatic amines its metabolic index has been correlated with risks of cancers possibly of dietary origin. A larger proportion of population having colon cancers are rapid metabolisers as compared to those being slow metabolisers although the difference is fairly modest.[26]

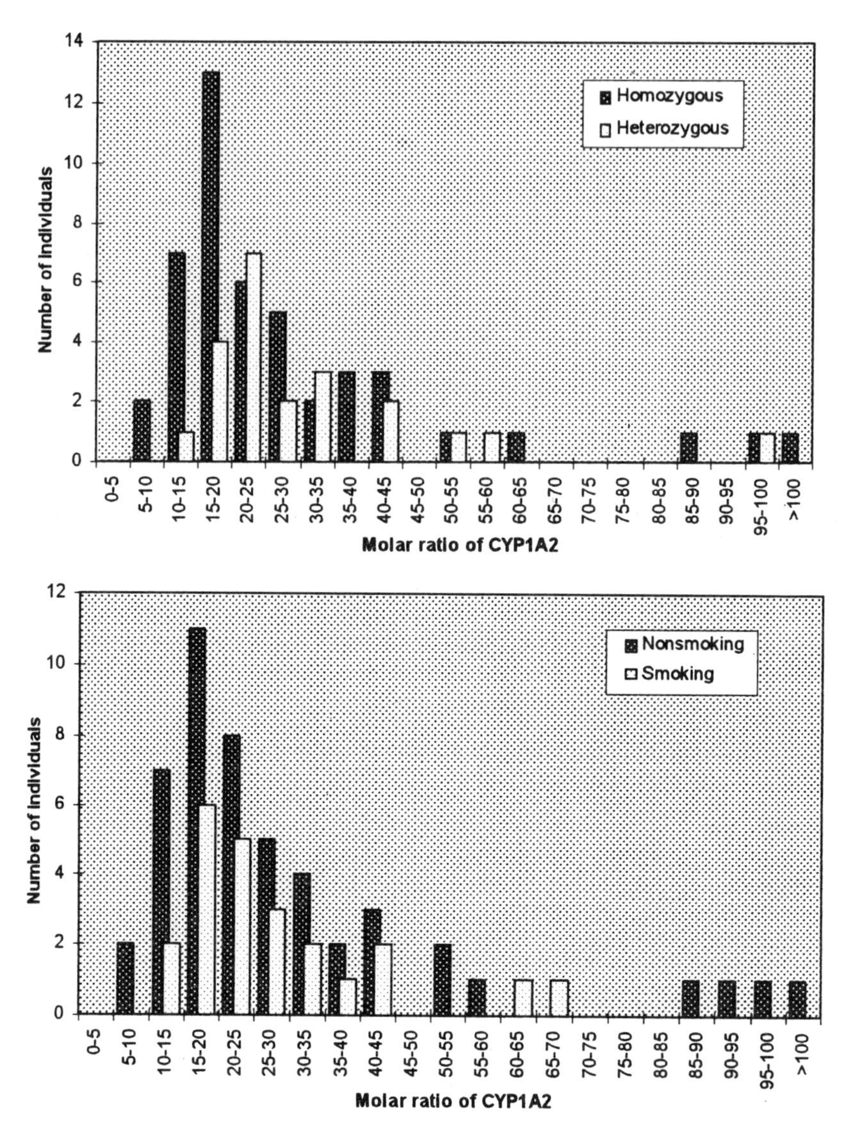

Figure 2. Frequency distribution of the urinary molar ratio of CYP1A2 in Finnish twin population of non-smokers and smokers.

CYP2D6 AND CANCER

Several studies have demonstrated that debrisoquine-metabolic rate seems to be involved with lung cancer risk in a way that those being extensive metabolisers are at a higher risk than those being poor metabolisers (PM), these data being quite reassuring when using genotyping or phenotyping assaying debrisoquine metabolism.[27,28] Numerous studies have demonstrated an association of *CYP2D6* gentoype and lung cancer risk in a way that those persons classified as poor metabolisers either by genotyping or phenotyping them have a lower risk than extensive or intermediate metabolisers (Table 2).[29,30] The association of PM genotype of *CYP2D6* with increased risk of lung adenocarcinoma has been speculated to be associated with the possibility of

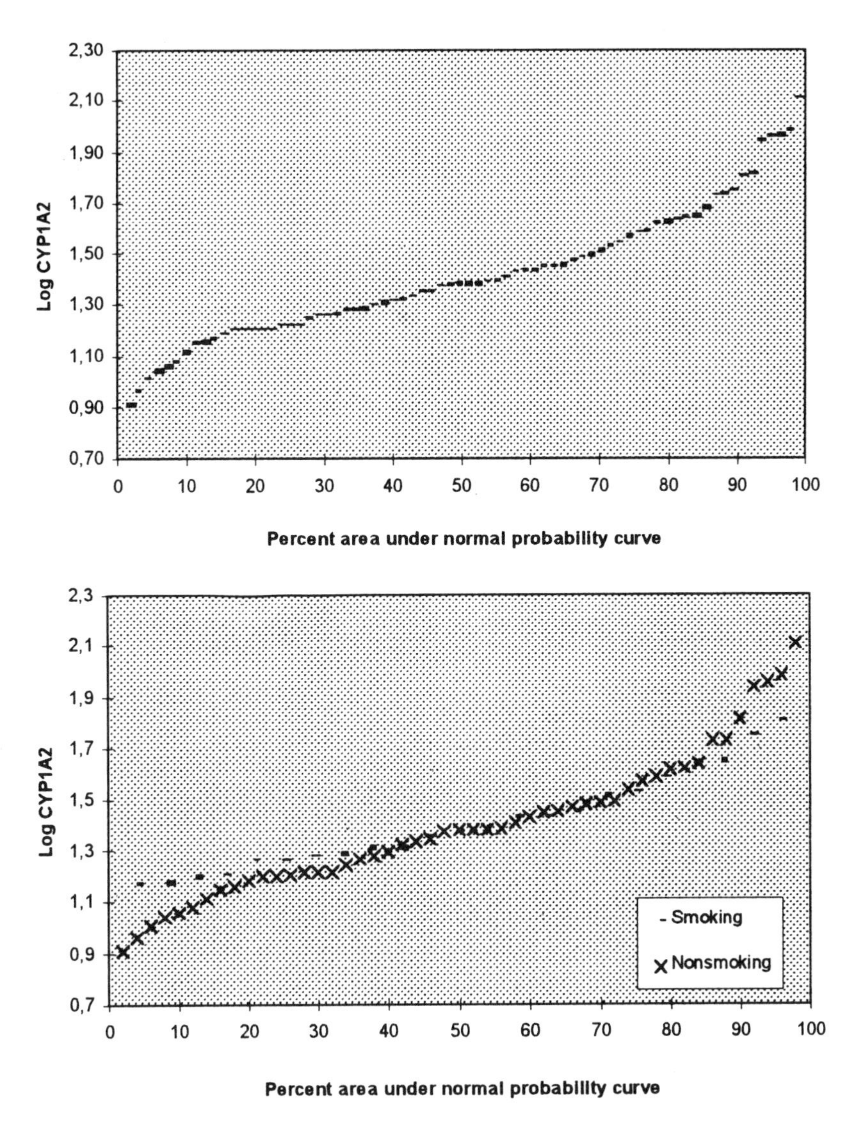

Figure 3. Probit analysis of the urinary molar ratio of CYP1A2 in 72 Finnish twins in the figure A (top) and divided according to their smoking habits to smokers and nonsmokers in the figure B (bottom).

the metabolism of cigarette smoke nitrosamines (NNK) by this cytochrome isozyme.[31] In a study by Bouchardy et al.[30] the increased risk of lung cancer was present in extensive metabolizers (as assayed phenotypically) of debrisoquine which risk was dose-dependent with smoking while in the persons with low or medium CYP2D6 activity no such an increase was found even at very high smoking levels (Table 2). The increased risk (OR 4.95) in extensive metabolisers became evident first at the tobacco consumption level of >30 g per day (as compared to smokers <20 g per day). This might be associated with the increasing level of the metabolism of cigarette-smoke containing carcinogen, 4-(methylnitrosamino)-1-(3-pyridyl)-1-butanone. CYP2D6 gene mutations may be involved also in increased risk to liver cancer in a way that those subjects with homozygous functional genes ("extensive metabolizers") seem to have 6.4-fold risk of primary liver cancer in comparison with those having inactivating mutations.[32]

Table 1. CYP1A1 (MspI) genotype and exposure to cigarette smoke (number of cigarettes × 10^4 in lifetime) in lung cancer (Adenocarcinoma/Squamous cell carcinoma, N = 41/78)[25]

	CYP1A1 genotype distribution (%)		
Smoking	m1m1	m1m2	m2m2
0	34/100	21/0	4/0
<30	26/28	57/38	17/34
>30	42/44	47/44	11/12
Total	35/40	53/40	13/21
All patients	36	47	16
Healthy controls (historical)	44	45	11

Table 2. CYP2D6 phenotypes in lung cancer patients and their smoking habits[30]

Tobacco consumption, g/day	Phenotype classification Low	Medium	High
<20	1.0	1.0	1.0
21–30	0.43 (0.14–0.37)	1.04 (0.30–3.60)	1.43 (0.42–4.85)
>30	0.40 (0.13–1.15)	0.43 (0.14–1.34)	4.95 (1.61–15.23)

CYP2E1 AND CANCER SUSCEPTIBILITY

In a recent study the risk of hepatocellular carcinoma was studied in a Taiwanese population most of whom being also hepatitis B positive.[33] In this study the significance of cytochrome P450E and *GSTM1* genotypes was studied. A dose-dependent increase in the hepatocellular cancer risk according to cigarette smoking was present in homozygous (c_1/c_1) *CYP2E* genotypes (Table 3). Furthermore, alcohol consumption enhanced the risk in smokers about two-fold. Tsutsumi et al.[34] found in a Japanese population that the prevalence of alcoholic liver disease was much lower in *CYP2E1* genotype c_1/c_1 (16% of cases in heavy drinkers) than in heterozygous or homozygous type c_2/c_2 (altogether 84% of cases), in control population the prevalence of c_1/c_1 was 62–68%.

Table 3. CYP2E1 at-risk genotype in hepatocellular cancer patients and their drinking and smoking habits[33]

Alcohol/Smoking	CYP2E1c_1/ c_1	Odds ratio (CI)	Cases/Controls
No/No	–	1.0	5/55
No/No	+	2.1 (0.7–6.4)	11/58
Yes/No	+	1.4 (0.1–13.3)	1/8
No/Yes	+	3.9 (1.1–13.5)	7/20
Yes/Yes	+	7.3 (1.8–29.2)	6/9

Table 4. GSTM1 genotype and smoking (in pack-years) in various lung cancers[38]

	Smoking, odds ratio and confidence limits (95%)		
Cancer type	<40	40–60	>60
Squamous cell carcinoma, N = 31	1.22 (0.36–4.14)	1.49 (0.53–4.22)	3.18 (0.98–10.9)
Small cell carcinoma, N = 30	0.87 (0.23–3.26)	2.44 (0.72–8.73)	3.66 (0.64–27.3)
Adenocarcinoma, N = 31	1.32 (0.54–3.24)	1.49 (0.53–4.22)	1.22 (0.33–4.55)

GSTM1 GENOTYPES AND CANCER RISK

The genetic polymorphism of the *GSTM1* gene that encodes the glutathione S-transferase μ1 enzyme is a result of a homozygous deletion of the entire GSTM1 gene locus. The *GSTM1* gene locus contains three alleles, i.e. the *GSTM1A* and *GSTM1B* alleles, which differ by a single amino acid, and a deficient *GSTM1* null allele. About 50% of the Caucasian population are known to inherit two deficient alleles, i.e. are homozygous for the null allele of the gene, and thus to be devoid of GSTM1 activity. In genotyping studies using PCR assay, no association was found between *GSTM1* null genotype and lung adenocarcinoma, but a tendency for an association between the *GSTM1* genotype and squamous cell carcinoma has been reported.[35–37] In a study on *GSTM1* genotypes and lung cancer the GSTM1 null genotype was associated with increased lung cancer risk (OR 1.87) even though no association with adenocarcinoma was found.[38] In this study there was a dose-dependent increase in lung cancer risk only in *GSTM* null genotype excluding adenocarcinoma suggesting even that GSTM1 gene might be a crucial in determining lung cancer risk (Table 4).

Lack of *GSTM1* gene seems to predispose to human bladder cancer. In the study by Bell et al.[39] relatively higher increase in urinary bladder cancer risk ratio was found in the *GSTM1* null genotype than GSTM1 homo- or heterozygous genotype although no dose-related "threshold" could be identified, yet there was roughly (<50 pack-years vs. >50 pack-years) some dose-dependence (Table 5). Brockmöller et al.[40] could not find dose-dependence in bladder cancer risk according to smoking in GSTM1 null genotype; if any there was a reverse relationship in the risk. In total *GSTM1* null genotype seemed to increase bladder cancer risk 1.40-fold.

Table 5. *GSTM1* genotype and smoking in bladder cancer patients[39]

Smoking, pack-years	GSTM1 wild type	GSTM1 null
Nonsmoker (38/81, cases/controls)	1	1.3
Smokers (175/118)		
1–50	2.2	4.3
>50	3.5	5.9

Table 6. The metabolic and genotype parameters in lung cancer according to smoking habits and GSTM1 genotype[43]

Parameter	Smokers, 46.6 pack-years		Ex-smokers		Nonsmokers	
GSTM1 genotype	Wild	Null	Wild	Null	Wild	Null
DNA (adducts $\times 10^{-8}$ nucleotide)	8.7 ± 4.7	9.9 ± 6.1	1.4 ± 0.9	3.4 ± 1.4	1.6 ± 0.9	1.6 ± 1.0
CYP1A2 (metabolic index)	10.8 ± 4.2	17.1 ± 5.3	14.7 ± 8.4	15.4 ± 6.6	—	—

NAT1 AND *NAT2* GENOTYPES AND CANCER

Although both N-acetyltransferase 1 and 2 catalyse the activation (O-acetylation) and inactivation (N-acetylation) of arylamine carcinogens only N-acetyltransferase 2 gene locus, *NAT2*, shows acetylation polymorphism and *NAT1* gene encodes a different acetyltransferase isozyme. Rapid acetylators (*NAT2*) may be at risk for colorectal cancers[41] yet a study by Bell et al.[42] in a British population could not confirm this but an increase risk (OR 1.9) was found to be associated with *NAT*10* allele of *NAT1*. Yet in this study the increased risk associated with NAT1 variant allele was most apparent in *NAT2* rapid acetylators (odds ratio 2.8) proposing a gene-to-gene interaction.

CLUSTERING OF CANCER RISK

Although a single mutation may have little risk increasing effect there are examples when there is a cluster of risk genotypes or phenotypes which might potentiate the cancer risk. When measuring CYP1A2 phenotype using caffeine as a probe drug in lung cancer patients among smokers a difference was found between GSTM1 null and positive genotypes suggesting a higher activity in GSTM1 null individuals (Table 6).[43] In a Japanese study combining the mutations of CYP1A1 and GSTM1 demonstrated en enhanced risk of lung cancer in a way that those having susceptible MspI and Ile-Val genotypes of P4501A1 contracted lung cancer after smoking fewer cigarettes than those with other genotypes; furthermore, this susceptibility was increased if these patients had GSTM1 null genotype also (OR 16-41).[24] When both CYP1A2 and NAT2 rapid metaboliser phenotypes were combined an odds ratio of 2.8 for the incidence of colorectal cancer was present (Table 7).[26]

Table 7. Clustering of NAT2 and CYP1A2 phenotypes in colorectal cancer and polyp patients according to their dietary habits[26]

NAT2-CYP1A2 phenotype and cooking preference	Odds ratio
Slow-slow, rare-medium/well done	1.00/2.06
Rapid-slow	0.91/1.87
Slow-rapid	1.39/2.86
Rapid-rapid	3.13/6.15

CONCLUSIONS

The genetic susceptibility of certain environmental-exposure related cancers and the formation of molecular adducts has now been demonstrated in many case-control and cohort studies. In some of the studies e.g. in the case of lung cancer conflicting results or not-confirming results have been obtained. It is only recently and in a very few studies which have shown that the dose might as well be important in determining the significance of genetic susceptibility. This seems to be dependent also on the organ concerned as well as on the dose even in the way that not always the role of dose is to the same direction but that in one organ the genetic susceptibility may become evident at low exposure doses while in other cases at the high exposure doses. In cases when only at high doses genetic susceptibility would come out the case might be that in the organ concerned repair-mechanisms at low doses would protect efficiently against carcinogenicity masking genetic susceptibility. It is also possible although outside this discussion that other, protective, factors might be involved masking genetic susceptibility. This would be the case e.g. when different dietary habits would exist although this is unlikely in certain studies on homogenous populations.

ACKNOWLEDGMENTS

The original studies by the author have been supported by Juho Vainio Fdn (Helsinki).

REFERENCES

1. Raunio, H., Husgafvel-Pursiainen, K., Anttila, S., Hietanen, E., Hirvonen, A., and Pelkonen, O. Diagnosis of polymorphisms in carcinogen-activaitng and inactivating enzymes and cancer susceptibility-a review. Gene 159:113–121 (1995).
2. Swift, M., Morrell, D., Massey, R.B., and Chase, C.L. Incidence of cancer in 161 families affected by ataxia-teleangiectasia. N. Engl. J. Med. 325:1831–1836 (1991).
3. Vainio, H. Biomarkers in metabolic subtyping-Relevance for environmental cancer control. Arch. Toxicol. (Suppl. 20):303–310 (1998).
4. Kleihues, P., Schauble, B., zur Hausen, A., Esteve, J., and Ohkagi, H. Tumors associated with p53 germline mutations: a synopsis of 91 families. Am. J. Pathol. 150:1–13 (1997).
5. Li, F.P. The 4th American Cancer Society Award for Research Excellence in Cancer Epidemiology and Prevention. Phenotypes, genotypes, and interventions for hereditary cancers. Cancer Epidem. Biomarkers Prev. 4:579–582 (1995).
6. Malkin, D., Jolly, K.W., Barbier, N., Look, A.T., Friend, S.H., Gebhardt, M.C., Andersen, T.L., Borresen, A.L., Li, F.P., Garber, J., and al. Germline mutations of the p53 tumor-suppressor gene in children and young adults with second malignant neoplasms. N. Engl. J. Med. 326:1309–1315 (1992).
7. Nebert, D.W., McKinnon, R.A., and Puga, A. Human drug-metabolizing enzyme polymorphisms: effects on risk of toxicity and cancer. DNA Cell Biol. 15:273–280 (1996).
8. Gooderham, N.J., Murray, S., Lynch, A.M., Edwards, R.J., Yadollahi-Farsani, M., Bratt, C., Rich, K.J., Zhao, K., Murray, B.P., Bhadresa, S., Crosbie, S.J., Boobis, A.R., and Davies, D.S. Heterocyclic amines: evaluation of their role in dietassociated human cancer. Br. J. Clin. Pharmacol. 42:91–98 (1996).
9. Catteau, A., Bechtel, Y.C., Poisson, N., Bechtel, P.R., and Bonaiti-Pellie, C. A population and family study of CYP1A2 using caffeine urinary metabolites. Eur. J. Clin. Pharmacol. 47:423–430 (1995).
10. Kadlubar, F.F. Biochemical individuality and its implications for drug and carcinogen metabolism: Recent insights from acetyltransferase and cytochrome P4501A2 phenotyping and genotyping in humans. Drug. Metab. Disp. 26:37–46 (1994).

11. MacLeod, S.L., Tang, Y.-M., Yokoi, T., Kamataki, T., Doublin, S., Lawson, B., Massengill, J., Kadlubar, F.F., and Lang, N.P. The role of recently discovered genetic polymorphism in the regulation of the human CYP1A2 gene. Proc. Amer Assoc. Cancer. Res. 396:•• (1998).
12. Tang, B.K., Zubovits, T., and Kalow, W. Determination of acetylated caffeine metabolites by high-performance exclusion chromatography. J. Chromatogr. 375:170–173 (1986).
13. Buters, J.T.M., Tang, B.-K., Pineau, T., Gelboin, H.V., Kimura, S., and Gonzalez, F.J. Role of CYP1A2 in caffeine pharmacokinetics and metabolism: studies using mice deficient in CYP1A2. Pharmacokinetics 6:291–296 (1996).
14. Petersen, D.D., McKinney, C.E., Ikeya, K., Smith, H.H., Bale, A.E., McBride, O.W., and Nebert, D.W. Human CYP1A1 gene: cosegregation of the enzyme inducibility phenotype and an RFLP. Am. J. Hum. Genet. 48:720–725 (1990).
15. Wedlund, P.J., Kimura, S., Gonzales, F.J., and Nebert, D.W. 1462 mutation in the human CYP1A1 allele gene: lack of correlation with either the MspI 1.9 kb (M2) allele or CYP1A1 inducibility in a three-generation family of East Mediterraean descent. Pharmacogenetics 4:21–26 (1994).
16. Crofts, F., Taioli, E., Trachman, J., Cosma, G.N., Currie, D., Toniolo, P., and Garte, S.J. Functional significance of different human CYP1A1 genotype, mRNA expression, and enzymatic activity in humans. Pharmacokinetics 4:242–246 (1994).
17. Landi, M.T., Bertazzi, P.A., Shields, P.G., Clark, G., Lucier, G.W., Garte, S.J., Cosma, G., and Caporaso, N.E. Association between CYP1A1 genotype, mRNA expression and enzymatic activity in humans. Pharmacokinetics 4:242–246 (1994).
18. Kawajiri, K., Nakachi, K., Imai, K., Yoshii, A., Shinoda, N., and Watanabe, J. Identification of genetically high risk individuals to lung cancer by DNA polymorphisms of the cytochrome P450IA1 gene. FEBS Lett. 263:131–133 (1990).
19. Nakachi, K., Imai, K., Hayashi, S., Watanabe, S., and Kawajiri, K. Genetic susceptibility of squamous cell carcinoma of the lung in relation to cigarette smoking dose. Cancer Res. 51:5177–5189 (1991).
20. Hayashi, S.I., Watanabe, J., Nakachi, K., and Kawajiri, K. Genetic linkage of lung cancer-associated MspI polymorhisms with amino acid replacement in the heme binding region of the human cytochrome P450IA1 gene. J. Biochem. 110:407–411 (1991).
21. Shields, P.G., Sugimura, H., Caporaso, N.E., Petruzzelli, S.F., Bowman, E.D., Trump, B.F., Weston, A., and Harris, C.C. Polycyclic aromatic hydrocarbon-DNA adducts and the CYP1A1 restriction fragment length polymorphism. Environ. Health Perspect. 98:191–194 (1992).
22. Tefre, T., Ryberg, D., Haugen, A., Nebert, D.W., Skaug, V., Brogger, A., and Borresen, A.L. Human CYP1A1 (cytochrome P1450) gene: lack of association between the MspI restriction fragment length polymorphism and incidence of lung cancer in a Norwegian population. Pharmacogenetics 1:20–25 (1991).
23. Hirvonen, A., Husgafvel-Pursiainen, K., Karjalainen, A., Anttila, S., and Vainio, H. Point-mutational MspI and Ile-Val polymorphisms closely linked in the CYP1A1 gene: Lack of association with susceptibility to lung cancer in a Finnish Study population. Cancer Epidem. Biomarkers Prevention 1:485–489 (1992).
24. Nakachi, K., Imai, K., Hayashi, S., and Kawajiri, K. Polymorphisms of the CYP1A1 and glutathione S-transferase genes associated with susceptibility to lung cancer in relation to cigarette dose in a Japanese population. Cancer Res. 53:2994–2999 (1993).
25. Okada, T., Kawashima, K., Fukushi, S., Minakuchi, T., and Nishimura, S. Association between a cytochrome P450 *CYP1A1* genotype and incidence of lung cancer. Pharmacogenetics 4:333–340 (1994).
26. Lang, N.P., Butler, M.A., Massengill, J., Lawson, M., Stotts, R.C., Hauer-jensen, M., and Kadlubar, F.F. Rapid metabolic phenotypes for acetyltransferase and cytochrome P4501A2 and putative exposure to food-borne heterocyclic amines increase the risk for colorectal cancer or polyps. Cancer Epidemiol. Biomarkers & Prev. 3:675–682 (1994).
27. London, S.J., Daly, A.K., Thomas, D.C., Caporaso, N.E., and Idle, J.R. Methodological issues in the interpretation of studies of the *CYP2D6* genotype in relation to lung cancer risk. Pharmacogenetics 4:107–108 (1994).
28. Caporaso, N.E., Tucker, M.A., Hoover, R.N., Hayes, R.B., Pickle, L.W., Issaq, H.J., Muschik, G.M., Green-Gallo, L., Buivys, D., Aisner, S., Resau, J.H., Trump, B.F., Tollerud, D., Weston, A., and Harris, C.C. Lung cancer and the debrisoquine metabolic phenotype. J. Natl. Cancer Inst. 82:1264–1272 (1990).
29. Stücker, I., Cosme, J., Laurent, Ph., Cenée, S., Beaune, Ph., Bignon, J., Depierre, A., Milleron, B., and Hémon, D. CYP2D6 genotype and lung cancer risk according to histologic type and tobacco exposure. Carcinogenesis 16:2759–2764 (1995).

30. Bouchardy, C., Benhamou, S., and Dayer, P. The effect of tobacco on lung cancer risk depends on CYP2D6 activity. Cancer. Res. 56:251–253 (1996).
31. Wynder, E.L. and Hoffmann, D. Smoking and lung cancer: scientific challenges and opprotunities. Cancer Res. 54:5284–5295 (1994).
32. Agundez, J.A.G., Ledesma, M.C., Benítez, J., Ladero, J.M., Rodríguez-Lescure, A., Díaz-Rubio, E., and Díaz-Rubio, M. CYP2D6 genes and risk of liver cancer. Lancet 345:830–831 (1995).
33. Yu, M.-W., Gladek-Yarborough, A., Chiamprasert, S., Santella, R.M., Liaw, Y.-F., and Chen, C.-J. Cytochrome P450 2E1 and glutathione S-transferase M1 polymorphisms and susceptibility to hepatocelluar carcinoma. Gastroenterology 109:1266–1273 (1995).
34. Tsutsumi, M., Takada, A., and Wang, J.-S. Genetic polymorphisms of cytochrome P4502E1 related to the development of alcoholic liver disease. Gastroenterology 107:1430–1435 (1994).
35. Brockmöller, J., Kerb, R., Drakoulis, N., Nitz, M., and Roots, I. Genotype and phenotype of glutathione S-transferase class m isoenzymes m and y in lung cancer patients and controls. Cancer Res. 53:1004–1011 (1993).
36. Hirvonen, A., Husgafvel-Pursiainen, K., Anttila, S., and Vainio, H. The GSTM1 null genotype as a potential risk modifier for squamous cell carcinoma of the lung. Carcinogenesis 14:1479–1481 (1993).
37. Zhong, S., Howie, A.F., Ketterer, B., Taylor, J., Hayes, J.D., Beckett, G.J., Wathen, C.G., Wolf, C.R., and Spurr, N.K. Glutathione S-transferase m locus: use of genotyping and phenotyping assays to assess association with lung cancer susceptibility. Carcinogenesis 12:1533–1537 (1991).
38. Kihara, M., Kihara, M., and Noda, K. Lung cancer risk of GSTM1 null genotype is dependent on the extent of tobacco smoke exposure. Carcinogenesis 15:415–418 (1994).
39. Bell, D.A., Taylor, J.A., Paulson, D.F., Robertson, C.N., Mohler, J.L., and Lucier, G.W. Genetic risk and carcinogen exposure: a common inherited defect of the carcinogen-metabolism gene glutathione S-transferase M1 (GSTM1) that increases susceptibility to bladder cancer. J. Natl. Cancer Inst. 85:1159–1164 (1993).
40. Brockmöller, J., Kerb, R., Drakoulis, N., Staffeldt, B., and Roots, I. Glutathione S-transferase M1 and its variants A and B as host factors of bladder cancer susceptibility: A case-control study. Cancer Res. 54:4103–4111 (1994).
41. Ilett, K.F., David, B., Dethcon, P., Castleden, W., and Kwa, R. Acetylator phenotype in colorectal carcinoma. Cancer Res. 47:1466–1469 (1991).
42. Bell, D.A., Stephens, E.A., Castranio, T., Umbach, D.M., Watson, M., Deakin, M., Elder, M., Henrickse, C., Duncan, H., and Strange, R.C. Polyadenylation polymorphism in the acetyltransferase 1 gene (NAT1) increases risk of colorectal cancer. Cancer Res. 55:3537–3542 (1995).
43. Bartsch, H. and Hietanen, E. The role of individual susceptibility in cancer burden related to environmental exposure. Environ. Health Perspect. 104(Suppl 3):569–577 (1996).

22

DNA REPAIR PATHWAYS AND CANCER PREVENTION

Anthony E. Pegg

Department of Cellular and Molecular Physiology
Pennsylvania State University College of Medicine
The Milton S. Hershey Medical Center
500 University Drive
Hershey Pennsylvania 17033

1. ABSTRACT

This article describes the five main classes of DNA repair processes that occur in humans with respect to their mechanism of action, major substrates, and role in protection against endogenous and environmental DNA damaging agents. The importance of all of these processes in protection from the initiation of neoplastic growth has been established either in studies of inheritable diseases affecting DNA repair or experiments with transgenic animals or both. The capacity of DNA repair pathways to deal with DNA damage is therefore a critical factor in the cellular response to environmental, and dietary carcinogens. DNA repair activity and factors affecting this activity either directly or indirectly must be taken into account in risk assessment.

2. INTRODUCTION

DNA repair is an essential function to maintain the genome from mutations resulting from DNA damage or DNA polymerase errors. DNA damage can be caused by the inherent instability of DNA, by its reaction with endogenous agents and by the reaction with exogenous agents. It is well documented that cellular DNA is continuously undergoing alteration as a result of these processes. Multiple defense mechanisms exist to prevent such changes from leading to mutations that may initiate or promote neoplastic growth. These include: detoxification of the potentially damaging agent; DNA repair pathways; and the induction of death of damaged cells by apoptosis. Factors that produce cell-cycle arrest after DNA damage can allow the DNA repair

Advances in Nutrition and Cancer 2, edited by Zappia *et al.*
Kluwer Academic / Plenum Publishers, New York, 1999.

machinery additional time to act before DNA replication can act to fix mutations and, therefore, are also a critical part of the response to DNA damage. The capacity for DNA repair is thus a major factor in the sensitivity to carcinogenic stimuli. Genetic, environmental and dietary influences that determine this repair capacity or DNA synthesis rates via checkpoint controls must be considered in risk assessment.

Five main classes of DNA repair occur in humans. These are nucleotide excision repair (NER), base excision repair (BER), mismatch repair (MMR), repair of double strand breaks (DSBR), and direct reversal by alkyltransferase (AGT). The importance of all of these processes in protection from the initiation of neoplastic growth has been established both by laboratory studies using experimental animals including transgenic mice in which DNA repair pathways are enhanced or eliminated and by studies of humans with inherited diseases affecting DNA repair capacity. The ability of DNA repair pathways to deal with DNA damage is therefore an integral component in the cellular response to environmental and dietary carcinogens. It should be stressed that the conversion of DNA damage into mutations requires both cell replication and survival. Thus, factors (e.g. the content of wild type p53) that promote death of damaged cells or arrest the cell cycle to allow increased time for DNA repair are also critical determinants of the response to DNA damage. Dietary components influencing the level of DNA damage, the activity of the DNA repair pathways and the activity of checkpoint controls or apoptotic responses would all be expected to influence the rate of mutations.

In this brief review, an outline of the five major pathways of DNA repair is provided along with a few references to recent review articles and papers describing these processes in more detail. The enormous topic of DNA repair is very well covered in the literature with many review articles and books as well as a vast and ever increasing literature of original papers. Space considerations prevent a more detailed citation of the literature and this can be found in many of the listed references.

3. NUCLEOTIDE EXCISION REPAIR (NER)

NER protects against damage by a wide variety of chemicals and ultraviolet light. The repair is accomplished by the combined action of a substantial number of polypeptides and more than 20 gene products are involved in this process.[1,2] The key step is the removal of a segment of DNA about 30 nucleotides in length containing the damaged site. The section is then filled in by DNA polymerase δ or ε acting in concert with proliferating-cell nuclear antigen (PCNA) and replication factor C (RFC) and the gap sealed by DNA ligase I.

Multiple proteins are needed to bind to the DNA lesion, unwind the DNA at this site and then cut the damaged DNA strand. The broad steps of this process are now well understood at the biochemical level although the exact order of some of the additions to the repair complex remains unclear. Until very recently, it was thought that the first step in NER is the recognition of DNA damage and the binding to it of the protein XPA which also binds replication protein A (RPA).[3] However, it now appears that the dimer between the proteins XPC and HR23B may actually act first and precede the binding of XPA/RPA to the lesion site.[4] The XPA/RPA complex then recruits the transcription factor TFIIH, which is a multi-subunit protein containing the helicases XPB and XPD. These proteins and HR23B unwind the damaged DNA and the DNA strand containing the lesion is cut on the 3′ side by XPG and on the 5′ side by a protein made up of XPF and ERCC1. The oligodeoxynucleotide containing the lesion and most of the proteins responsible for its excision are then displaced and the gap filled in as described above.

It is noteworthy that in many cases, NER occurs more rapidly when damage is present in the transcribed strand of active genes.[5] This transcription coupling of the NER process requires the proteins CS-A and CS-B. It now appears that RNA polymerase II may be able to replace the XPC/HR23B complex in damage detection and, in connection with CS-A and CS-B, facilitate the binding of XPA and the rest of the repair complex to the lesion.[4] Thus, XPC may be needed only for global genome repair and not required for transcription-coupled repair.

NER is able to recognize and bring about the repair of a wide variety of DNA lesions. This is probably due to the versatile nature of the interaction of XPA with DNA damage.[3] This 40-kDa protein contains a zinc finger structure and can interact with small single stranded regions caused by helix-destabilizing lesions as well as several of the other proteins needed for excision repair including RPA, TFIIH, and ERCC1. NER is therefore able to bring about the repair of DNA containing bulky DNA adducts such as the cyclobutane pyrimidine dimers and 6,4-pyrimidine-pyrimidone photoproducts produced by UV light and a variety of adducts produced by carcinogens such a polycyclic hydrocarbons, and acetylaminofluorene. Even within these classes of adducts, there are considerable variations in the rates of repair. For example, pyrimidine dimers are repaired more rapidly than 6,4-photoproducts. It is possible that transcription coupling of repair is important only for those lesions that are relatively poorly recognized by global genome repair.

Smaller adducts such as O^6-ethylguanine which produce less distortion of the DNA structure are also recognized by NER but repair is slow and AGT provides an alternative route. It has been claimed that there are no known covalent base adducts that cannot be substrates for NER.[1,6]. This may well be correct in experiments *in vitro* in which large amounts of recombinant proteins can be used in prolonged incubations in reconstituted systems. However, adducts such as O^6-methylguanine (m^6G) and 8-oxoguanine, which produce very little distortion of the DNA, are repaired *in vivo* much more rapidly by other pathways described below. NER is likely to act only as an inefficient back-up system when these other pathways are inactive or saturated.

The importance of NER in protecting against ubiquitous environmental carcinogenic insults is demonstrated clearly by a large body of work on patients suffering from the inherited disease xeroderma pigmenosum (XP). The 7 complementation groups of this disease correspond to partial or complete defects in the 7 proteins in the NER pathway designated XPA to XPG respectively. Patients with XP have a greatly increased risk of developing skin cancer such that the incidence is almost 100% before the age of 20.[7,8] XP patients also have a significant increase in the incidence of primary tumors at other sites. Interestingly, patients with Cockayne's syndrome in which either CS-A or CS-B protein are defective resulting in the loss of transcription-coupled NER do not have an increased incidence of cancer. This suggests this coupling is not likely to be critical in the prevention of mutations even though it provides substantial protection against the acute toxic effects of DNA damage.

Results with transgenic mice are consistent with these conclusions. Knockouts of either the *XPA* or the *XPC* genes lead to a phenotype in which the mice are highly susceptible to carcinogenesis by UV light, polycyclic hydrocarbons or 4-nitroquinoline N-oxide.[9] Knockouts of *CS-A* or *CS-B* are have a variety of growth defects and are sensitive to the toxic effects of UV and polycyclic hydrocarbons but have not been shown to be cancer-prone. Complete deletions of the gene products for any one of several of the other components in the NER pathway (such as ERCC1, XPB, XPD, or XPG) are lethal at early stages during mouse development. This is consistent with the role of these proteins in other cellular processes such as recombination and transcription. More subtle changes in the proteins such as point mutations or small deletions that better mimic the changes in XP patients have been shown to produce viable mice but tumor incidence has not yet been fully investigated.

Overall, it is clear that NER is a major factor in resistance to DNA damage caused by UV light and many exogenous carcinogens. It is possible that minor changes in the rate of NER may alter the individual sensitivity to such agents and many attempts to demonstrate such correlations by measuring unscheduled DNA synthesis or other indirect assessments of NER capacity have been reported (see below). The absence of clear information of the rate limiting factors in the NER pathway and the overlapping specificities of DNA repair pathways renders such studies very difficult to interpret.

4. BASE EXCISION REPAIR (BER)

The BER pathway is the major route for the repair of a variety of small lesions that do not produce substantial helix distortion. The pathway is started by the action of DNA glycosylases that cleave the N-glycosidic bond between the target base and the deoxyribose.[10] The resulting AP site in DNA is then processed in one of two ways. In the simplest pathway ("short-patch BER"), it is cleaved on the 5′ side by a 5′-endonuclease (in humans, called APE) and on the 3′ side by phosphodiesterase (in humans, DNA polymerase β provides this activity) to form a single nucleotide gap. This is filled in by DNA polymerase β and ligated by DNA ligase III. The protein XRCC1 assists in the polymerase and ligation steps. Some glycosylases have an additional enzymatic activity that allows them to remove the deoxyribose via a β-lyase mechanism that produces a 3′ nick and then eliminates the sugar by δ-elimination. The alternative pathway for dealing with the AP site also involves the 5′ AP lyase activity of APE but a short patch of DNA is then excised via displacement by DNA polymerase δ or ε combined with PCNA and the resulting overhang is removed by the flap-endonuclease FEN-1. This "long-patch BER" may be necessary for those sites that are not good substrates for the phosphodiesterase activity of polymerase β.[10]

Thus, a DNA lesion can only be a substrate for BER if it is recognized by a glycosylase. Only a limited number of these enzymes are known. These can remove a range of potentially mutagenic lesions including those produced by deamination of bases, oxidative damage, ionizing radiation and methylating agents. Some are highly specific such as uracil-DNA glycosylase. Others, such as methylpurine-DNA glycosylase (MPG) that removes a variety of methylation products and thymine-glycol-DNA glycosylase (also called Endo III) that removes a number of oxidation products including urea and other fragments, have a much broader range of specificity. Other noteworthy glycosylases are the 8-oxoguanine-DNA glycosylase (OGG); the MutY glycosylase that removes adenine when paired incorrectly with guanine or 8-oxoguanine; and thymine-mismatch DNA glycosylase that repairs thymine or uracil paired incorrectly with guanine.

Several of the glycosylases mentioned above are involved in the repair of oxidative damage. The extent of such damage may be influenced by dietary factors providing anti-oxidants and by other detoxification pathways. Oxidative damage may also be minimized by lowering exposure to conditions including ionizing radiation that favor its production. However, despite such precautions and protective influences, the amount of unavoidable oxidative damage due mainly to endogenous sources is considerable. BER pathways may play a central role in the repair of such DNA damage. The broad-spectrum thymine-glycol-DNA glycosylase removes a number of oxidized and ring-fragmented bases. The OGG and MutY glycosylases counteract the effects of the formation of oxidation products from purines. 8-Oxoguanine and the ring opened product, 2,6-diamino-4-hydroxy-5-N-methylformamidopyrimidine, are removed by OGG while MutY counteracts the mutagenic potential of 8-oxoguanine by removing adenine from mismatches presumably formed by the incorrect copying of 8-oxoguanine by DNA polymerases.

Many studies in microorganisms have shown the importance of BER pathways in protection from endogenous and environmental agents. It is highly likely that BER protects from carcinogenesis by these agents but no compelling direct studies of the relationship between BER pathways and cancer have yet been reported. Several of the genes involved in BER have been studied using knockout mutations in mice.[9,11] The lack of APE, DNA polymerase β, FEN1, XRCC1, or DNA ligase I are all lethal at embryonic stages. This could relate to the multiple functions of these proteins or to the loss of a critical BER pathway.

Of the glycosylases, so far only the MPG has been eliminated in mice. These animals show normal development despite the absence of any ability to repair 3-methyladenine, 1,N^6-ethenoadenine, and hypoxanthine. These results suggest that the daily burden of these lesions contributed by endogenous processes has little effect in either toxicity or mutagenesis. Furthermore, there was no increase in toxicity towards bone marrow cells from these mice in response to methylating or chloroethylating agents. However, there was an increase in mutations attributable to 3-methyladenine and 3- or 7-methylguanine after treatment of the mice with methyl methanesulfonate[12] and it remains to be seen how these mice will respond to a variety of exogenous exposures to alkylating agents. Furthermore, in contrast to the studies of hematopoietic cells described above, the lack of MPG in embryonic stem cells did result in increased sensitivity to killing by methylating and chloroethylating agents.[13]

Overexpression of gene products involved in base excision repair has not been shown to result in increased protection although down-regulation or inactivation of these genes may increase sensitivity to genotoxic stimuli.[14] In a number of cases, increased production of only one component of the BER pathway actually has detrimental effects. For example, studies in which the MPG (or a more specific 3-methyladenine-DNA glycosylase from microorganisms) were over expressed in mammalian cells or yeast show an increase in toxicity of methyl methanesulfonate and an increase in the spontaneous mutation rate.[15] These observations suggest that the coordinated expression of the steps in the BER is necessary in order to get an optimal repair rate and that, in some cases, the presence of unrepaired AP sites may be more harmful than the initial lesion. Thus, factors that alter the balance between these steps may of major importance. Although the results are highly preliminary, the findings of an increased expression of MPG in breast cancer are of potential interest this context.[16]

The glycosylase steps of BER are the best understood reactions in DNA repair at the structural level. Elegant experiments by X-ray crystallography have shown the nature of the interaction between the enzyme and the DNA.[10,17] These studies indicate that the enzymes use a nucleoside flipping mechanisms to bind the lesion into a specificity pocket. Very recently, it has been shown that the binding of uracil DNA-glycosylase to the

AP site continues until the enzyme is displaced by the AP-endonuclease.[18] This is likely to protect the cell from the inherent toxicity of the AP site and is entirely consistent with studies described above showing that the unbalanced expression of glycosylases and AP-endonucleases may increase toxicity and mutation rates.

At present, no inherited conditions involving a deficiency in BER have been reported and none of the glycosylases involved in oxidative damage repair has been eliminated in mice. In view of the likely endogenous exposure to such damage, these alterations may have much more profound effects that the elimination of alkylpurine repair. It has been reported that mutations in the *OGG1* gene occurred in 3 out of 40 tumors but were not present in 15 samples of normal tissue. It was therefore suggested that OGG1 mutations could play a role in the carcinogenic process.[19] Although these numbers are very small and the significance is not fully established, this is an interesting hypothesis in view of the wide spared prevalence of oxidative damage and the key role played in response to is by the 8-oxoguanine DNA glycosylase. However, a more detailed comparison of gastric cancer patients with normal controls showed only a very slight [and not statistically significant difference] in the presence of a polymorphism reducing OGG1 activity.[20]

5. MISMATCH REPAIR (MMR)

MMR reverses the effects of errors made by DNA polymerase. These errors consist of mispaired bases and small loops or deletions. This process is very well understood in *E. coli* and appears to be similar in most basic respects in mammalian cells, although considerably more complicated with many more proteins being involved.[21,22] In human cells, mismatches and single displaced nucleotides are recognized by hMutSα, which is a dimer of MSH2 and MSH6. This interacts with hMutLα, which is a dimer of PMS2 and MLH1. Under the influence of these proteins and other factors, the newly synthesized strand containing the incorrect base and surrounding sequences of up to 1000 nucleotides is degraded [probably by the exonuclease Hex1[23]] and replaced by PCNA and DNA polymerase δ using the parent strand as a template. Larger displaced loops are repaired in the same way but the recognition involves hMutSβ, which is a dimer of MSH2 and MSH3. It is also noteworthy that MLH1 may mediate G_2-M cell-cycle delays in response to DNA damage and, in this way, allow adequate time for the damage to be repaired.[24]

The critical role of MMR in protecting against carcinogenesis has been demonstrated by studies of hereditary nonpolyposis colorectal cancer (HNPCC).[25] This is a relatively common condition (having an incidence of about 1 in 300) which there is a high incidence of colorectal cancer and is associated with germ line deficiencies in the components of MutS and MutL. The most common mutations are in MSH2 (45%) and MLH1 (49%) with rare cases (6%) in PMS2. HNPCC is also associated with microsatellite instability that occurs due to uncorrected errors in the copying of short repeated sequences of DNA. Such sequences are frequent sites of errors by DNA polymerase and the absence of effective mismatch repair allows these errors to be propagated. Microsatellite instability is also found in up to 20% of the cases of colon cancer not associated with a family history. In these cases, there is either somatic mutation of MMR genes or epigenetic inactivation frequently associated with methylation of the *MLH1* promoter.[26] The Muir-Torre syndrome is similar to HNPCC in having a MMR defect (usually in MSH2) but is characterized by both sebaceous skin tumors and internal malignancies. Several other malignancies are also associated with MMR deficiency including cancers of the endometrium, ovary, stomach, pancreas, and skin. The *HEX1* gene may be deleted in some tumors and the lack of this exonuclease could provide another form of MMR deficiency.[23]

Transgenic mice lacking MSH2, MLH1, or PMS2 all show a mutation and cancer prone phenotype and microsatellite instability.[9,27] Knockouts of the *MSH6* gene are also cancer prone but do not have microsatellite instability since hMutSβ is not affected. Deletions of the *MLH1* and *PMS2* genes also have effects on fertility consistent with an essential role for these proteins in meiosis.

The importance of coordinated expression of the proteins forming MMR is emphasized by recent studies showing that a hPMS2 mutation can confer a dominant negative mutator phenotype[28] and that overexpression of the *MSH3* gene can actually can reduce mismatch repair.[29] The latter effect is likely to be due to the presence of multiple partners able to form heterodimers which each have specific functions. If one of these is present in excess, then it may bind all of its partner and prevent the formation of other dimers involving this partner. Thus, when MSH3 levels are artificially increased, all of the MSH2 is bound to it and hMutSα levels become deficient.[29]

It is also noteworthy that MLH1 may mediate G_2-M cell-cycle delays in response to DNA damage and in this way allow adequate time for the damage to be repaired.[24]

6. REPAIR OF DOUBLE STRAND BREAKS (DSBR)

The repair of double strand breaks presents a particular challenge because the absence of a complementary sequence to use as a template makes correct alignment difficult. Accurate DSBR is extremely important. If such breaks are left unrepaired, they may lead to cell death; but, if repaired improperly, they may lead to mutations, rearrangements and chromosome translocations that can all initiate neoplastic transformation. Several methods for repair of double strand breaks are known including homologous recombination, which may play a greater role in humans than previously realized[30] but it appears that the major mechanism in mammalian cells is nonhomologous end joining.[31]

The details of this process are not fully understood but several of the genes essential for DSBR have been determined and in some cases, the biochemistry of their gene product has been clarified. Two key proteins are Ku and DNA-PK. The Ku protein binds to the broken ends protecting them while it recruits and activates the kinase activity of the DNA-PK complex. The activated DNA-PK_{CS} phosphorylates itself, Ku, and probably other proteins, which may also play critical roles. Ku then acquires a helicase activity and unwinds the DNA ends so that exposed areas of homology can anneal. This process would require an endonuclease such as FEN-1 to remove the unpaired DNA regions and a DNA polymerase and ligase to fill in the gaps and seal the nicks. The critical importance of DNA-PK is demonstrated by the phenotype of severely combined immunodeficient (*scid*) mice. These fail to develop lymphocytes because of a mutation in DNA-PK_{CS} that prevents V(D)J recombination. These mice have a major defect in DSBR.

It is highly likely that several other proteins play critical roles in DSBR. Such gene products have been identified in studies in yeast and are known to have mammalian counterparts (see below). It also appears that cell-cycle checkpoint controls may be of particular importance in allowing adequate time for DSBR to occur.

There are two heritable diseases [ataxia telangiectasia[32] and Nijmegen breakage syndrome[33]] that may indicate a relationship in humans between an impaired ability to carry out DSBR and carcinogenesis. Both are autosomal recessive conditions in which patients are highly sensitive to ionizing radiation, show chromosomal instability and a predisposition to cancer. Patients with ataxia telangiectasia have a minor impairment in DSBR and have an increased risk of T-cell lymphomas, leukamias and some other tumors. Patients with Nijmegen breakage syndrome have reduced DSBR and an elevated occurrence of hematological malignancies and neuroblastomas. In both cases, the affected genes have been cloned. The *ATM* gene encodes a phosphatidylinositol-3 kinase family member and is widely thought to play a role in cell-cycle checkpoint control. However, recent data indicates that cells lacking ATM activity have distinct repair and checkpoint defects.[32] The kinase activity of the ATM protein may therefore play a more central role in DSBR. Transgenic mice lacking the *ATM* gene are cancer prone developing thymic lymphomas at an early age and are also infertile and sensitive to ionizing radiation.[9,32]

The *NBS* gene has recently been identified and its product cloned.[33] This protein is a component of a complex with proteins Mre11 and Rad50, which are known from yeast studies to be involved in DSBR and in meiotic recombination.[34] A speculative model that may explain the role of NBS in DSBR suggests that the NBS/hMre11/hRad50 may be recruited to the sites of breaks by the activity of DNA-PK. It could then function either to aid in repair or to signal to the cell-cycle checkpoint machinery perhaps by increasing p53 levels. The NBS protein contains a domain similar to the carboxyl terminus of the breast cancer susceptibility protein BRCA1 that is present in several other proteins involved in DNA repair including DNA ligases III and IV and PARP. This domain is thought to be involved in protein:protein interactions.

7. DIRECT REVERSAL BY ALKYLTRANSFERASE (AGT)

The simplest DNA repair systems are those that restore the original DNA structure by direct removal of the adduct itself. Two such systems are known but one, a photolyase for the reversal of pyrimidine dimers, does not occur in humans. The other is a very widespread system for the repair of O^6-alkylguanine. This is brought about by a protein termed O^6-alkylguanine-DNA alkyltransferase (AGT).[35] AGT contains a cysteine acceptor site, which is located in the amino sequence—Pro-Cys-His-Arg-. It transfers the alkyl group to this site forming S-alkylcysteine and restoring the altered guanine in the DNA without any other alterations. The S-alkylcysteine formed this reaction is not converted back to cysteine so each AGT molecule can act only once and AGT is not really an enzyme. It is however more than an alkyl acceptor protein as it overcomes a significant energy barrier to bring about the alkyl transfer reaction.

AGT genes have been identified in many different species including *Eubacteria, Archaea*, and *Eukarya*. The very widespread distribution of AGT probably relates to the ubiquitous endogenous formation of m^6G. It appears highly likely that there is an endogenous source for the formation of this adduct and several possibilities including *S*-adenosylmethionine and *N*-methyl-*N*-nitroso-compounds have been suggested. It is clear that m^6G is a highly dangerous DNA lesion that needs to be removed rapidly because it is very readily copied incorrectly by DNA polymerase causing G:C to A:T transition mutations. AGT acts very rapidly to prevent such mutations. It is noteworthy that AGT is not specific for the removal of the methyl groups. It can also act quickly on ethyl and 2-chloroethyl-adducts and more slowly on a variety of other aliphatic alkyl groups.[35] Recent studies show that *O*^{6}-benzylguanine is actually a preferred substrate for the human AGT[36] and pyridyloxobutyl adducts formed by the tobacco carcinogen NNK are also rapidly repaired.[37] It is therefore likely that AGT can protect against some other potentially carcinogenic agents but more bulky adducts are also good substrates for NER suggesting that AGT exists to provide the first line of defense against m^{6}G, which is a very poor substrate for NER.

The mechanism of action of AGT is not fully understood. It appears that the cysteine acceptor site is activated by surrounding amino acids to form a thiolate anion that attacks the alkyl group. Recognition of the m^{6}G probably occurs by a nucleoside flipping mechanism. Models based on the crystal structure of the *E. coli* Ada-C AGT are consistent with this[38] but, at present, no direct experimental data for the structure of AGT bound to a substrate or the crystal structure of any AGT except that of Ada-C are available.

It is interesting to speculate why m^{6}G is repaired in this unique way rather than via a glycosylase. Major advantages of the AGT mechanism are that there are no transient breaks in DNA and the repair can proceed without any other proteins such as APE and DNA polymerase β being involved. The disadvantage is that the AGT reaction is not catalytic and the AGT protein can only be used once. However, this may not be a serious disadvantage for protection against low levels of endogenous damage. Also, the alkylation of the active site of the AGT protein causes a conformational change that leads to the rapid degradation of the AGT. In addition to removing unwanted inactive AGT from the nucleus, this may act as signal for the synthesis of additional AGT protein. There is evidence that AGT is inducible in response to agents that cause DNA damage but DNA breaks may be more potent than *O*^{6}-alkylguanine in this respect.[35,39]

Studies with AGTs from bacteria and recombinant mammalian AGTs show that *O*^{4}-methylthymine (m^{4}T) is also a substrate for repair. However, the repair of m^{4}T is very slow and recent experiments in which human AGT or its inactive C145A mutant are expressed in *E. coli* strains lacking endogenous AGT show that AGT actually increases the A:T to G:C transition mutations due to m^{4}T.[40] [The inactive C145A mutant AGT also increases the frequency of mutations caused by m^{6}G, which are very effectively prevented by wild type AGT.] The increases are abolished when experiments are carried out in an *E. coli* strain lacking NER indicating that AGT binds to m^{4}T without repairing it and that this binding prevents repair by NER. These studies show that interference between overlapping DNA repair pathways may actually retard the removal of adducts.

Some polymorphisms in the human *AGT* gene have been reported and, in some cases, the *AGT* gene may be silenced, possibly by promoter methylation.[35,41] These phenomena are the subject of ongoing experimentation but at present there are no clear reports of AGT deficiencies leading to malignant disease. However, there is a wealth of experimental evidence indicating that the AGT status affects sensitivity to alkylating agents.[35] Since the AGT acts alone, it is relatively easy to construct transgenic mice that have increased AGT expression derived from strong and tissue specific promoters. These mice show a strong resistance to carcinogens such as N-methyl-N-nitrosourea.[9,42,43] Conversely, transgenic mice in which the *AGT* gene has been inactivated show an increased sensitivity to dimethylnitrosamine and *N*-methyl-*N*-nitrosourea.[44,45]

The AGT DNA repair pathway is unique in that it apparently only involves one protein and the content of active AGT is depleted by exposure to alkylating agents. Studies of the level of AGT and its alkylated form may provide information on exposure to alkylation damage. All of the other DNA repair pathways involve multiple proteins and the inappropriate expression of any one of the proteins in these pathways may influence its activity and thus the response to DNA damage.

The stoichiometric nature of the repair of m^6G by AGT produces an inverse relationship between residual AGT levels and the extent of methylation damage after exposure to a methylating agent such as dimethylnitrosamine.[46] Significant levels of m^6G have been found in many human DNA samples suggesting that there may be substantial exposure to methylating agents.[47] It has been suggested that monitoring the content of the alkylated inactive form of AGT by use of selective protease digestion of alkylated AGT or reaction with a specific antibody would provide a method for quantitating human exposure to alkylating agents.[48] However, the imprecision of these assays and the rapid degradation of the alkylated AGT protein renders this approach problematic.

8. AGT AND MMR IN RESPONSE TO METHYLATION DAMAGE

AGT prevents the toxicity and mutagenic potential of m^6G by repairing this adduct. If the extent of methylation of the genome exceeds the capacity of AGT for repair, cell death or mutagenesis may occur. The killing of cells by m^6G in their genome is actually brought about by MMR. This recognizes the m^6G:thymine pair (and possibility also the m^6G:cytosine pair) as a mismatch causing the section of DNA containing the thymine to be degraded. However, it is the non-lesion containing strand that is degraded. DNA polymerase then re-copies this strand forming the same m^6G:thymine pair and triggering further mismatch correction. The repeated cycles of synthesis and degradation lead to apoptosis.[49] If MMR is absent, this toxicity does not occur and thus the lack of MMR leads to a tolerance of m^6G in the genome.[9,45] However, the mutagenic potential of the m^6G lesions is not lessened by the lack of MMR but is enhanced by the increased likelihood that cells with m^6G lesions will survive. Thus, the absence of both AGT and MMR provides a particularly high risk of mutagenesis in response to methylating agents.

9. BRCA1 AND BRCA2

Two genes causing a predisposition to familial early-onset breast and ovarian cancer have been identified and are referred to *BRCA1* and *BRCA2*. The functions of these gene products are currently still unclear but may involve DNA repair. Complete knockouts of either gene in transgenic mice are lethal at early embryonic stages[9] but a mutated form of the *BRCA2* gene gave some mice that survived to adulthood with a variety of abnormalities including an increased development of thymic lymphomas.[50] Cells with such deficiencies in the BRCA2 protein show an increased sensitivity to genotoxic agents perhaps via cell-cycle arresting mechanisms.[50,51] The BRCA1 protein is now known to associate with the known DNA repair protein Rad51 as well as RNA polymerase II and transcription factors such as TFIIH. *BRCA1* deficient embryonic stem cells are hypersensitive to ionizing radiation, and to oxidative damage and a defect in the transcription-coupled repair of oxidative damage has been reported.[52] Thus, it appears that DNA repair deficiencies may be linked to susceptibility to breast cancer. Preliminary data supporting this hypothesis has been published based on the screening of lymphocytes using indirect measurements of DNA repair capacity such as the frequency of chromosomal breakages induced by exposure to bleomycin or radiation.[53,54]

10. POLY(ADP-RIBOSE) POLYMERASE (PARP)

PARP is a major component of the nuclear matrix. It is a chromatin-associated enzyme that catalyzes the transfer of the ADP-ribosyl- component from NAD to acceptor proteins forming branched polymers of these units. These are rapidly degraded by a PARP-associated glycohydrolase. Acceptor proteins include PARP itself. The ribosylation reaction is a rapid but transient response to DNA strand breaks and PARP is likely to be involved in survival and DNA repair after such damage. PARP binds to strand breaks triggering its autoribosylation activity. Several different PARP deficient strains of transgenic mice have been derived and experiments with some of these strains confirm that PARP is an important part of the DNA damage response since these mice are highly sensitive to γ-irradiation and to alkylating agents.[55,56] This is in agreement with observations that PARP inhibitors lead to hypersensitivity to alkylating agents, ionizing radiation and reactive oxygen species. However, it has not been possible to demonstrate a direct role for PARP in repair and one attractive hypothesis is that PARP functions to prevent strand breaks from leading to aberrant recombination. PARP is therefore possibly a component of both the BER and the DSBR pathways. The absence of PARP ribosylating activity in response to radiation in leukocytes of patients with familial adenomatous polyposis, which imparts a genetic predisposition to colon cancer provides an additional preliminary link between PARP and carcinogenesis.[50]

11. OTHER HEREDITARY CANCER-PRONE SYNDROMES THAT MAY BE LINKED TO DNA REPAIR DEFICIENCIES

Several other heritable human diseases may be associated with DNA repair defects. However, in these cases, it has not been fully established that the deficiency resides definitely in a DNA repair gene. These conditions include Fanconi's anemia in which there is a high sensitivity to DNA-cross linking agents and Bloom's syndrome in which some (but not all patients) have high sensitivity to UV light and some chemical mutagens. Both conditions show major increases in chromosome breaks and recombinational defects. The gene for Bloom's syndrome has been cloned and is known to be a helicase that could play a role in DNA replication and/or repair.[58] There are now up to 8 complementation groups reported for Fanconi's anemia mostly of unknown function and the conditions may result from alterations in either DNA damage recognition, response to DNA damage or cell-cycle checkpoint controls.[59] The profound effect of the latter on the sensitivity to tumor development is well illustrated by the effects of the loss of p53 function (the Li-Fraumeni syndrome when present in the germ line). This leads to a propensity for the development of genomic instability and a wide variety of neoplasms.[60] The extent to which the loss of well documented abilities of p53 to induce apoptosis, to induce DNA repair responses or to slow cell-cycle progression [reviewed by[61]] relates to the susceptibility to endogenous and exogenous carcinogens is not well understood but it is likely that all contribute. The importance of the content of wild type p53 is well illustrated by studies with transgenic animals.[9] Even the heterozygous *p53* deficient state has a major effect; p53 (+/–) mice show an increase in spontaneous tumors and in susceptibility to genotoxic carcinogens.

12. CONCLUSIONS AND DNA REPAIR AND MOLECULAR EPIDEMIOLOGY

Many studies have shown by indirect methods that DNA repair rates are reduced or absent in cells such as peripheral lymphocytes or skin fibroblasts from patients with known deficiencies in repair pathways or cell-cycle checkpoints [for example[62]]. These studies use unscheduled DNA synthesis, host cell reactivation assays of viral DNA treated with DNA damaging agents, or restoration of activity of reporter genes damaged by reaction with chemical carcinogens to provide measurements of DNA repair responses. Considerable individual variation in repair capacities have been revealed in this type of assay. Attempts to correlate low levels in such assays with susceptibility to neoplastic disease have been made including recent studies using peripheral blood lymphocytes from lung cancer patients[63] or head and neck cancer patients.[64] These approaches, and those described in the section on *BRCA* genes above, do apparently show statistically significant correlations in which reduced DNA repair was linked to the occurrence of neoplastic disease. These pioneering studies require much more experimental work before the conclusions can be fully accepted but they are an interesting approach to move studies of DNA repair into risk assessment and molecular epidemiology. The profound complications in DNA repair, which are described briefly in the sections above, render it unlikely that it will be easy to come up with a standardized test that will allow a simple determination of the relevant DNA repair capacity. These complications include: cell and tissue specificity or repair; induction of repair pathways; induction of apoptosis or cell-cycle blockages in cells with DNA damage; the overlapping nature of DNA repair pathways; the multifunctional aspects of many DNA repair pathways; and the complexities of the interactions between the multiple protein components that make up many DNA repair pathways.

ACKNOWLEDGMENTS

Research in the author's laboratory on DNA repair is supported by grants CA-18137, CA-57725, and CA-71976 from the National Cancer Institute, NIH, Bethesda, MD, USA.

REFERENCES

1. Sancar, A. DNA Excision Repair, Annu. Rev. Biochem. *65*:43–81, 1996.
2. Wood, R.D. Nucleotide excision repair in mammalian cells, J. Biol. Chem. *272*:23465–23468, 1997.
3. Cleaver, J.E. and States, J.C. The DNA-damage-recognition problem in human and other eukaryotic cells: the XPA damage binding protein, Biochem. J. *328*:1–12, 1997.

4. Sugasawa, K., Ng, J.M.Y., Masutani, C., Iwai, S., van der Spek, P.J., Eker, A.P.M., Hanaoka, F., Bootsma, D., and Hoeijmakers, J.H.J. Xeroderma pigmentosum group C protein complex is the initiator of global genome nucleotide excision repair, Molecular Cell. *2*:223–232, 1998.
5. Hanawalt, P.C. Genomic instability: environmental invasion and the enemies within, Mutation Res. *400*:117–125, 1998.
6. Sancar, A. DNA repair in humans, Annu. Rev. Genetics. *29*:69–105, 1995.
7. Kraemer, K.H. Sunlight and skin cancer: Another link revealed, Proc. Natl. Acad. Sci. USA. *94*:11–14, 1997.
8. Ford, J.M. and Hanawalt, P.C. Role of DNA Excision Repair Gene Defects in the Otiology of Cancer. *In*: M.B. Kastan (ed.) Current Topics in Microbiology and Immunology, Vol. 221, pp. 47–70. Heidelberg: Springer-Verlag, 1997.
9. Friedberg, E.C., Meira, L.B., and Cheo, D.L. Database of mouse strains carrying targeted mutations in genes affecting cellular responses to DNA damage. Version 2, Mutation Res. *407*:217–226, 1998.
10. Krokan, H.E., Standahl, R., and Slupphaug, G. DNA glycosylases in the base excision repair of DNA, Biochem. J. *325*:1–16, 1997.
11. Wilson III, D.M. and Thompson, L.H. Life without DNA repair, Proc. Natl. Acad. Sci. USA. *94*:12754–12757, 1997.
12. Elder, R.H., Jansen, J.G., Weeks, R.J., Willington, M.A., Deans, B., Watson, A.J., Mynett, K.J., Bailey, J.A., and Margison, G.P. Alkylpurine-DNA-*N*-glycosylase knockout mice show increased susceptibility to induction of mutations by methyl methanesulfonate, Mol. Cell. Biol. *18*:5828–5837, 1998.
13. Allan, J.A., Engelward, B.P., Dreslin, A.J., Wyatt, M.D., Tomasz, M., and Samson, L.D. Mammalian 3-methyladenine DNA glycosylase protects against the toxicity and clastogenicity of certain chemotherapeutic DNA cross-linking agents, Cancer Res. *58*:3965–3973, 1998.
14. Kaina, B. Critical steps in alkylation-induced aberration formation, Mutation Res. *404*:119–124, 1998.
15. Glassner, B.J., Rasmussen, L.J., Najarian, M.T., Posnick, L.M., and Samson, L.D. Generation of a strong mutator phenotype in yeast by imbalanced base excision repair, Proc. Natl. Acad. Sci. USA. *95*:9997–10002, 1998.
16. Cerda, S.R., Turk, P.W., Thor, A.D., and Weitzman, S.A. Altered expression of the DNA repair protein, *N*-methylpurine-DNA glycosylase (MPG) in breast cancer, FEBS Lett. *431*:12–18, 1998.
17. Parikh, S.S., Mol, C.D., and Tainer, J.A. Base excision repair enzyme family portrait: integrating the structure and chemistry of an entire DNA repair pathway, Structure. *5*:1543–1550, 1997.
18. Parikh, S.S., Mol, C.D., Slupphaug, G., Bharati, S., Krokan, H.E., and Tainer, J.A. Base excision repair initiation revealed by crystal structures and binding kinetics of human uracil-DNA glycosylase with DNA, EMBO J. *17*:5214–5226, 1998.
19. Chevillard, S., Radicella, J.P., Levalois, C., Lebeau, J., Poupon, M.-F., Oudard, S., Dutrillaux, B., and Boiteux, S. Mutations in *OGG1*, a gene involved in the repair of oxidative DNA damage, are found in human lung and kidney tumors, Oncogene. *16*:3083–3086, 1998.
20. Shinmura, K., Kohno, T., Kasai, H., Koda, K., Sugimura, H., and Yokota, J. Infrequent mutations of the hOGG1 gene that is involved in the excision of 8-hydroxyguanine in damaged DNA, in human gastric cancer, Jpn. J. Cancer Res. *89*:825–828, 1998.
21. Modrich, P. and Lahue, R. Mismatch repair in replication fidelity, genetic recombination, and cancer biology, Ann. Rev. Biochem. *65*:101–133, 1996.
22. Tindall, K.R., Glaab, W.E., Umar, A., Risinger, J.I., Koi, M., Barrett, J.C., and Kunkel, T.A. Complementation of mismatch repair gene defects by chromosome transfer, Mutation Res. *402*:15–22, 1998.
23. Wilson III, D.M., Carney, J.P., Coleman, M.A., Adamson, A.W., Christensen, M., and Lamerdin, J.E. Hex1: a new human Rad2 nuclease family member with homology to yeast exonuclease 1, Nucleic Acid Res. *26*:3762–3768, 1998.
24. Davis, T.W., Wilson-Van Patten, C., Meyers, M., Kunugi, K.A., Cuthil, S., Reznikoff, C., Garces, C., Boland, C.R., Kinsella, T.J., Fishel, R., and Boothman, D.A. Defective expression of the DNA mismatch reapir protein, MLH1, alters G2-M cell cycle checkpoint arrest following ionizing radiation, Cancer Res. *58*:767–778, 1998.
25. Toft, N.J. and Arends, M.J. DNA Mismatch Repair and Colorectal Cancer, J. Pathol. *185*:123–129, 1998.
26. Veigl, M.L., Kasturi, L., Olechnowicz, J., Ma, A., Lutterbaugh, J.D., Periyasamy, S., Modrich, P., and Markowitz, S.D. Biallelic inactivation of *hMLH1* by epigenetic gene silencing, a novel mechanism causing human MSI cancers, Proc. Natl. Acad. Sci. USA. *95*:8698–8702, 1998.
27. Prolla, T. DNA mismatch repair and cancer, Curr. Opinion in Cell Biol. *10*:311–316, 1998.
28. Nicolaides, N.C., Littman, S.J.P.M., Kinzler, K.W., and Vogelstein, B. A naturally occurring *hPMS2* mutation can confer a dominant negative mutator phenotype, Mol. Cell. Biol. *18*:1635–1641, 1998.

29. Marra, G., Iaccarino, I., Lettieri, T., Roscilli, G., Delmastro, P., and Jiricny, J. Mismatch repair deficiency associated with overexpression of the *MSH3* gene, Proc. Natl. Acad. Sci. USA. *95*:8568–8573, 1998.
30. Liang, F., Han, M., Romanienko, P.J., and Jasin, M. Homology-directed repair is a major double-strand break repair pathway in mammalian cells, Proc. Natl. Acad. Sci. USA. *95*:5172–5177, 1998.
31. Chu, G. Double strand break repair, J. Biol. Chem. *272*:24097–24100, 1997.
32. Jeggo, P.A., Carr, A.M., and Lehmann, A.R. Splitting the ATM: distinct repair and checkpoint defects in ataxia-telangiectasia, Trends in Genet. *14*:312–316, 1998.
33. Featherstone, C. and Jackson, S.P. DNA repair: The Nijmegen breakage syndrome protein, Current Biol. *8*:R622–R625, 1998.
34. Carney, J.P., Maser, R.S., Olivares, H., Davis, E.M., Le Beau, M., Yates, J.R., III, and Petrini, J.H.J. The hMre11/hRad50 protein complex and Nijmegen breakage syndrome: Linkage of double-strand break repair to the cellular DNA damage response, Cell. *93*:477–486, 1998.
35. Pegg, A.E., Dolan, M.E., and Moschel, R.C. Structure, function and inhibition of O^6-alkylguanine-DNA alkyltransferase, Progr. Nucleic Acid Res. Mol. Biol. *51*:167–223, 1995.
36. Goodtzova, K., Kanugula, S., Edara, S., Pauly, G.T., Moschel, R.C., and Pegg, A.E. Repair of O^6-benzylguanine by the *Escherichia coli* Ada and Ogt and the human O^6-alkylguanine-DNA alkyltransferase., J. Biol. Chem. *272*:8332–8339, 1997.
37. Wang, L., Spratt, T.E., X.-L., L., Hecht, S.S., Pegg, A.E., and Peterson, L.A. Pyridyloxobutyl adduct, O^6-[4-oxo-4-(3-pyridyl)butyl]guanine, is present in 4-(acetoxymethylnitrosamino-1-(3-pyridyl)-1-butanone-treated DNA and is a substrate for O^6-alkylguanine-DNA alkyltransferase., Chem. Res. Toxicol. *10*:562–567, 1997.
38. Vora, R., Pegg, A.E., and Ealick, S.E. A new model for how O^6-methylguanine-DNA methyltransferase binds DNA, Proteins. *32*:3–6, 1998.
39. Boldogh, I., Ramana, C.V., Chen, Z., Biswas, T., Hazra, T.K., Grösch, S., Grombacher, T., Mitra, S., and Kaina, B. Regulation of expression of the DNA repair gene O^6-methylguanine-DNA methyltransferase via protein kinase C-mediated signaling, Cancer Res. *58*:3950–3956, 1998.
40. Edara, S., Kanugula, S., and Pegg, A.E. Expression of the inactive C145A mutant human O^6-alkylguanine-DNA alkyltransferase in *E. coli* increases cell killing and mutations by *N*-methyl-*N'*-nitro-*N*-nitrosoguanidine., Carcinogenesis in press, 1998.
41. Qian, X.C. and Brent, T.P. Methylation hot spots in the 5′ flanking region denote silencing of the O^6-methylguanine-DNA methyltransferase gene, Cancer Res. *57*:3672–3677, 1997.
42. Gerson, S.L., Zaidi, N.H., Dumenco, L.L., Allay, E., Fan, C.Y., Liu, L., and O'Connor, P.J. Alkyltransferase transgenic mice: probes of chemical carcinogenesis, Mutation Res. *307*:541–555, 1994.
43. Kaina, B., Fritz, G., Ochs, K., Haas, S., Grombacher, T., Dosch, J., Christmann, M., Lund, P., Gregel, C.M., and Becker, K. Transgenic systems in studies on genotoxicity of alkylating agents: critical lesions, thresholds, and defense mechanisms, Mutation Res. *405*:179–191, 1998.
44. Iwakuma, T., Sakumi, K., Nakatsuru, Y., Kawate, H., Igarashi, H., Shiraishi, A., Tsuzuki, T., Ishikawa, T., and Sekiguchi, M. High incidence of nitrosamine-induced tumorigenesis in mice lacking DNA repair methyltransferase, Carcinogenesis. *18*:1631–1635, 1997.
45. Kawate, H., Sakumi, K., Tsuzuki, T., Nakatsuru, Y., Ishikawa, T., Takahashi, S., Takano, H., Noda, T., and Sekiguchi, M. Separation of killing and tumorigenic effects of an alkylating agent in mice defective in two of the DNA repair genes, Proc. Natl. Acad. Sci. USA. *95*:5116–5120, 1998.
46. Pegg, A.E. Mammalian O^6-alkylguanine-DNA alkyltransferase: regulation and importance in response to alkylating carcinogenesis and therapeutic agents, Cancer Res. *50*:6119–6129, 1990.
47. Kyrtopoulos, S.A. DNA adducts in humans after exposure to methylating agents, Mutation Res. *405*:135–143, 1998.
48. Oh, H.-K., Teo, A.K.-C., Ali, R.B., Lim, A., Ayi, T.-C., Yarosh, D.B., and Li, B.F.-L. Conformational change in human DNA repair enzyme O^6-methylguanine-DNA methyltransferase upon alkylation of its active site by SN1 (indirect-acting) and SN2 (direct acting) alkylating agents: breaking a "salt-link"?, Biochemistry. *35*:12259–12266, 1996.
49. Karran, P. and Hampson, R. Genomic instability and tolerance to alkylating agents, Cancer Surveys. *28*:69–85, 1996.
50. Connor, F., Bertwistle, D., Mee, P.J., Ross, G.M., Swift, S., Grigorieva, E., Tybulewicz, V.L.J., and Ashworth, A. Tumorigenesis and a DNA repair defect in mice with a truncating *Brca2* mutation, Nature Genetics. *17*:423–430, 1997.
51. Patel, K.J., Yu, V.P.C.C., Lee, H., Corcoran, A., Thistlewaite, F.C., Evans, M.J., and Colledge, W.H. Involvement of Brca2 in DNA repair, Mol. Cell. *1*:347–357, 1998.
52. Gowen, L.C., Avrutskaya, A.V., Latour, A.M., Koller, B.H., and Leadon, S.A. BRCA1 required for transcription-coupled repair of oxidative DNA damage, Science. *281*:1009–1012, 1998.

53. Parshad, R., Bohr, V.A., Cowans, K.H., Zujewski, J.A., and Sanford, K.K. Deficient DNA repair capacity, a predisposing factor in breast cancer, Br. J. Cancer. *74*:1–5, 1996.
54. Jyothish, B., Ankathil, R., Chandini, R., Vinodkumar, B., Nayar, G.S., Roy, D.D., Madhavan, J., and Nair, M.K. DNA repair proficiency: a potential marker for identification of high risk members in breast cancer families, Cancer Lett. *124*:9–13, 1998.
55. Jeggo, P.A. DNA Repair: PARP—another guardian angel?, Current Biol. *8*:R49–R51, 1998.
56. Le Rhun, Y., Kirkland, J.B., and Shah, G.M. Cellular responses to DNA damage in the absence of poly(ADP-ribose) polymerase, Biochem. Biophys. Res. Comm. *245*:1–10, 1998.
57. Cristovao, L., Lechner, M.C., Leitão, C.N., Mira, F.C., and Rueff, J. Absence of stimulation of poly(ADP-ribose) polymerase activity in patients predisposed to colon cancer, Br. J. Cancer. *77*:1628–1632, 1998.
58. Ellis, N.A., Groden, J., Ye, T.-Z., Straughen, J., Lennon, D.J., Ciocci, S., Proytcheva, M., and German, J. The Bloom's syndrome gene product is homologous to RecQ helicases, Cell. *83*:655–666, 1995.
59. Buchwald, M. and Moustacchi, E. Is Fanconi anemia caused by a defect in the processing of DNA damage?, Mutation Res. *408*:75–90, 1998.
60. Gottleib, T.M. and Oren, M. p53 in growth control and neoplasia, Biochim. Biophys. Acta. *1996*:77–102, 1996.
61. Wang, X.W. and Harris, C.C. *TP53* tumour suppressor gene: clues to molecular carcinogenesis and cancer therapy, Cancer Surveys. *28*:169–196, 1996.
62. Abrahams, P.J., Houweling, A., Cornelissen-Steijger, P.D.M., Jaspers, N.G.J., Darroudi, F., Meijers, C.M., Mullenders, L.H.F., Filon, R., Arwert, F., Pinedo, H.M., Natarajan, A.P.T., Terleth, C., Van Zeeland, A.A., and van der Eb, A.J. Impaired DNA repair capacity in skin fibroblasts from various hereditary cancer-prone syndromes, Mutation Res. *407*:189–201, 1998.
63. Wei, Q. and Spitz, M.R. The role of DNA repair capacity in susceptibility to lung cancer: a review, Cancer and Metastasis Reviews. *16*:295–307, 1997.
64. Cheng, L., Eicher, S.A., Guo, Z., Hong, W.K., Spitz, M.R., and Wei, Q. Reduced DNA repair capacity in head and neck cancer patients, Cancer Epidemiology, Biomarkers and Prevention. *7*:465–468, 1998.

23

CEREALS, FIBER, AND CANCER PREVENTION

Attilio Giacosa[1] and Michael J. Hill[2]
on behalf of the ECP Consensus Panel* (see end of chapter)

[1]National Cancer Institute, 16132 Genova, Italy
[2]ECP (UK) Headquarters, Wexham Park Hospital, Slough, Berks SL2 4HL, United Kingdom

All plant foods contain plant cell walls which contain dietary fibre and a range of other anticarcinogenic agents including vitamins, antioxidants, tannins, polyphenolics, and flavonoids. In general, vegetables contain relatively modest amounts of dietary fibre but are rich in a wide array of anticarcinogens, the amounts and classes of which vary between vegetable type. Cereals are relatively rich in dietary fibre and also contain phytate and a range of anticarcinogens. However, these latter are partly removed with the husk during milling. Fruit contains the least dietary fibre but contains an array of anticarcinogens which differ from those in cereals and vegetables.

Current hypotheses suggest that fruit and vegetables provide protection against cancer mainly through the action of their anticarcinogens. In contrast, cereals have been assumed in the past to act mainly through the action of dietary fibre.

In this Consensus Statement "cereals fibre" will imply unrefined or high-extraction cereal, with its husk (and the accompanying anticarcinogens) largely intact. In Europe, cereals may be consumed as breakfast cereals which are often rich in dietary fibre and also rich in B vitamins and anticarcinogens. At other time of day, cereals are usually eaten as breads, pasta, rice, pastries, etc. These are usually made from low-extraction cereals which contain lower levels of dietary fibre and anticarcinogens; wholemeal breads and products are richer in both, however.

Different cereals contain different amounts of dietary fibre and anticarcinogens (rice has least and wheat and rye have most of both). Further, rice is almost always eaten in polished and refined form and so contains even less dietary fibre and anticarcinogen than usual. The cereals which are most often consumed in unrefined and high-extraction form are wheat and rye.

The postulated mechanisms of action indicate that the protective action will be greater in the unrefined cereal than in that in which the husk has been removed by

Advances in Nutrition and Cancer 2, edited by Zappia *et al.*
Kluwer Academic / Plenum Publishers, New York, 1999.

milling. In most epidemiological studies the cereals are primarily low-extraction products and so are low in dietary fibre and anticarconogens. A major recommendation was that in future, questionnaires should be framed to distinguish between low-extraction and high-extraction cereals.

On 1997 the European Organization for Cancer Prevention (ECP) held in S. Margherita Ligure (Italy) a Consensus Meeting on the role of cereals, fibre, and cancer prevention. The ECP Panel (reported at the end of this paper) achieved a consensus statement that is reported in the following pages.[1]

COLORECTAL CANCER

A diet rich in high-fibre cereal is associated with a reduced risk of colorectal cancer. In support of this we cite the review of 58 previous studies of diet and colon cancer, in only 19 of which cereal fibre was measured. Of these, 16 reported an inverse association between cereal fibre and colon cancer risk and the other three showed no relationship 2, 3 (Hill, 1997, 1998). In addition, the review of Food and Agriculture Organization data by Caygill et al.[4] showed that there is an inverse relationship between the risk of colorectal and of breast cancer and cereal and vegetable disappearance, no relationship with fruit and starchy root intake and a positive correlation with total energy intake.

These data are consistent with those from the Italian study;[5] in the context of the Italian diet, high consumption of refined cereal was shown to be a major contributor to high total energy intake and was a risk factor for cancers of the colon and breast. This suggested that the real association was with total energy intake.

This consensus reaffirms and extends the consensus reached by the Colon Group at the World Health Organization (WHO) Consensus Conference in 1996,[6] and with the Committeee on Medical Aspects of Food Policy (COMA) recommendations in the United Kingdom.

A variety of mechanisms has been proposed for the protective effect of cereal fibre. Burkitt[7] popularised the idea that a diet high in fibre-rich foods could influence the course of colorectal carcinogenesis: He proposed that it was fermentation of the fibre itself that gave the protection through (1) increased faecal weight; (2) increased frequency of defecation; (3) decreased transit time; and (4) dilution of the colonic contents. The evidence is strongest for (1) and (4) being important, although there is evidence against all four mechanisms. In addition he proposed that fibre metabolism influenced microbial growth in the colon, an area we know very little about.

More recently, mechanisms involving the metabolic consequences of fibre metabolism have been proposed including (5) alteration of energy metabolism. It is now generally accepted that energy restriction will inhibit carcinogenesis and a fibre-rich diet may make a contribution to overall energy management; (6) influence bile acid metabolism, a theory that appears to refuse to go away; (7) production of short-chain fatty acids, which may inhibit carcinogenesis through its effects on colonic pH, and through the supply of butyrate. This latter has been shown in vitro to promote apoptosis, and cell differentiation, both of which are central to the carcinogenesis process. In vivo verification of these actions is still awaited.

BREAST CANCER

There is suggestive evidence that cereal fibre provides protection against breast cancer. Although many epidemiological studies have shown that cereal fibre has a protective effect, others have shown no effect and there is insufficient evidence to reach a definitive conclusion. In Stuttgart, the WHO Consensus Group on Breast Cancer concluded that the epidemiological evidence was suggestive of a protective effect (as did we) and recommended that cereal fibre consumption should be increased.

It is generally accepted that high levels of circulating oestrogens and insulin growth factors represent major risks for the development of breast cancer. Diets low in fat and rich in cereal fibre reduce levels of plasma oestrogens, in particular by interfering with their enterohepatic circulation and so increasing the rate of faecal excretion. Such diets also contain phytoestrogens, which have been proposed to be protective. Rose et al.[8] and Woods et al.[9] have shown that diets low in fat and high in wheat bran fibre significantly reduce plasma levels of oestradiol and oestradiol suphate. Fibre intakes have also been shown to be inversely related to total, subcutaneous and extra-abdominal fat and to lower insulin levels. These findings reflect the influence of fibre in controlling aspects of the insulin-resistance syndrome.

OTHER SITES

There is good reason to examine seriously the relationship between cereal fibre intake and cancer at other sites. The preliminary analyses reported by La Vecchia and Chatenoud[10] suggested that people who reported consuming whole grain cereals had a lower risk of cancer at a range of other sites in addition to the large bowel and breast. There were many potential confounding factors in these Italian data, and they need to be confirmed. However, there are good theoretical reasons for suspecting a general protective effect. If the mechanisms proposed to explain the protective effects against breast cancer are true, then we would expect them to apply also to other hormone-related cancer sites such as endometrium, ovary, and prostate. Carcinogen binding in the colon lumen might also give rise to a generalised protection, and the presence of anticarcinogens in the cereal husk would provide a mechanism similar to that proposed for vegetables and fruit. If such a generalised protection were to be confirmed it would, of course, strengthen the recommendation to increase intakes of high-fibre cereals.

GENERAL RECOMMENDATIONS

- Questionnaires need to be directed in future to the study of food groups (e.g. cereals) rather than nutrients or anutrients (e.g. dietary fibre), since the latter are highly heterogeneous and not necessarily well quantitated.
- In view of the data presented in the review by Hill,[2] meta-analyses of the case-control and the cohort studies should be carried out.
- Many of the effects of dietary fibre that provide protection against colorectal and breast cancer are concerned with events in the caecum and proximal colon. We need to understand much more about the ecology of this important but experimentally inaccessible subsite of the large bowel.

REFERENCES

1. ECP Consensus Panel (1998). Consensus statement on cereals, fibre, and colorectal and breast cancers. Eur J Cancer Prev 7 (suppl 2):S1–S3.
2. Hill M.J. (1997). Cereals, cereal fibre, and colorectal cancer risk: a review of the epidemiological literature. Eur J Cancer Prev 6:219–225.
3. Hill M.J. (1998). Composition and control of ileal contents. Eur J Cancer Prev 7 (suppl 2):S75–S78.
4. Caygill C.P.J., Charlett A., and Hill M.J. (1998). Relationship between the intake of high-fibre foods and energy and the risk of cancer of the large bowel and breast. Eur J Cancer Prev 7 (suppl 2):S11–S17.
5. Franceschi S., Favero A., Parpinel M., Giacosa A., and La Vecchia C. (1998). Italian study on colorectal cancer with emphasis on the influence of cereals. Eur J Cancer Prev 7 (suppl 2):S19–S23.
6. Biesalski H.K. and Fürst P. (1997): WHO consensus conference on diet and cancer: 28–30 November 1996. Eur J Cancer Prev 6:315.
7. Burkitt D.P. (1971). Some neglected leads to cancer causation. J Natl Cancer Inst 47:913–916.
8. Rose D.P., Goldman M., Connelly J.M., and Strong L.Z. (1997). High fiber diet reduces serum estrogen concentrations in premenopausal women. Am J Clin Nutr 54:520–525.
9. Woods M.N., Gorbach S.L., Longcope C., Goldin B.R., Dwyer J.T., and Morrill-LaBrode A. (1989). Low-fat, high-fiber diet and serum estrone sulfate in premenopausal women. Am J Clin Nutr 49:1179–1183.
10. La Vecchia C. and Chatenoud L. (1998). Fibres, whole-grain foods and breast and other cancers. Eur J Cancer Prev 7 (suppl 2):S25–S28.

* Signatories to the ECP Consensus Statement

Dr. Michael Hill (Chairman of ECP)	Slough, UK
Dr. Attilio Giacosa (Scientific Coordinator of ECP)	Genoa, Italy
Dr. David Beckley	Plymouth, UK
Dr. Christine PJ Caygill	Slough, UK
Dr. Paula Chaves	Lisbon, Portugal
Dr. David Evans	London, UK
Prof. Jean Faivre	Dijon, France
Dr. Fabio Farinati	Padua, Italy
Prof. Silvia Franceschi	Aviano, Italy
Dr. Miquel Gassull	Badalona, Spain
Dr. Mariette Gerber	Montpelier, France
Dr. Ian T Johnson	Norwich, UK
Dr. David Kritchevsky	Philadelphia, USA
Prof. Carlo La Vecchia	Milan, Italy
Prof. Paul Mainguet	Brussels, Belgium
Dr. Alain Maskens	Brussels, Belgium
Dr. Robert W Owen	Heidelburg, Germany
Dr. Joseph Rafter	Stockholm, Sweden
Prof. Ian N Rolwland	Irish United Nutr. Assoc.
Dr. David Southgate	Norwich, UK
Prof. Reinhold Stocckbrugger	Maastricht, The Netherlands

24

CARNITINE SYSTEM AND TUMOR

Menotti Calvani,[1] Raffaela Nicolai,[1] Alfonso Barbarisi,[2] Emilia Reda,[1] Paola Benatti,[1] and Gianfranco Peluso[3]

[1] Scientific Department, Sigma Tau S.p.A., Via Pontina Km 30,400, Pomezia, Rome, Italy
[2] Institute of Clinical Surgery, Faculty of Medicine, 2 University of Naples, Italy
[3] CNR, Via Toiano 6, Arco Felice (Naples), 2 University of Naples, Italy

1. INTRODUCING CARNITINE

Carnitine, a name derived from the Latin *carnis* (flesh), was isolated from meat extracts in 1905[1] and early its chemical formula ($C_7H_{15}NO_3$) was proposed. Its structure, a trimethylbetaine of γ-amino-β-hydroxybutyric acid, was correctly identified and published about twenty years later.[2] Initially, some circumstances led to consider carnitine as a vitamin. By about 1945, all of the important vitamins of the B group had been identified, but the interest in the discovery of still missing B-vitamins, their lack being possibly correlated with anemia, was tremendous. In those years Fraenkel and coworkers observed that the mealworm *Tenebrio molitor* required for normal growth and survival, in addition to at least eight of the known B-vitamins, also folic acid and a new factor contained in brewers yeast or in liver extract, which they tentatively named vitamin-B_T ($_T$ for Tenebrio).[3] The unfavorable properties of this factor (it was hygroscopic and extremely water soluble, thus, hard to crystallize) made its isolation difficult but, finally, the missing vitamin-B_T was identified as carnitine.[4] The widespread distribution of carnitine was established in microorganisms, lower animals, and in all organs of mammals, and in plants too.[5]

But soon after, the finding that microorganisms as well as higher animals were also able to synthesize carnitine by themselves, came to light.[6–8] Hence, the assumption upon which carnitine was included among vitamins failed.

The physiological role of carnitine in microorganisms has not been elucidated for a long time. To date it is known that the role of carnitine in growth stimulation and metabolism in microorganisms varies depending on species and living conditions. For example in *Escherichia coli*, carnitine and other quaternary compounds, such

Advances in Nutrition and Cancer 2, edited by Zappia *et al.*
Kluwer Academic / Plenum Publishers, New York, 1999.

Table 1. Carnitine content in food stuffs

	mg/100 g
Meat	
Sheep	210
Lamb	80
Beef	60
Pork	30
Rabbit	20
Chicken	7.5
Other food stuffs	
Yeast	2.4
Milk	2
Egg	0.8
Peanut	0.1
Corn products	
Wheat germ	1.0
Bread	0.2
Vegetable	
Avocado	1.3
Cauliflower	0.1
Potato	0
Orange	0

from: Neumann 1996

as crotonobetaine and γ-butyrobetaine, may serve as osmoprotectants,[9] may be an ancestral conserved function. Other microorganisms use carnitine as both a carbon and nitrogen source for aerobic growth, and, under anaerobic conditions and in absence of preferred substrates, some bacteria use carnitine as an electron acceptor.[10]

In higher organisms carnitine has specific functions in intermediary metabolism. Its first role to be discovered was to promote mitochondrial oxidation of long-chain fatty acids,[11] that has been widely studied successively. Today, carnitine implications in the utilization of substrates for energy production are well understood and, in the next paragraphs, will be discussed in more detail.

1.1. Carnitine Daily Needs

The human organism synthesizes only 25% of carnitine for itself, the rate of biosynthesis being 1.2 μmole/kg body weight per day. The other 75% has to be assumed by diet and a carnitine daily supply of about 200 mg is required. The main source of carnitine is meat but other food stuffs, such as milk, eggs and, to a lesser extent, vegetables contain this nutrient (Table 1).[12]

Because carnitine is contained primarily in meats and dairy products, vegetarian diets provide a model for assessing the impact of prolonged low carnitine intake on carnitine status. As reported by Lombard in 1989 in adults, plasma carnitine concentration and urinary carnitine excretion after strict vegetarian and lactoovovegetarian diet are significantly lower than those observed after mixed diet. The differences in plasma carnitine concentrations are greater in children than in adults, possibly reflecting the effects of growth and tissue deposition. For this reason, vegetarian children are at greater risk for overt carnitine deficiency.[13]

Carnitine is present in breast milk, as well as in milk of other mammals. A study has reported that in normal fullterm infants fed soy milk, plasma and urine free carnitine and carnitine esters (acylcarnitines) concentrations, at one, two, and three months of age, were significantly lower than in breast-fed infants. Adding carnitine to soy milk, in an amount comparable to the average found in human breast milk, plasma and urine free carnitine and carnitine esters concentrations were restored to normal values.[14]

Results of kinetic and pharmacokinetic studies have suggested that approximately 54–87% of orally introduced carnitine is absorbed by gastrointestinal tract in humans, the remainder is degraded by bacteria in the large intestine and the metabolites are excreted in urine and feces.[15] Absorbed carnitine appears primarily in portal circulation and is extracted by the liver, prior to systemic distribution. A movement of carnitine into bile has been observed, but the total carnitine (free and esterified) concentrations found in bile are highly variable relatively to different metabolic conditions.[10]

Carnitine is a threshold substance with a rapid turnover in the kidneys, so it is likely that the level of carnitine in the blood is mainly under control of the kidney. Under physiological conditions, 90 to 95% of the carnitine that undergoes renal glomerular filtration is subsequently reabsorbed by the tubules, tubular reabsorption threshold of free carnitine being 20 and 74 µmol/L in children and adults, respectively. The remaining proportion is excreted with the urine either as free carnitine or, predominantly, in acetylated (ester) form. Since the carnitine esters mainly undergo secretion at the tubular site, the carnitine esters clearance results 4 to 8 times higher than the clearance rate for free carnitine.[16]

Carnitine synthesized or assumed with diet is extensively transported and redistributed to the different tissues. The normal concentration in blood plasma is 25–50 µM, while most tissues have carnitine concentrations 10-100-fold higher, most of carnitine found in skeletal muscle (3 mM) but the highest concentrations being in epididymis (up to 80 mM).[17,18] It is clear that an active uptake of carnitine from blood must take place, at different rates.

Once carnitine is transported into the cell, carnitine intracellular machineries are switched on, all the components of this system acting in a concerted way to maintain the physiological role of carnitine. The carnitine system consists of free carnitine and acylcarnitines, and the cellular enzymes required for their metabolism and transport.[19]

In summary, to provide optimal function, carnitine concentrations are preserved by a modest rate of endogenous synthesis and by a strong absorption from dietary sources; mechanisms of transport present in most tissues establish and maintain substantial concentration gradients between intracellular and extracellular carnitine pools; finally, intracellular machineries are involved in the development of carnitine metabolic role.

Hence, synthesis, transport and intracellular machineries represent a conceivable paradigm to approach the complexity of carnitine system in health and disease.

2. CARNITINE SYSTEM PARADIGM

2.1. Carnitine Synthesis

Carnitine is synthesized in mammals from the essential amino acids lysine and methionine to form trimethyllysine. In humans, most tissue can convert trimethyllysine to butyrobetaine, but the last step, the hydroxylation of butyrobetaine to carnitine, is limited to the kidney (presenting with the major activity), the liver, the brain and the testes. The availability of trimethyllysine seems to regulate the rate of carnitine biosynthesis[20] and no further regulation has been observed to date.

2.2. Carnitine Transport

Since carnitine is a water soluble, polarized molecule, a saturable carrier-mediated transport across cellular membranes has been postulated. As mentioned above, carnitine concentrations in various tissues are 10-100-fold higher than those in blood, and an active uptake of carnitine is strongly believed. Moreover, the properties of this transport are supposed to be different in a tissue-specific manner.

Recently, a sodium ion-dependent, high affinity human carnitine transporter, has been identified. Because of its similarity (75.8%) to the organic cation transporter OCTN1, carnitine transporter was named OCTN2.[21] In adult tissues OCTN2 is strongly expressed in kidney, skeletal muscle, placenta, heart, prostate and thyroid, and weakly expressed in pancreas, liver, lung, brain, small intestine, uterus, thymus, adrenal gland, trachea, and spinal cord. In fetal tissues OCTN2 is expressed strongly in kidney and weakly in liver, lung and brain.[21]

2.3. Carnitine Intracellular Machineries

Carnitine intracellular machineries exist within each cell of mammal tissues. Specific enzymes, widely distributed within the cell, and their carnitine substrates join in concerted activities developing the physiological role of carnitine, that is following described.[19,22,23]

2.3.1. Transport of Long-Chain Acyl Fatty Acids across the Mitochondrial Membrane. Long-chain acyl fatty acids are transported into mitochondrial matrix through a mechanism involving three enzymes located on the mitochondrial outer and inner membranes[23,24] (Fig. 1). The enzyme carnitine palmitoyltransferase I (CPT I), located on the outer mitochondrial membrane, catalyzes the transfer of acyl groups from acyl-CoA to carnitine to produce acylcarnitine.[25–27] The enzyme acylcarnitine/carnitine translocase (CT), located in the inner mitochondrial membrane, exchanges cytoplasmic acylcarnitine for mitochondrial free carnitine.[28,29] The presence of cardiolipin in the mitochondrial membrane is also required for this transport.[31] Finally, the enzyme carnitine palmitoyltransferase II (CPT II), located in the matrix side of the inner mitochondrial membrane, catalyzes a reaction that is the reverse of that of CPT I, reconverting acylcarnitine to acyl-CoA conveyed to β-oxidation.[26,27]

2.3.2. Modulation of the Mitochondrial Acetyl-CoA/CoA Ratio to Regulate Pyruvate Utilization. The end product of fatty acid β-oxidation and glycolysis is acetyl-CoA, entering the Krebs cycle. In particular, the conversion of pyruvate (formed from glycolysis) to acetyl-CoA is catalysed by the enzyme pyruvate dehydrogenase (PDH) located in the inner mitochondrial membrane. Under normal diet, this reaction permits the use of glucose (and glucose-related fuels) for fatty acid synthesis in adipose tissue, and fatty acid and cholesterol synthesis in liver. The extent of PDH activity is dependent on fatty acid-produced increase in acetyl-CoA (mitochondrial increased acetyl-CoA/CoA ratio), which inhibits PDH reaction. This crucial step in the interplay between fatty acids and glucose utilization obviously relies on the modulation of the acetyl-CoA/CoA ratio. One enzyme of the carnitine system, carnitine acetyltransferase (CAT), located in the matrix side of the inner mitochondrial membrane[32–34] works in this sense: it transfers acetyl group from acetyl-CoA to carnitine, forming acetylcarnitine that can be exported from mitochondria probably by a one-way transport operated by CT.[30] Hence, carnitine, CAT and CT work as a buffer system to maintain the acetyl-CoA/CoA ratio. The removing of intramitochondrial acetyl-CoA leads to the release of PDH inhibition, and consequent pyruvate utilization (Fig. 1). Once in the cytosol, acetyl-CoA forms malonyl-CoA (reaction catalyzed by the enzyme acetyl-CoA carboxylase) which inhibits CPT I activity, that is, carnitine-dependent β-oxidation of long-chain fatty acid[35] (Fig. 1). Thus, the malonyl-CoA/CPT I interaction is the other crucial step in the interplay between fatty acids and glucose utilization.

2.3.3. Peroxisomal Fatty Acid Oxidation, Intracellular Communication. It is now striking that intracellular organelles, other than mitochondria, contain CPTs, CT, and CAT, their activity being demonstrated in the membranes of peroxisomes, microsomes (endoplasmic reticulum), and nucleus.[23,34,36] Although there is no general agreement on the function of CPTs in peroxisome, carnitine seems to be needed to transport long- and medium-chain fatty acid into peroxisome for β-oxidation.[37] Instead, there is direct evidence[38] that through transfer of short-chain acyl-CoA (propionyl-, acetyl-) produced by peroxisomal β-oxidation, and branched-chain acyl-CoA produced by peroxisomal branched-chain amino acids oxidation to mitochondria depends on carnitine system (shuttle of shortened acylcarnitines to mitochondria to further and complete oxidation).

2.3.4. Branched-Chain Amino Acid Metabolism. Carnitine is involved in the metabolism of branched chain amino acids. In fact, the formation of branched-chain acylcarnitines from branched chain amino acids was detected in mitochondria and in peroxisomes, and branched-chain acylcarnitines oxidation has been demonstrated in both organelles, being higher in the latter.[39,40]

2.3.5. Scavenger System for Acyl Groups. Inside the mitochondria, CAT and CT permit transfer of the excess acyl groups to carnitine in the cytosol,[41] to assure mitochondrial metabolism relieved from disturbances (Fig. 1). In the liver, cytosolic carnitine and even more acylcarnitines equilibrate very rapidly with blood (very slowly in muscle); subsequently they are released to the urine by the kidney. Under some conditions (e.g. starvation; acute exercise; inborn errors; drug therapy) acetyl-CoA and other short-CoA esters (e.g., propionylcarnitine; pivaloylcarnitine due to pivalic acid therapy) can accumulate in mitochondria with a parallel accumulation in the cytosol.[42] Thus, carnitine scavenger system represents not only a route for detoxifying cells of exceeding acyl groups but also a mechanism for carnitine loss from the body.

2.3.6. Membrane Stabilization and Repair (Acyl Trafficking). CPTs have been also found in erythrocytes membranes[43–45] and in neurons.[46] Carnitine system, either in erythrocytes and in neurons, is assumed to be an important partner in the pathway of phospholipid and triglyceride fatty acid turnover.

2.3.7. Re-Esterification of Triacyl-Glycerol before Secretion of Very Low Density Lipoproteins (VLDL) Occurring in Microsomes. In the liver, cytosolic triacylglycerol

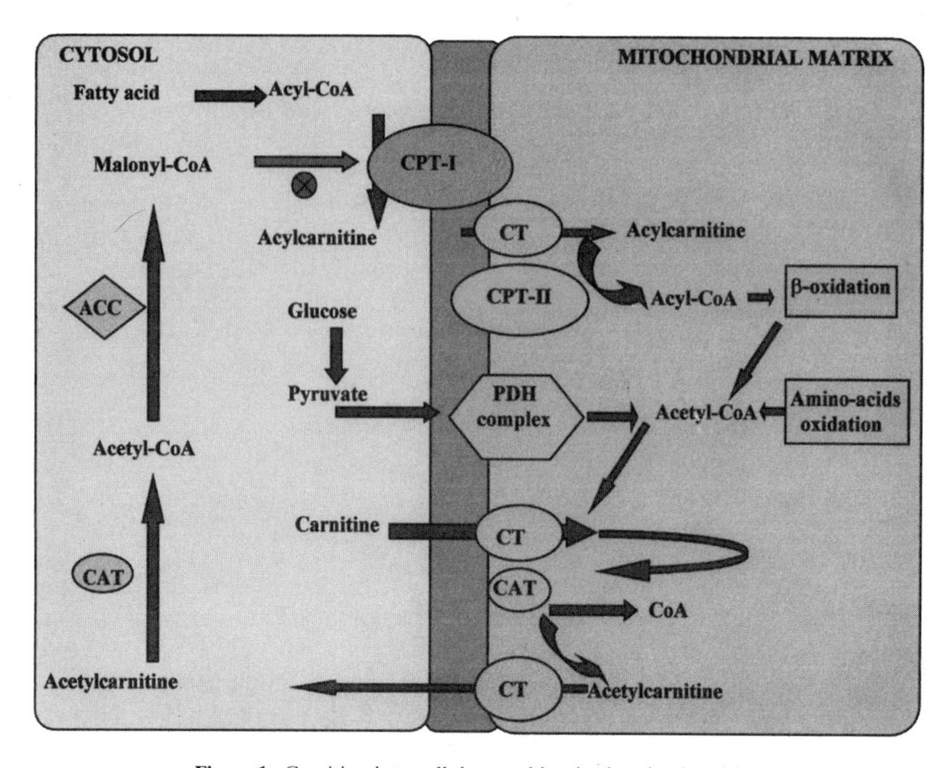

Figure 1. Carnitine intracellular machineries in mitochondria.

is not secreted directly into the blood, but a breakdown by lipolysis and re-esterification is needed. This takes place as part of the transport into the lumen of microsomes (endoplasmic reticulum) before secretion as VLDL by the Golgi apparatus. The microsomal CPTs may take part in this process.[47]

2.3.8. Nuclear Transcriptional Control. Nuclear specific forms of CPTs may have a role in the regulation of acyl-CoA levels intranuclearly and thereby affect the transcriptional regulation of certain genes involved in the control of adipogenesis, gluconeogenesis, and glycolysis.[48,49]

3. KEY POINTS ON TUMOR METABOLISM

3.1. Tumor Cells Utilize Glucose as a Major Energy Source

Tumor cells accumulate deoxyglucose and, as a consequence, an increased number of glycolytic enzymes is expressed. Since, in tumor cells, inhibitors of both mRNA and protein synthesis and glycoprotein transport suppress the increase of deoxyglucose accumulation to control levels, it can be postulated that the transiently elevated glucose metabolism occurs via processes at the levels of gene expression.[50]

3.2. Glucose Utilization in Tumors Is Characterized by Increased Aerobic and Anaerobic Glucose Metabolism

In particular, tumor cells show an accelerated glycolysis and a low O_2 dependence, which are metabolic modifications involved in the resistance of many tumor cell lines to radiation. Thus, a strategy to enhance the radiosensitivity could be the transformation of the glycolytic metabolism of tumor cells into an oxidative type of metabolism, i.e., to induce the ATP supply to depend solely on oxidative phosphorylation.[51]

3.3. Overexpression of Glucose Transporter 1 (GLUT 1)

Increased cellular glucose uptake and enhanced glycolytic metabolism observed in tumor cells is mediated by the overexpression of GLUT 1 (either by constitutive activation of GLUT 1 gene or by decreased degradation of GLUT 1 mRNA), other than key regulatory glycolytic enzymes.[52]

3.4. Low Utilization of Fatty Acids and Ketone Bodies Production

Impaired long-chain fatty acid oxidation and ketogenesis observed in FAO hepatoma cells result from an exquisite sensitivity of CPT I to malonyl-CoA inhibition and unexpressed mitochondrial hydroxymethylglutaryl-CoA synthase gene, respectively.[53] This metabolic characteristic could be a useful way for preferentially directing long-chain fatty acids into phospholipids synthesis in tumor tissues to cover the needs of rapid membrane turnover.

3.5. Increased Pentose Phosphate Pathway (PPP)

PPP are considered important in tumor proliferation processes because of their role in supplying tumor cells with reduced NADP and carbons for intracellular anabolic processes. Direct involvement of PPP in tumor DNA/RNA synthesis is not considered as significant as in lipid and protein synthesis. Instead, recent data indicate that PPP are directly involved in ribose synthesis in tumor cells, through oxidative steps and transketolase reactions. For this reason, control of both oxidative and nonoxidative PPP may be critical in the treatment of cancer.[54]

3.6. Carnitine System Abnormalities

Carnitine system abnormalities in tumor is discussed in the next paragraph.

4. CARNITINE SYSTEM ABNORMALITIES IN TUMOR

Carnitine system in tumor has been studied in different experimental and clinical models. Although much information is still lacking, a landscape appears in that carnitine system is otherwise modified, abnormalities in the modulation and expression of the components of the system being present differently in the various models of tumor studied.

A framework of the findings joined up to date is given, taking into account the paradigm "synthesis, transport, and intracellular machineries" to approach the complexity of the carnitine system.

4.1. Experimental Models of Tumor

4.1.1. Synthesis. No data.

4.1.2. Transport. No data.

4.1.3. Intracellular Machineries. Walker 256 carcinosarcoma bearing rat is an experimental model of tumor where carnitine system abnormalities have been more studied, both in tumor tissue and in non-tumor tissue.

Tumor tissue. Walker 256 tumor tissue contains detectable CPT I and II activities, thereby demonstrating that tumor tissue in vivo has the capacity for the processing of fatty acyl-CoA in the mitochondrion.[55,56] Moreover, CPT I and II activities are not modified in tumor tissue in comparison with that of liver of control rats, hence, fatty acids oxidation in Walker 256 tumor cells proceeds at rates comparable with those found for healthy rat liver.

The possible regulation of tumor CPT I and II was investigated using the hormone insulin, that normally decreases both CPT I activity and CPT I mRNA transcription. Controversially, in the Walker 256 rat tumor in vivo, insulin was without effect on CPT I activity but was found to increase both CPT II activity and the level of CPT II protein expression. Moreover, a different isoform of CPT I has been suggested by the finding that an antibody directed to the rat liver CPT I does not recognize the enzyme from Walker 256 tumor mitochondria.[57]

Rats bearing Walker 256 carcinosarcoma when fed with soya oil (rich in linoleic acid) diet have tumor CPT I activity markedly inhibited, and CPT I mRNA expression markedly increased. Thus, soya oil can modulate, in vivo, the β-oxidative pathway of tumor tissue and further supports the hypothesis of tumor CPT I regulation by polyunsaturated fatty acids.[58]

Nontumor tissue. In Walker 256 tumor-bearing rats, liver fatty acid oxidation and ketone body synthesis have been demonstrated to be reduced, while plasma and liver fatty acids concentrations increased. As non-esterified fatty acid uptake and intracytoplasmic transport are not impaired in the livers of tumor-bearing animals,[59] that is, the metabolic defect appears to occur in the transport of fatty acids into the mitochondria. Contrary to the expectation that the regulation of CPT I activity (by malonyl-CoA) would fulfill the role of controlling the rate of fatty acids oxidation in all different metabolic conditions, neither the activity nor the expression of CPT I were affected in the liver mitochondria of Walker 256 tumor-bearing rats. Instead, CPT II activity was reduced by half compared with that of the controls and the presence of an additional CPT II isoform, different from that of control livers, was detected.[60] The possibility arises that a less active form of CPT II is expressed in these livers, either as the product of a different gene or through post-transcriptional/post-translational modifications, since an occurrence of tumor-induced expression of different enzymes isoforms is known to occur.[61]

It has been suggested that PGE2 may play a role in the control of CPT II expression in the liver of tumor-bearing rats. In fact, indomethacin, a prostaglandin synthesis inhibitor, increased CPT II activity in the liver of tumor bearing rats to levels higher than those found in control animals. Indomethacin did not affect CPTs activities of the mitochondria isolated from tumor tissue.[60]

In rats bearing the Walker 256 carcinosarcoma, lymphocytes from mesenteric lymph nodes preferably oxidize intracellular lipids, and this capacity is greatly enhanced by factors circulating in the serum of tumor-bearing rats. CPT I activity is slightly increased as compared with the control rats, while CPT II is demonstrable in the lymphocytes of tumor-bearing rats and is not detected in the control.[62]

In Fisher 344 rats with methylcholanthrene-induced sarcoma CPT activity is not altered in liver, even in the presence of a large tumor burden. Fatty acid oxidation, reflected by CPT activity, is not enhanced in spite of an ample supply of NEFAs to the liver from the peripheral tissues.[63]

In rats bearing Morris hepatoma 7777 and 7800 has been observed a decrease in liver fatty acids oxidation and low or absent CPT activity.[55]

In vitro studies using rat FAO hepatoma cells have demonstrated that long-chain fatty acids oxdation and keton bodies production are impaired. Neither the activities of CPT I and II nor the CPT II protein amount are affected, whereas a high malonyl-CoA concentration and a CPT I very sensitive to malonyl-CoA inhibition are present.[53]

4.2. Human Models of Tumor

4.2.1. Synthesis. No data.

4.2.2. Transport. Sodium ion-dependent, high affinity human carnitine transporter OCTN2 is strongly expressed in tumor cells[21] such as: Melanoma G361, Lung Carcinoma A549, Colorectal Carcinoma SW480, Chronic Myelogenous Leukemia K562, Cervix Carcinoma Hela S3.

4.2.3. Intracellular Machineries. Little information is known about carnitine intracellular machinery abnormalities in human cancers, both in tumor tissue and in non tumor tissue. Nevertheless, data available confirm that carnitine system is affected in different forms of human cancer in a specific manner. In particular the ratio between free carnitine and carnitine esters concentrations appears modified in different tissues of tumor patients compared with healthy controls. This suggests that a dysmetabolic syndrome, extended to the whole organism, may be occurring. In this regard, Table 2 synoptically reports the more representative findings obtained to date.

5. ANTINEOPLASTIC DRUGS FACING CARNITINE SYSTEM

Carnitine system undergoes several modifications in tumor diseases, as mentioned above. It is noteworthy that some antineoplastic drugs contribute, as side effects, to the dysregulation of this system. It is the case of cysplatin, ifosfamide and, in particular, adriamycin which represents a prototype of antitumor drug affecting carnitine system.

Table 2. Carnitine system status in patients with various forms of tumor

Populations	Free Carnitine	Long-chain Acylcarnitine	Short-chain Acylcarnitine	Total Carnitine	CPT Levels
Pediatric patients with various forms of cancer[64]	= plasma	= urine	= urine	= plasma	
54 patients with various forms of cancer[65]	= plasma	= plasma		= plasma	
10 patients with esophageal carcinoma[66]	↑ plasma	↓ muscle	= muscle	↑ plasma	
21 patients with metastatic disease[67]	↓ plasma ↑ urine	↓ plasma ↑ urine ↓ renal clearance ↓ renal reabsorption	↓ plasma ↑ urine ↑ renal clearance	↓ plasma ↑ urine	
6 patients with colon cancer[68]		↑ tumor tissue	↑ tumor tissue	↑ tumor tissue	↑ tumor tissue
52 women with early breast tumors[69]	↓ plasma				

=: not different from healthy volunteer.
↑: increased compared to healthy volunteer.
↓: decreased compared to healthy volunteer.

5.1. Ifosfamide

After treatment with ifosfamide dysfunction of the tricarboxylic acid (TCA) cycle is observed, in that TCA intermediates in the urine are severely decreased. The almost complete reabsorption of TCA cycle intermediates by renal brush border may reflect a dysfunction of TCA cycle.[70]

Metabolic pathway of ifosfamide leads to formation of chloroacetyl-CoA, with subsequent depression of CoASH level, an indispensable activator in most of the energy providing systems (TCA cycle, fatty acids oxidation). Carnitine is known to detoxify excess amounts of CoA-bound moieties with formation of acylcarnitines and a subsequent release of free CoA. In fact, after treatment with ifosfamide the presence of chloroacetyl-carnitine is detected in urine. In this way formation of chloroacetaldehyde (a non-active metabolite of the drug) responsible for the observed ifosfamide-induced neuro and nephrotoxicity, is thought to be prevented.[70]

5.2. Cisplatin

Cisplatin can lead to reduction in glomerular filtration and to tubular damage. Since carnitine is absorbed proximal to the tubular level, patients treated with cisplatin may run into an increased loss of carnitine through the kidney. In fact, it has been observed that during the course of therapy with cisplatin, total carnitine clearance increases by a factor of 8, while plasma carnitine slightly increases. The increased renal excretion of carnitine (likely due to inhibition of carnitine reabsorption) may be considered an early marker of tubular damage due to cisplatin.[71,72]

5.3. Adriamycin

Adriamycin (Doxorubicin), an anthracycline antibiotic, is among the most important antitumor agents. A number of important biochemical effects have been described for anthracyclines any one or all of which having a role in the therapeutic or toxic effects of such drugs.

Adriamycin can intercalate with DNA, affecting both DNA and RNA synthesis. DNA single and double-strand-breaks occur by activation of topoisomerase II or by generation of free radicals, the latter being also highly destructive to cells.

The clinical value of adriamycin is limited by an unusual cardiomyopathy, the occurrence of which is related to the total dose of the drug, that is often irreversible.

5.3.1. Adriamycin and Carnitine System Paradigm. No data are available concerning adriamycin induced modifications both in carnitine synthesis and transport. Otherwise, ample evidence has demonstrated that carnitine intracellular machineries are deeply influenced by adriamycin treatment.

Adriamycin Binding to Cardiolipin. Much evidence suggests that the mitochondrial membrane could be the target responsible for adriamycin cardiotoxicity. The formation of a very stable complex between adriamycin and cardiolipin, a phospholipid specific to the inner mitochondrial membrane, has been shown to inhibit several mitochondrial membrane enzymes, whose activities depend on the presence of cardiolipin.[73]

Cytochrome c oxidase usually contains some molecules of tightly bound cardiolipin per cytochrome aa3 complex. Without the tightly bound cardiolipin, cytochrome c oxidase activity is decreased by a half.[74]

Besides, it has been demonstrated that the interaction of adriamycin with the cardiolipin-dependent cytochrome c oxidase determines a progressive and irreversible loss of oxidase activity.[75] This uncoupling effect of adriamycin on mitochondrial oxidative phosphorylation, leads to consequent free radicals formation which affects mitochondrial membrane integrity.

Cardiolipin is also essential for mitochondrial carnitine acylcarnitine translocase (CT) activity (see "Carnitine intracellular machineries", above). It is supported by the finding that in intact mitochondria of rat liver and heart, the CT activity is markedly inhibited by micromolar concentrations of adriamycin, through binding to cardiolipin.[31]

Adriamycin Inhibition of Phosphatidyletanolamine N-Methyltransferase. In rat heart adriamycin has been observed to inhibit phosphatidylethanolamine N-methylation, resulting in decreased production of methylated intermediates, phosphatidyl-N-monomethylethanolamine and phosphatidyl-N,N-dimethylethanolamine as well as phosphatidylcholine.[76] Since these phospholipids are involved in cardiolipin synthesis, it is conceivable that adriamycin-induced phosphatidyletanolamine N-methyltransferase inhibition leads to a decrease in cardiolipin content of the inner mitochondrial membrane. It could represent another route for decreasing CT activity.

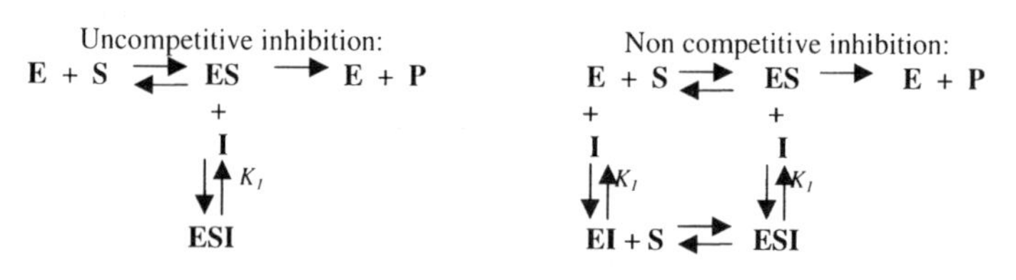

Figure 2. Schematic diagram showing mechanisms of enzyme inhibition.

Adriamycin Binding to Carnitine Palmitoyltransferase (CPT). Adriamycin has been found to inhibit rat heart and liver carnitine palmitoyltransferases of both mitochondrial outer (CPT I) and inner (CPT II) membranes. CPT I was more sensitive than CPT II to inhibition by this drug. Moreover, the cardiac mitochondrial carnitine palmitoyltransferases seemed to be more sensitive to the inhibitory effects of adriamycin than the liver enzyme (useful to remind that CPT I is present in heart and liver in two different isoforms), this giving account of the toxic effects of the drug observed in heart but not in liver.[77,78]

Enzymatic kinetic studies have revealed that adriamycin behaves as an uncompetitive inhibitor with respect to palmitoyl-CoA and as a noncompetitive inhibitor with respect to carnitine for both mitochondrial outer and inner membrane enzymes.[78] A schematic representation of uncompetitive and non competitive inhibition is given in Fig. 2.

Adriamycin causes a concentration- and time-dependent inhibition of CPT I dependent long-chain fatty acid oxidation, whereas acute or chronic administration of carnitine completely abolished adriamycin inhibition. Interestingly, medium- and short-chain fatty acid oxidation, which are independent of CPT I, are also inhibited acutely by adriamycin and could be reversed by carnitine. These data suggest that adriamycin inhibits fatty acid oxidation in part secondary to inhibition of CPT I and/or depletion of its substrate, L-carnitine, in cardiac tissue.[79]

5.3.2. Adriamycin-Induced Cardiomyopathy: Effects of Carnitine Treatment. Many studies have reported that carnitine treatment significantly prevents or decreases both acute and delayed forms of cardiomyopathy induced by adriamycin, either in animal models[80–83] and in patients.[84–88] The characteristics of adriamycin-induced cardiomyopathy, and the effects of carnitine treatment are summarized in Table 3.

It is noteworthy to mention that carnitine protective effects on adriamycin toxicity are accomplished without either decreasing adriamycin antitumor activity or promoting tumor growth.

In vivo studies, in experimental models, have been demonstrated that: a) Carnitine pretreatment increased survival time and did not affect adriamycin's inhibition of leukemic colony formation in mice;[89] b) Carnitine administration did not promote tumor growth, neither in mice bearing osteosarcoma nor in mice with implanted mammary carcinoma.[90]

In vitro studies have demonstrated that: Carnitine addition to three different human cell lines derived from pancreatic tumors had no effect on the number of cells incubated in the presence or absence of adriamycin.[91,92]

Table 3. Adriamycin-induced cardiomyopathy: Effects of carnitine treatment

Acute Form Manifested by ECG Abnormalities	Delayed Form Manifested by Congestive Heart Failure (CHF) Unresponsive to Digitalis
Occurring few hours after a single administration of drug. This is brief and rarely a serious problem.	Occurring after chronic, cumulative drug administration (1–6 months, or more, after the last dose of drug). Serious risk of cardiomyopathy is 1% to 10% at total dose below 450 mg/m^2, and increases to >20% at total doses higher than 550 mg/m^2.
▲ • ST-T wave alterations ▲ • Arrhythmias	▲ • Signs and symptoms • Histological features Light microscopy: ●—Edema ●—Sarcoplasmic alterations ●—Myofibrillar fragmentation ●—Myocyte damage and loss Electron microscopy: ●—sarcoplasmic vacuolation ●—mitochondrial degeneration ●—myofibrillar lysis
Carnitine prevention or improvement in: ● Experimental models	▲ Patients

6. PALMITOYLCARNITINE IN TUMOR PREVENTION: A NATURAL ANTI-TUMOR PROMOTING MOLECULE WITH ANTI-DRUG RESISTANCE ACTION

Palmitoylcarnitine is a natural ester of carnitine. It is the metabolic product of the reaction catalized by CPT I but the effects of palmitoylcarnitine do not appear to be mediated by CPT I.

During the early 1960's, palmitoylcarnitine has been investigated as an antitumoral substance in ehrlich's ascites tumor cells, exhibiting an inhibitory effect on respiration and glycolysis in these cells.[93]

The main feature of palmitoylcarnitine came to light when it was tested in experimental conditions using a tumor promoter such as the phorbol esters, known to alter the normal regulation of cellular growth control and gene expression. Tumor promoters cause disruption or abnormal activation of intracellular signal transduction pathways, which control proliferation and differentiation. Palmitoylcarnitine can be located somewhere in the normal control pathways.

Evidence has been accumulated that the phorbol esters receptor, in intact cells, is a Ca^{++}-phospholipid-dependent protein kinase C (PKC). However, it is not certain whether the various actions of the phorbol esters on cultured cells are all mediated by the activation of PKC. Palmitoylcarnitine has been recognized as a PKC inhibitor.[94]

In human promyelocytic leukemia cells (HL-60) the phorbol ester 12-O-tetradecanoyl-phorbol-13-acetate (TPA) is shown to cause cell adhesion concomitant with morphological changes. Palmitoylcarnitine is able to inhibit the TPA-induced cell adhesion and differentiation. Neither palmitic acid nor carnitine possess this action.[95]

The tumor promoter TPA stimulates a rapid increase in ornithine decarboxylase (ODC) activity in different target cells. ODC is widely associated with the aberrant growth regulation of cancer cells and an elevated expression of ODC protein is an early event in chemical carcinogenesis in the skin, liver, and other tissues. High levels of ODC activity are also found in human cancer.

It has been observed that stimulation of ODC mRNA by TPA is blocked by palmitoylcarnitine (and other PKC inhibitors) in rat hepatoma cells, consistent with a requirement for PKC activation in the induction mechanism. However, additional kinases may be involved in the intracellular signalling process.[96]

The effect of palmitoylcarnitine on ODC induction has also been evaluated after topical applications of TPA on mouse epidermis. Two applications of TPA at 12-hour intervals caused a marked suppression of ODC induction by the 2nd application of TPA (refractory state); instead, at 96-hour interval between the 1st and 2nd application, the ODC activity induced by the 2nd application of TPA is greater than that induced by the single 1st application (enhanced state). Changes in the PKC activity are hypothesized to mediate TPA-induced refractory and/or enhanced state for ODC induction. Palmitoylcarnitine (applied concurrently with the 1st TPA application) is shown to restore the ODC induction. Acetylcarnitine and palmitic acid were not effective, and steroylcarnitine less effective than palmitoylcarnitine.[97]

Studies concerning the inhibitory effect of adriamycin and daunomycin in combination with verapamil or palmitoylcarnitine on skin mouse tumor promotion by TPA have been conducted. Adriamycin and daunomycin alone were unable to inhibit the promotion of skin papillomas by repeated applications of TPA. Moreover, preatreatment with adriamycin also failed to prevent the tumor-promoting activities of smaller doses of TPA. By combining verapamil (simultaneously with each promotion TPA treatment) with the anthracycline antibiotics, the number of papillomas/mouse is inhibited by 47–50%. Verapamil alone results in a reduction of 26%. By combining palmitoylcarnitine with the anthracycline antibiotics, the number of papillomas/mouse is inhibited by 78–86%. Palmitoylcarnitine alone results in a reduction of 44%.

Thus, anthracyclines need verapamil and palmitoylcarnitine to exert anti-cancer activity, and palmitoylcarnitine is shown to be the most efficacious substance in circumventing drug resistance.[98]

Palmitoylcarnitine is found to inhibit melanoma cell growth in a dose-dependent manner. This effect in part is related to the molecular as aspects linking PKC activity and the ionic events in the initiation of cell growth.[99]

7. CONCLUSIONS

a) No data are available on synthesis and/or transport defects in tumor models, either animal or human.
b) The OCTN2, sodium ion-dependent high affinity human carnitine transporter, is strongly expressed in tumor cells.
c) Carnitine intracellular machineries appear abnormally expressed in tumor, in such a way to greatly reduce fatty acid β-oxidation.
d) Yet, carnitine system in tumor appears to respond to pharmacological agents.
e) Some important antitumor drugs contribute to dysfunction of carnitine system.

f) Adriamycin cardiotoxicity seems to be related to CPTs and CT inhibition, and is reversed by carnitine treatment, without affecting adriamycin antitumor therapeutic efficacy.
g) Early studies suggest a tumor prevention role of palmitoylcarnitine and constitute the basis of ongoing research of this carnitine ester as a natural antitumoral compound.

REFERENCES

1. Gulewitsh W. and Krimberg R. Zur Kenntnis der Extraktivstoffe der Muskeln. Z. Physiol. Chem. 45:326–330, 1905.
2. Tomita M. and Sendju Y. Uber die Oxyaminoverbindungen, welche die Biuretreaktion zeigen, Z. Physiol. Chem. 169:263, 1927.
3. Fraenkel G., Blewett M., and Coles M. B_T, a new vitamin of the B-group and its relation to the folic acid group, and other anti-anaemia factors. Nature 161(4103):981–983, 1948.
4. Carter H.E., Bhattacharyya P.K., Weidman K.R., and Fraenkel G. Chemical studies on vitamin B_T. Isolation and characterization as *carnitine*, Arch. Biochem. Biophys. 38:405–416, 1952.
5. Fraenkel G. and Friedman S. *Carnitine*. Vitam. Horm.15:73–118, 1957.
6. Lindstedt G. and Lindstedt S. On the biosynthesis and degradation of *carnitine*, Biochem. Biophys. Res. Commun. 6(5):319–323, 1961.
7. Bremer J. *Carnitine* precursors in the rat, Biochim. Biophys. Acta 57:327–335, 1962.
8. Horne D.W., Tanphaichitr V., and Broquist H.P. Role of lysine in *carnitine* biosynthesis in Neurospora crassa, J. Biol. Chem. 246(13):4373–4375, 1971.
9. Jung H., Jung K., and Kleber H.P. L-*carnitine* metabolization and osmotic stress response in Escherichia coli, J. Basic Microbiol. 30(6):409–413, 1990.
10. Rebouche C.J. and Seim H. *Carnitine metabolism* and its regulation in microorganisms and mammals, Annu. Rev. Nutr. 18:39–61, 1998.
11. Fritz I.B. The effect of muscle extracts on the oxidation of palmitic acid by liver slices and homogenates, Acta Physiol. Scand. 34:367, 1955.
12. Neumann G. Effect of L-*carnitine* on athletic performance. In: *Carnitine*. Pathobiochemical Basics and Clin. Applications, pp. 61–71, 1996.
13. Lombard K.A., Olson A.L., Nelson S.E., and Rebouche C.J. *Carnitine* status of lactoovovegetarians and strict vegetarian adults and children, Am. J. Clin. Nutr. 50(2):301–306, 1989.
14. Novak M. The role and importance of L-*carnitine* in infants nutrition. International symposium on infant nutrition, Beijing, China, 21–23/9/1993.
15. Rebouche C.J. and Chenard C.A. Metabolic fate of dietary *carnitine* in human adults: identification and quantification of urinary and fecal metabolites, J. Nutr. 121(4):539–546, 1991.
16. Bulla M., Glögger A., Rößle C., and Fürst P. Dysregulation of *carnitine metabolism* in renal insufficiency: a summary of findings in adults and children. In: *Carnitine*. Pathobiochemical Basics and Clin Applications. H. Seim and H Löster, editors; Ponte Press, pp. 177–194, 1996.
17. Rebouche C.J. and Engel A.G. Kinetic compartmental analysis of *carnitine metabolism* in the human *carnitine* deficiency syndromes. Evidence for alterations in tissue *carnitine* transport, J. Clin. Invest. 73(3):857–867, 1984.
18. Brass E.P. *Carnitine* transport. In: L-*Carnitine* and its role in Medicine: from function to therapy. R. Ferrari, S. DiMauro, and G. Sherwood, editors; Academic Press, pp. 21–36, 1992.
19. Scholte H.R., Boonman A.M.C., Hussaarts-Odijk L.M., Ross J.D., Van Oudheusden L.J., Rodrigues Pereira R., and Wallenburg H.C.S. New aspects of the biochemical regulation of the *carnitine* system and mitochondrial fatty acid oxidation. In: *Carnitine*. Pathobiochemical basics and clinical applications. Seim H. and Löster H. editors, Ponte Press Verlags-GmbH, pp. 11–34, 1996.
20. Rebouche C.J., Lehman L.J., and Olson L. Epsilon-N-trimethyllysine availability regulates the rate of *carnitine* biosynthesis in the growing rat, J. Nutr. 116(5):751–759, 1986.
21. Tamai I., Ohashi R., Nezu J., Yabuuchi H., Oku A., Shimane M., Sai Y., and Tsuji A. Molecular and functional identification of sodium ion-dependent, high affinity human *carnitine* transporter OCTN2, J. Biol. Chem. 273(32):20378–20382, 1998.
22. Hoppel C. The physiological role of *carnitine*. In: L-*Carnitine* and its role in medicine: From function to therapy. Ferrari R, DiMauro S, and Sherwood G editors; Academic Press Limited, pp. 5–19, 1992.

23. Bremer J. The role of *carnitine* in cell *metabolism*. In: *Carnitine* Today, C. de Simone and G. Famularo, editors; Springer-Verlag, Heidelberg, pp. 1–37, 1997.
24. McGarry J.D. and Brown N.E. The mitochondrial *carnitine* palmitoyltransferase system. From concept to molecular analysis, Eur. J. Biochem. 244(1):1–14, 1997.
25. Murthy M.S.R. and Pande S.V. Malonyl-CoA binding site and the overt *carnitine* palmitoyltransferase activity reside on the opposite sides of the outer mitochondrial membrane, Proc.Natl. Acad. Sci. USA 84(2):378–382, 1987.
26. Declercq P.E., Falck J.R., Kuwajima M., Tyminski H., Foster D.W., and McGarry J.D. Characterization of the mitochondrial *carnitine* palmitoyltransferase enzyme system. I. Use of inhibitors, J. Biol. Chem. 262(20):9812–9821, 1987.
27. Woeltje K.F., Kuwajima M., Foster D.W., and McGarry J.D. Characterization of the mitochondrial *carnitine* palmitoyltransferase enzyme system. II. Use of detergents and antibodies, J. Biol. Chem. 262:9822–9827, 1987.
28. Pande S.V. A mitochondrial *carnitine* acyl*carnitine* translocase system, Proc. Natl. Acad. Sci. USA 72(3):883–887, 1975.
29. Indiveri C., Tonazzi A., and Palmieri F. Identification and purification of *carnitine* carrier from rat liver mitochondria, Biochi. Biophys. Acta 1020(1):81–86, 1990.
30. Indiveri C., Tonazzi A., and Palmieri F. Characterization of the unidirectional transport of *carnitine* catalized by the reconstituted *carnitine* carrier from rat liver mitochondria, Biochim. Biophys Acta 1069(1):110–116, 1991.
31. Noël H. and Pande S.V. An essential requirement of cardiolipin for mitochondrial *carnitine* acyl*carnitine* translocase activity. Lipid requirement of *carnitine* acyl*carnitine* translocase, Eur. J. Biochem. 155(1):99–102, 1986.
32. Edwards Y.H., Chase J.F.A., Edwards M.R., and Tubbs P.K. *Carnitine* acetyltransferase: the question of multiple forms, Eur. J. Biochem. 46(1):209–215, 1974.
33. Bremer J. *Carnitine Metabolism* and functions, Physiol. Rev. 63(4):1420–1480, 1983.
34. Bieber L.L. *Carnitine*. Annu. Rev. Biochem. 57:261–283, 1988.
35. Murthy M.S.R. and Pande S.V. Characterization of a solubilized malonyl-CoA sensitive *carnitine* palmitoyltransferase from the mitochondrial outer membrane as a protein distinct from the malonyl-CoA insensitive enzyme of the inner membrane. Biochem. J. 268(3):599–604, 1990.
36. Murthy M.S.R. and Pande S.V. Molecular biology of *carnitine* palmitoyltransferases and role of *carnitine* in gene transcription, In: *Carnitine* Today, C. de Simone and G. Famularo, editors; Springer-Verlag, Heidelberg, pp. 39–70, 1997.
37. Buechler K.F. and Lowenstein J.M. The involvement of *carnitine* intermediates in peroxisomal fatty acid oxidation: a study with 2-bromofatty acids, Arch. Biochem. Biophys. 281(2):233–238, 1990.
38. Jakobs B.S. and Wanders R.J. Fatty acid beta-oxidation in peroxisomes and mitochondria: the first, unequivocal evidence for the involvement of *carnitine* in shuttling propionyl-CoA from peroxisomes to mitochondria, Biochem. Biophys. Res. Commun. 213(3):1035–1041, 1995.
39. Solberg H.E. and Bremer J. Formation of branched chain acyl*carnitines* in mitochondria, Biochim. Biophys. Acta 222(2):372–380, 1970.
40. Singh H., Beckman K., and Poulos A. Peroxisomal beta-oxidation of branched chain fatty acids in rat liver. Evidence that *carnitine* palmitoyltransferase I prevents transport of branched chain fatty acids into mitochondria, J. Biol. Chem. 269(13):9514–9520, 1994.
41. Lysiak W., Lilly K., Di Lisa F., Toth P.P., and Bieber L.L. Quantitation of the effect of L-*carnitine* on the levels of acid- soluble short-chain acyl-Coa and CoASH in rat heart and liver mitochondria, J. Biol. Chem. 263(3):1151–1156, 1988.
42. Diep Q.N. and Bohmer T. Increased pivaloyl*carnitine* in the liver of the sodium pivalate treated rat exposed to clofibrate, Biochim. Biophys. Acta 1256(2):245–247, 1995.
43. Wittels B. and Hochstein P. The identification of *carnitine* palmityltransferase in erythrocyte membranes, J. Biol. Chem. 242(1):126–130, 1967.
44. Ramsay R.R., Mancinelli G., and Arduini A. *Carnitine* palmitoyltransferase in human erythrocyte membrane. Properties and malonyl-CoA sensitivity, Biochem. J. 275(Pt3):685–688, 1991.
45. Arduini A., Mancinelli G., Radatti G.L., Dottori S., Molajoni F., and Ramsay R.R. Role of *carnitine* and *carnitine* palmitoyltransferase as integral components of the pathway for membrane phospholipid fatty acid turnover in intact human erythrocytes, J. Biol. Chem. 267(18):12673–12681, 1992.
46. Arduini A., Denisova N., Virmani A., Avrova N., Federici G., and Arrigoni Martelli E. Evidence for the involvement of *carnitine*-dependent long-chain acyltransferases in neuronal triglyceride and phospholipid fatty acid turnover, J. Neurochem. 62(4):1530–1538, 1994.

47. Broadway N.M. and Saggerson E.D. Microsomal *carnitine* acyltransferases. Biochem. Soc. Trans. 23(3):490–494, 1995.
48. Tomaszewski K.E. and Melnick R.L. In vitro evidence for involvement of CoA thioesters in peroxisome proliferation and hypolipidaemia, Biochim. Biophys. Acta 1220(2):118–124, 1994.
49. Nishimaki-Mogami T., Takahashi A., and Hayashi Y. Activation of a peroxisome-proliferating catabolite of cholic acid to its CoA ester, Biochem. J. 296 (Pt 1):265–270, 1993.
50. Fujibayashi Y., Waki A., Sakahara H., Konishi J., Yonekura Y., Ishii Y., and Yokoyama A. Transient increase in glycolytic *metabolism* in cultured *tumor* cells immediately after exposure to ionizing radiation: from gene expression to deoxyglucose uptake, Radiat. Res. 147(6):729–734, 1997.
51. Rodriguez-Enriquez S. and Moreno-Sanchez R. Intermediary *metabolism* of fast-growth *tumor* cells, Arch. Med. Res. 29(1):1–12, 1998.
52. Reske S.N., Grillenberger K.G., Glatting G., Port M., Hildebrandt M., Gansauge F., and Beger H.G. Overexpression of glucose transporter 1 and increased FDG uptake in pancreatic carcinoma, J. Nucl. Med. 38(9):1344–1348, 1997.
53. Prip Buus C., Bouthillier Voisin A.C., Kohl C., Demaugre F., Girard J., and Pegorier J.P. Evidence for an impaired long-chain fatty acid oxidation and ketogenesis in FAO hepatoma cells, Eur. J. Biochem. 209(1):291–298, 1992.
54. Boros L.G., Lee P.W., Brandes J.L., Cascante M., Muscarella P., Schirmer W.J., Melvin W.S., and Ellison E.C. Nonoxidative pentose phosphate pathways and their direct role in ribose synthesis in *tumors*: is cancer a disease of cellular glucose *metabolism*?, Med. Hypotheses 50(1):55–59, 1998.
55. Fields A.L., Wolman S.L., Cheema-Dhadli S., Morris H.P., and Halperin M.L. Regulation of energy *metabolism* in Morris hepatoma 7777 and 7800, Cancer Res. 41(7):2762–2766, 1981.
56. Tisdale M.J. and Brennan R.A, Loss of acetoacetate coenzyme A transferase activity in tumours of peripheral tissues, Br. J. Cancer 47(2):293–297, 1983.
57. Colquhoun A. and Curi R. Human and rat tumour cells possess mitochondrial *carnitine* palmitoyltransferase I and II: effects of insulin, Biochem. Mol. Biol. Int. 37(4):599–605, 1995.
58. Colquhoun A., de Mello F.E., and Curi R. In vivo inhibition of Walker 256 tumour *carnitine* palmitoyltransferase I by soya oil dietary supplementation, Biochem. Mol. Biol. Int. 44(1):151–156, 1998.
59. Evans R.D. and Williamson D.H. Tissue-specific effects of rapid tumour growth on lipid *metabolism* in the rat during lactation and on litter removal, Biochem. J. 252(1):65–72, 1988.
60. Seelaender M.C., Curi R., Colquhoun A., Williams J.F., and Zammit V.A. *Carnitine* palmitoyltransferase II activity is decreased in liver mitochondria of cachectic rats bearing the Walker 256 carcinosarcoma: effect of indomethacin treatment, Biochem. Mol. Biol. Int. 44(1):185–193, 1998.
61. Siddiqui R.A. and Williams J.F. The regulation of fatty acid and branched-chain amino acid oxidation in cancer cachectic rats: a proposed role for a cytokine, eicosanoid, and hormone trilogy, Biochem. Med. Met. Biol. 42(1):71–86, 1989.
62. Seelaender M.C., Costa-Rosa L.F., and Curi R. Fatty acid oxidation in lymphocytes from Walker 256 *tumor*-bearing rats, Braz. J. Med. Biol. Res. 29(4):445–451, 1996.
63. Noguchi Y., Vydelingum N.A., and Brennan M.F. *Tumor*-induced alterations in hepatic malic enzyme and *carnitine* palmitoyltransferase activity, J. Surg. Res. 55(4):357–363, 1993.
64. Yazdanpanah M., Luo X., Lau R., Greenberg M., Fisher L.J., and Lehotay D.C. Cytotoxic aldehydes as possible markers for childhood cancer, Free Radic. Biol. Med. 23(6):870–878, 1997.
65. Sachan D.S. and Dodson W.L. The serum *carnitine* status of cancer patients, J. Am. Coll. Nutr. 6(2):145–150, 1987.
66. Rössle C., Pichard C., Roulet M., Bergstrom J., and Furst P. Muscle *carnitine* pools in cancer patients, Clin. Nutr. 8(6):341–346, 1989.
67. Dodson W.L., Sachan D.S., Krauss S., and Hanna W. Alterations of serum and urinary *carnitine* profiles in cancer patients: hypothesis of possible significance, J. Am. Coll. Nutr. 8(2):133–142, 1989.
68. Willson J., Weese J., Wolberg W., and Shug A. Differences between normal and cancerous human colon in *carnitine* (C) and CoA levels, AACR Annual Meeting, San Diego, California 25–28/5/1983.
69. De la Morena E., Montero C., and De la Vieja J. Low levo-*carnitine* levels in serum of women with early breast *tumors*, International Conference Predictive Drug Testing on Human *Tumor* Cells. Zurich, 20–23/7/1983.
70. Schlenzig J.S., Charpentier C., Rabier D., Kamoun P., Sewell A.C., and Harpey J.P. L-*Carnitine*: a way to decrease cellular toxicity of *ifosfamide*?, Eur. J. Pediatr. 154(8):686–687, 1995.
71. Berardi S., Heuberger W., Jacky E., and Krahenbuhl S. Renal *carnitine* excretion is a marker for *cisplatin*-induced tubular nephrotoxicity, FASEB J. 10(3):A470, 1996.
72. Heuberger W., Berardi S., Jacky E., Pey P., and Krahenbuhl S. Increased urinary excretion of *carnitine* in patients treated with *cisplatin*, Eur. J. Clin. Pharmacol. 54(7):503–508, 1998.

73. Goormaghtigh E., Brasseur R., Huart P., and Ruysschaert J.M. Study of the *adriamycin*-cardiolipin complex structure using attenuated total reflection infrared spectroscopy, Biochemistry 26(6):1789–1794, 1987.
74. Robinson N.C. Functional binding of cardiolipin to cytochrome c oxidase, J. Bioenerg. Biomembr 25(2):153–163, 1993.
75. Demant E.J. Inactivation of cytochrome c oxidase activity in mitochondrial membranes during redox cycling of doxorubicin, Biochem. Pharmacol. 41(4):543–552, 1991.
76. Iliskovic N., Panagia V., Slezak J., Kumar D., Li T., and Singal P.K. *Adriamycin* depresses in vivo and in vitro phosphatidylethanolamine N-methylation in rat heart sarcolemma, Mol. Cell. Biochem. 176(1–2):235–240, 1997.
77. Brady L.J. and Brady P.S. Hepatic and cardia *carnitine* palmitoyltransferase activity. Effects of *adriamycin* and galactosamine, Biochem. Pharmacol. 36(20):3419–3423, 1987.
78. Kashfi K., Israel M., Sweatman T.W., Seshadri R., and Cook G.A. Inhibition of mitochondrial *carnitine* palmitoyltransferases by *adriamycin* and *adriamycin* analogues, Biochem. Pharmacol. 40(7):1441–1448, 1990.
79. Abdel-aleem S., el-Merzabani M.M., Sayed-Ahmed M., Taylor D.A., and Lowe J.E. Acute and chronic effects of *adriamycin* on fatty acid oxidation in isolated cardiac myocytes, J. Mol. Cell. Cardiol. 29(2):789–797, 1997.
80. Payne C.M. A quantitative analysis of leptomeric fibrils in an *adriamycin/carnitine* chronic mouse model, J. Submicrosc. Cytol. 14(2):337–45, 1982.
81. McFalls E.O., Paulson D.J., Gilbert E.F., and Shug A.L. *Carnitine* protection against *adriamycin*-induced cardiomyopathy in rats, Life Sci. 38(6):497–505, 1986.
82. Torresi U., Miseria S., Piga A., Cellerino R., Quacci D., Dell'orbo C., and Murer B. An ultrastructural study of the protective effect of L-*carnitine* against cardiotoxicity of anthracyclines in experimental animals, Clin. Trials. J. 27(2):128–140, 1990.
83. Vick J.A., De Felice S.L., and Barranco I.S. Prevention of adrianycin induced cardiac toxicity with *carnitine*: a study in primates, Pharmacodynamics & Therapeutics (Life Sci Adv) 9:1–5, 1990.
84. Neri B., Comparini T., Miliani A., and Torcia M. Protective effects of L-*carnitine* on acute *adriamycin* and daunomycin cardiotoxicity in cancer patients. A preliminary report, Clin. Trials J. 20(2):98–103, 1983.
85. Gulizia M., Cardillo R., Olivieri M., Raciti S., Tosto A., Valadà F., and Circo A. Valutazione funzionale cardiaca in pazienti oncologici in trattamento con adriamicina, Clin. Europea 2:1–40, 1984.
86. Durante C., Ghio R., Ratti M., Dototero D., Dejana A., Gatti A., Minale P., and Boccaccio P. Valutazione della cardiotossicità da antraciclinici mediante test enzimatico: possibile ruolo protettivo della L-carnitina, XXX Congresso Nazionale della Società Italiana di Ematologia, Palermo 6–11/10/1985.
87. Anselmi G., Alvarez M., Strauss M., Gonzales M.I., Gomez J.R., Lopez J.R., Suarez C., Mathison Y., and Horvat D. Indicaciones de la L-carnitina en cardiologia pediatrica, Rev. Latina Cardiol. 12(3):137–145, 1991.
88. Anselmi G., Chazzim G., Eleizalde G., Machado H.I., Mathison Y., Alvarez M., and Strauss M. Prevention of *adriamycin* (ADM) cardiotoxicity with L-*carnitine*. Results in 100 children treated for different types of *tumors*. World Congress of Pediatric Cardiology and Cardiac Surgery. Paris, 21–25/6/1993, pp. 24–51.
89. Alberts D.S., Peng Y.M., Moon T.E., and Bressler R. *Carnitine* prevention of *adriamycin* toxicity in mice, Biomedicine 29(8):265–268, 1978.
90. Senekowitsch R., Lohninger A., Kriegel H., Staniek H., Krieglsteiner H., and Kaiser E. Protective effects of *carnitine* on *adriamycin* toxicity to heart, In: Kaiser E. "*Carnitine*—Its role in lung and heart disorders". Karger, Basel, pp. 126–137, 1987.
91. Culbreath A., Howard E.F., and Carter A.L. Lack of an effect of *carnitine* on the chemotherapeutic properties of *adriamycin* towards human pancreatic cells, FASEB Fed Am Soc Exp Biol J 3(4):A1264, 1989.
92. Carter A.L., Pierce R., Culbreath C., and Howard E. Conjunctive enhancement of *adriamycin* by *carnitine*. In: Carter AL "Current concepts in *carnitine* research". Crc Press, Inc, pp. 245–251, 1992.
93. Thomitzek W.D. and Strack E. The effect of palmitoyl*carnitine* on Ehrlich ascites *tumor* cells in vitro and in vivo. Acta Biol. Med. Ger. 17(2):145–159, 1966.
94. Nakadate T. and Blumberg P.M. Modulation by palmitoyl*carnitine* of protein kinase C Activation, Cancer Res. 47(24 Pt 1):6537–6542, 1987.
95. Nakaki T., Mita S., Yamamoto S., Nakadate T., and Kato R. Inhibition by palmitoyl*carnitine* of adhesion and morphological changes in HL-60 cells induced by 12-O-tetradecanoylphorbol-13-acetate, Cancer Res. 44(5):1908–1912, 1984.

96. Butler A.P., Mar P.K., McDonald F.F., and Ramsay R.L. Involvement of protein kinase C in the regulation of ornithine decarboxylase mRNA by phorbol esters in rat hepatoma cells, Exp. Cell Res. 194(1):56–61, 1991.
97. Aizu E., Yamamoto S., Nakadate T., Kiyoto I., and Kato R. Palmitoyl*carnitine* reverses 12-O-tetradecanoylphorbol 13-acetate-induced refractory state for the TPA-caused ornithine decarboxylase induction in mouse epidermis, Carcinogenesis 9(2):309–313, 1988.
98. Satyamoorthy K. and Perchellet J.P. Inhibition of mouse skin *tumor* promotion by *adriamycin* and daunomycin in combination with verapamil or palmitoyl*carnitine*, Cancer Lett. 55:135–142, 1990.
99. Vescovi G., Weber B., Matrat M., Ramacci C., Nabet P., and Kmemer B. Modulation by palmitoyl-*carnitine* of calcium activated, phospholipid-dependent protein kinase activity and inhibition of melanoma cell growth, Br. J. Dermatol. 119(2):171–178, 1988.

CONTRIBUTORS

Abbondanza, Ciro
Istituto di Patologia Generale ed Oncologia
Facoltà di Medicina e Chirurgia
Seconda Università di Napoli
Larghetto Sant'Aniello a Caponapoli 2
80138 Napoli, Italy

Abbruzzese, Alberto
Dipartimento di Biochimica e Biofisica "F. Cedrangolo"
Seconda Università di Napoli
Via Costantinopoli, 16
80138 Napoli, Italy
E-mail: abbruzze@cds.unina.it

Addeo, Francesco
Istituto di Scienze dell'Alimentazione
Consiglio Nazionale delle Ricerche
via Roma 52 A/C
83100 Avellino, Italy

Airoldi, Luisa
Istituto di Ricerche Farmacologiche Mario Negri
Via Eritrea 62,
20157 Milano, Italy
E-mail: airoldi@irfmn.mnegri.it

Armetta, Ignazio
Istituto di Patologia Generale ed Oncologia
Facoltà di Medicina e Chirurgia
Seconda Università di Napoli
Larghetto Sant'Aniello a Caponapoli 2
80138 Napoli, Italy

Barbarisi, Alfonso
Istituto di Chirugia Clinica
Facoltà di Medicina
Seconda Università di Napoli
Via Pansini 5
80131 Napoli, Italy
E-mail: abarbarisi@unina.it

Benatti, Paola
Sigma Tau S.p.A.
Via Pontina Km 30,400
00040 Pomezia (Roma), Italy

Berrino, Franco
Divisione di Epidemiologia
Istituto Nazionale per lo Studio e la Cura dei Tumori
Via Venezian, 1
20133 Milano, Italy

Bonithon-Kopp, Claude
Registre des Cancers Digestifs
(Equipe Associèe INSERM-DGS, CRI 95-05)
Facultè de Mèdecine
7 Boulevard Jeanne d'Arc
21033 Dijon Cèdex, France

Bontempo, Paola
Istituto di Patologia Generale ed Oncologia

Facoltà di Medicina e Chirurgia
Seconda Università di Napoli
Larghetto Sant'Aniello a Caponapoli 2
80138 Napoli, Italy

Borriello, Adriana
Istituto di Biochimica delle Macromolecule
Seconda Università di Napoli
Via Costantinopoli, 16
80138 Napoli, Italy

Bradlow, H. Leon
Strang Cancer Research Laboratory
Box 231, The Rockefeller University
1230 York Ave., Smith Hall
New York, NY 10021 USA
E-mail: hbradlow@ix.netcom.com

Budillon, Alfredo
Divisione di Oncologia Sperimentale
Istituto Nazionale dei Tumori
Fondazione "G. Pascale"
Via M. Semmola, 1
80131 Napoli, Italy

Calvani, Menotti
Direzione Scientifica
Sigma Tau S.p.A.
Via Pontina Km 30,400
00040 Pomezia (Roma), Italy
E-mail: menotti.calvani@sigma-tau.it

Caraglia, Michele
Dipartimento di Biochimica e Biofisica "F. Cedrangolo"
Seconda Università di Napoli
Via Costantinopoli, 16
80138 Napoli, Italy

Celentano, Egidio
Tumour/Tissues Central Bank; Senology Surgery
Istituto Nazionale dei Tumori
Fondazione "G. Pascale"
Via M. Semmola, 1
80131 Napoli, Italy

Cucciolla, Valeria
Istituto di Biochimica delle Macromolecule
Seconda Università di Napoli
Via Costantinopoli, 16
80138 Napoli, Italy

D'Amicis, Amleto
Statistics and Food Economics Unit
Istituto Nazionale della Nutrizione
Via Ardeatina, 546
00178 Roma, Italy
E-mail: damicis@inn.ingrm.it

D'Angelo, Stefania
Istituto di Biochimica delle Macromolecule
Seconda Università di Napoli
Via Costantinopoli, 16
80138 Napoli, Italy

D'Argenio, Giuseppe
Istituto di Gastroenterologia e Endoscopia
Facoltà di Medicina e Chirurgia
Università di Napoli "Federico II"
Via S. Pansini, 5
80131 Napoli, Italy

Della Pietra, Valentina
Istituto di Biochimica delle Macromolecule
Seconda Università di Napoli
Via Costantinopoli, 16
80138 Napoli, Italy

Della Ragione, Fulvio
Istituto di Biochimica delle Macromolecule
Seconda Università di Napoli
Via Costantinopoli, 16
80138 Napoli, Italy
E-mail: dellarag@cds.unina.it

Dello Iacovo, Rossano
Tumour/Tissues Central Bank; Senology Surgery
Istituto Nazionale dei Tumori
Fondazione "G. Pascale"
Via M. Semmola, 1

80131 Napoli, Italy
E-mail: jacow@iol.it

Faivre, Jean
Registre des Cancers Digestifs
(Equipe Associèe INSERM-DGS, CRI 95-05)
Facultè de Mèdecine
7 Boulevard Jeanne d'Arc
21033 Dijon Cèdex, France
E-mail: jean.faivre@u-bourgogne.fr

Fanelli, Roberto
Istituto di Ricerche Farmacologiche Mario Negri
Via Eritrea 62,
20157 Milano, Italy

Farchi, Sara
Statistics and Food Economics Unit
Istituto Nazionale della Nutrizione
Via Ardeatina, 546
00178 Roma, Italy

Favero, Adriano
Epidemiology Unit, Aviano Cancer Center
Centro di Riferimento Oncologico
Via Pedemontana Occ. 1
33081 Aviano (PN), Italy

Franceschi, Silvia
Epidemiology Unit, Aviano Cancer Center
Centro di Riferimento Oncologico
Via Pedemontana Occ. 1
33081 Aviano (PN), Italy
E-mail: franceschis@ets.it

Galletti, Patrizia
Istituto di Biochimica delle Macromolecule
Seconda Università di Napoli
Via Costantinopoli, 16
80138 Napoli, Italy

Garte, Seymour
EOHSI, UMDNJ
Robert Wood Johnson Medical School
Piscataway NJ 08855, USA

Giacosa, Attilio
Istituto Nazionale per la Ricerca sul Cancro
Largo Rosanna Benzi, 10
16132 Genova, Italy
E-mail: giacosa@hp380.ist.unige.it

Giovannucci, Edward
Channing Laboratory, Department of Medicine
Brigham and Women's Hospital and Harvard Medical School
Departments of Nutrition and Epidemiology
Harvard School of Public Health
Bldg II, Room 319
Boston, MA 02115 USA
E-mail: edward.giovannucci@channing.harvard.edu

Goldbohm, R. Alexandra
TNO Nutrition and Food Research Institute
Utrechtseweg, 48
3700 AJ Zeist, The Netherlands

Greenwald, Peter
Division of Cancer Prevention
National Cancer Institute
Building 31, Room 10A52
31 Center Drive, MSC-2580
National Institutes of Health
Bethesda, Maryland 20892-2580, USA
E-mail: pg37g@nih.gov

Hietanen, Eino
Department of Clinical Physiology
Turku University Hospital
FIN-20520 Turku, Finland
E-mail: eino.hietanen@utu.fi

Hill, Michael J.
ECP (UK) Headquarters
Wexham Park Hospital
Slough, Berks SL2 4HL, United Kingdom

Iacomino, Giuseppe
Istituto di Scienze dell'Alimentazione
Consiglio Nazionale delle Ricerche
via Roma 52 A/C
83100 Avellino, Italy

Iazzetta, Giacomo
Tumour/Tissues Central Bank; Senology Surgery
Istituto Nazionale dei Tumori
Fondazione "G. Pascale"
Via M. Semmola, 1
80131 Napoli, Italy

Iolascon, Achille
Istituto di Biochimica delle Macromolecule
Seconda Università di Napoli
Via Costantinopoli, 16
80138 Napoli, Italy

Jacob, Robert A.
Western Human Nutrition Research Center
U.S.D.A. Agricultural Research Service
University of California
2047 Wickson Hall
Davis, CA 95616 USA

Krogh, Vittorio
Divisione di Epidemiologia
Istituto Nazionale per lo Studio e la Cura dei Tumori
Via Venezian, 1
20133 Milano, Italy

Magagnotti, Cinzia
Istituto di Ricerche Farmacologiche Mario Negri
Via Eritrea 62,
20157 Milano, Italy

Manna, Caterina
Istituto di Biochimica delle Macromolecule
Seconda Università di Napoli
Via Costantinopoli, 16
80138 Napoli, Italy
E-mail: catmanna@unina.it

Mazzacca, Gabriele
Istituto di Gastroenterologia e Endoscopia
Facoltà di Medicina e Chirurgia
Università di Napoli "Federico II"
Via S. Pansini, 5
80131 Napoli, Italy

Mercurio, Ciro
Istituto di Biochimica delle Macromolecule
Seconda Università di Napoli
Via Costantinopoli, 16
80138 Napoli, Italy

Molinari, Anna Maria
Istituto di Patologia Generale ed Oncologia
Facoltà di Medicina e Chirurgia
Seconda Università di Napoli
Larghetto Sant'Aniello a Caponapoli 2
80138 Napoli, Italy

Montella, Maurizio
Epidemiology Unit
Istituto Nazionale dei Tumori
Fondazione "G. Pascale"
Via M. Semmola, 1
80131 Napoli, Italy

Nicolai, Raffaela
Sigma Tau S.p.A.
Via Pontina Km 30,400
00040 Pomezia (Roma), Italy

Nola, Ernesto
Istituto di Patologia Generale ed Oncologia
Facoltà di Medicina e Chirurgia
Seconda Università di Napoli
Larghetto Sant'Aniello a Caponapoli 2
80138 Napoli, Italy

Oliva, Adriana
Istituto di Biochimica delle Macromolecule
Seconda Università di Napoli
Via Costantinopoli, 16
80138 Napoli, Italy

Orsini, Fulvia
Dipartimento di Chimica Organica e Industriale
Università degli Studi di Milano
Via Venezian 21
20133 Milano, Italy
E-mail: orsiful@unimi.it

Osborne, Michael P.
Strang Cancer Research Laboratory
The Rockefeller University
1230 York Ave., Smith Hall
New York, NY 10021 USA

Palli, Domenico
Sezione Epidemiologia Analitica CSPO
Azienda Ospedaliera Careggi
Viale Volta 171
50131 Firenze, Italy
E-mail: md0632@mclink.it

Palumbo, Rosanna
Istituto di Scienze dell'Alimentazione
Consiglio Nazionale delle Ricerche
via Roma 52 A/C
83100 Avellino, Italy

Panico, Salvatore
Dipartimento di Medicina Clinica e Sperimentale
Università di Napoli "Federico II"
Via Pansini, 5
80131 Napoli, Italy

Parpinel, Maria
Epidemiology Unit
Aviano Cancer Center
Via Pedemontana Occ. 1
33081 Aviano (PN), Italy

Pastorelli, Roberta
Istituto di Ricerche Farmacologiche Mario Negri
Via Eritrea 62,
20157 Milano, Italy

Pegg, Anthony E.
Department of Cellular and Molecular Physiology
Pennsylvania State University College of Medicine
The Milton S. Hershey Medical Center
500 University Drive
Hershey, PA 17033 USA
E-mail: aep1@psu.edu

Peluso, Gianfranco
CNR, Istituto di Biochimica delle Proteine
Via Toiano 6
Seconda Università di Napoli
80072 Arco Felice (Napoli), Italy

Puca, Giovanni Alfredo
Istituto di Patologia Generale ed Oncologia
Facoltà di Medicina e Chirurgia
Seconda Università di Napoli
Larghetto Sant'Aniello a Caponapoli 2
80138 Napoli, Italy
E-mail: gapuca@tin.it

Randazzo, Giacomo
Biochimica, Scienza dell'Alimentazione
Università di Napoli "Federico II"
80100 Napoli, Italy

Reda, Emilia
Sigma Tau S.p.A.
Via Pontina Km 30,400
00040 Pomezia (Roma), Italy

Russo, Antonio
Sezione Epidemiologia Analitica CSPO
Azienda Ospedaliera Careggi
Viale Volta 171
50131 Firenze, Italy

Russo, Gian-Luigi
Istituto di Scienze dell'Alimentazione
Consiglio Nazionale delle Ricerche
via Roma 52 A/C
83100 Avellino, Italy
E-mail: glrusso@isa.av.cnr.it

Russo, Maria
Istituto di Scienze dell'Alimentazione
Consiglio Nazionale delle Ricerche
via Roma 52 A/C
83100 Avellino, Italy

Schiavone, Ettore Maria
2a Divisione di Ematologia
Ospedale Cardarelli
80131 Napoli, Italy

Sepkovic, Daniel W.
Strang Cancer Research Laboratory
The Rockefeller University
1230 York Ave., Smith Hall
New York, NY 10021 USA

Strollo, Anna Maria
Tumour/Tissues Central Bank; Senology Surgery
Istituto Nazionale dei Tumori
Fondazione "G. Pascale"
Via M. Semmola, 1
80131 Napoli, Italy

Tagliaferri, Pierosandro
Dip.to di Endocrinologia ed Oncologia Molecolare e Clinica
Università "Federico II" di Napoli
Present address: Dipartimento di Medicina Sperimentale e Clinica,
Università "Magna Graecia"
88100 Catanzaro, Italy

Taioli, Emanuela
Policlinico IRCCS
Università di Milano
Via F. Sforza 28
20122 Milano, Italy
E-mail: sget@iol.it

Telang, Nitin T.
Strang Cancer Research Laboratory
The Rockefeller University
1230 York Ave., Smith Hall
New York, NY 10021 USA

Tortora, Vincenzo
Istituto di Patologia Generale ed Oncologia
Facoltà di Medicina e Chirurgia
Seconda Università di Napoli
Larghetto Sant'Aniello a Caponapoli 2
80138 Napoli, Italy

Tosto, Mariarosaria
Stazione Zoologica "Anton Dohrn"
Viale Dohrn
80121 Napoli, Italy

Tumino, Rosario
Registro dei Tumori della Provincia di Ragusa
Azienda Ospedaliera Civile M.P. Arezzo
07100 Ragusa, Italy

Verotta, Luisella
Dipartimento di Chimica Organica e Industriale
Università degli Studi di Milano
Via Venezian 21
20133 Milano, Italy

Vineis, Paolo
Servizio di Epidemiologia dei Tumori
Dipartimento di Scienze Biomediche e Oncologia Umana
Università di Torino
10100 Torino, Italy

van Poppel, Geert
TNO Nutrition and Food Research Institute
Utrechtseweg, 48
3700 AJ Zeist, The Netherlands
E-mail: vanpoppel@voeding.tno.nl

Verdicchio, Mariantonietta
Istituto di Patologia Generale ed Oncologia
Facoltà di Medicina e Chirurgia
Seconda Università di Napoli
Larghetto Sant'Aniello a Caponapoli 2
80138 Napoli, Italy

Verhagen, Hans
TNO Nutrition and Food Research Institute

Utrechtseweg, 48
3700 AJ Zeist, The Netherlands

Verhoeven, Dorette T. H.
TNO Nutrition and Food Research Institute
Utrechtseweg, 48
3700 AJ Zeist, The Netherlands

Zappia, Vincenzo
Istituto di Biochimica delle Macromolecule
Seconda Università Napoli
Via Costantinopoli, 16
80138 Napoli, Italy
E-mail: zappia@unina.it

INDEX